Humic Matter in Soil
and the Environment

BOOKS IN SOILS, PLANTS, AND THE ENVIRONMENT

Editorial Board

Agricultural Engineering	Robert M. Peart, University of Florida, Gainesville
Animal Science	Harold Hafs, Rutgers University, New Brunswick, New Jersey
Crops	Mohammad Pessarakli, University of Arizona, Tucson
Irrigation and Hydrology	Donald R. Nielsen, University of California, Davis
Microbiology	Jan Dirk van Elsas, Research Institute for Plant Protection, Wageningen, The Netherlands
Plants	L. David Kuykendall, U.S. Department of Agriculture, Beltsville, Maryland
	Kenneth B. Marcum, Texas A&M University, El Paso, Texas
Soils	Jean-Marc Bollag, Pennsylvania State University, University Park, Pennsylvania
	Tsuyoshi Miyazaki, University of Tokyo

Soil Biochemistry, Volume 1, edited by A. D. McLaren and G. H. Peterson
Soil Biochemistry, Volume 2, edited by A. D. McLaren and J. Skujiņš
Soil Biochemistry, Volume 3, edited by E. A. Paul and A. D. McLaren
Soil Biochemistry, Volume 4, edited by E. A. Paul and A. D. McLaren
Soil Biochemistry, Volume 5, edited by E. A. Paul and J. N. Ladd
Soil Biochemistry, Volume 6, edited by Jean-Marc Bollag and G. Stotzky
Soil Biochemistry, Volume 7, edited by G. Stotzky and Jean-Marc Bollag
Soil Biochemistry, Volume 8, edited by Jean-Marc Bollag and G. Stotzky
Soil Biochemistry, Volume 9, edited by G. Stotzky and Jean-Marc Bollag
Soil Biochemistry, Volume 10, edited by Jean-Marc Bollag and G. Stotzky

Organic Chemicals in the Soil Environment, Volumes 1 and 2, edited by C. A. I. Goring and J. W. Hamaker
Humic Substances in the Environment, M. Schnitzer and S. U. Khan
Microbial Life in the Soil: An Introduction, T. Hattori
Principles of Soil Chemistry, Kim H. Tan
Soil Analysis: Instrumental Techniques and Related Procedures, edited by Keith A. Smith
Soil Reclamation Processes: Microbiological Analyses and Applications, edited by Robert L. Tate III and Donald A. Klein
Symbiotic Nitrogen Fixation Technology, edited by Gerald H. Elkan

Soil–Water Interactions: Mechanisms and Applications, Shingo Iwata and Toshio Tabuchi with Benno P. Warkentin
Soil Analysis: Modern Instrumental Techniques, Second Edition, edited by Keith A. Smith
Soil Analysis: Physical Methods, edited by Keith A. Smith and Chris E. Mullins
Growth and Mineral Nutrition of Field Crops, N. K. Fageria, V. C. Baligar, and Charles Allan Jones
Semiarid Lands and Deserts: Soil Resource and Reclamation, edited by J. Skujiņš
Plant Roots: The Hidden Half, edited by Yoav Waisel, Amram Eshel, and Uzi Kafkafi
Plant Biochemical Regulators, edited by Harold W. Gausman
Maximizing Crop Yields, N. K. Fageria
Transgenic Plants: Fundamentals and Applications, edited by Andrew Hiatt
Soil Microbial Ecology: Applications in Agricultural and Environmental Management, edited by F. Blaine Metting, Jr.
Principles of Soil Chemistry: Second Edition, Kim H. Tan
Water Flow in Soils, edited by Tsuyoshi Miyazaki
Handbook of Plant and Crop Stress, edited by Mohammad Pessarakli
Genetic Improvement of Field Crops, edited by Gustavo A. Slafer
Agricultural Field Experiments: Design and Analysis, Roger G. Petersen
Environmental Soil Science, Kim H. Tan
Mechanisms of Plant Growth and Improved Productivity: Modern Approaches, edited by Amarjit S. Basra
Selenium in the Environment, edited by W. T. Frankenberger, Jr., and Sally Benson
Plant–Environment Interactions, edited by Robert E. Wilkinson
Handbook of Plant and Crop Physiology, edited by Mohammad Pessarakli
Handbook of Phytoalexin Metabolism and Action, edited by M. Daniel and R. P. Purkayastha
Soil–Water Interactions: Mechanisms and Applications, Second Edition, Revised and Expanded, Shingo Iwata, Toshio Tabuchi, and Benno P. Warkentin
Stored-Grain Ecosystems, edited by Digvir S. Jayas, Noel D. G. White, and William E. Muir
Agrochemicals from Natural Products, edited by C. R. A. Godfrey
Seed Development and Germination, edited by Jaime Kigel and Gad Galili
Nitrogen Fertilization in the Environment, edited by Peter Edward Bacon
Phytohormones in Soils: Microbial Production and Function, William T. Frankenberger, Jr., and Muhammad Arshad
Handbook of Weed Management Systems, edited by Albert E. Smith
Soil Sampling, Preparation, and Analysis, Kim H. Tan
Soil Erosion, Conservation, and Rehabilitation, edited by Menachem Agassi
Plant Roots: The Hidden Half, Second Edition, Revised and Expanded, edited by Yoav Waisel, Amram Eshel, and Uzi Kafkafi
Photoassimilate Distribution in Plants and Crops: Source–Sink Relationships, edited by Eli Zamski and Arthur A. Schaffer

Mass Spectrometry of Soils, edited by Thomas W. Boutton and Shinichi Yamasaki
Handbook of Photosynthesis, edited by Mohammad Pessarakli
Chemical and Isotopic Groundwater Hydrology: The Applied Approach, Second Edition, Revised and Expanded, Emanuel Mazor
Fauna in Soil Ecosystems: Recycling Processes, Nutrient Fluxes, and Agricultural Production, edited by Gero Benckiser
Soil and Plant Analysis in Sustainable Agriculture and Environment, edited by Teresa Hood and J. Benton Jones, Jr.
Seeds Handbook: Biology, Production, Processing, and Storage, B. B. Desai, P. M. Kotecha, and D. K. Salunkhe
Modern Soil Microbiology, edited by J. D. van Elsas, J. T. Trevors, and E. M. H. Wellington
Growth and Mineral Nutrition of Field Crops: Second Edition, N. K. Fageria, V. C. Baligar, and Charles Allan Jones
Fungal Pathogenesis in Plants and Crops: Molecular Biology and Host Defense Mechanisms, P. Vidhyasekaran
Plant Pathogen Detection and Disease Diagnosis, P. Narayanasamy
Agricultural Systems Modeling and Simulation, edited by Robert M. Peart and R. Bruce Curry
Agricultural Biotechnology, edited by Arie Altman
Plant–Microbe Interactions and Biological Control, edited by Greg J. Boland and L. David Kuykendall
Handbook of Soil Conditioners: Substances That Enhance the Physical Properties of Soil, edited by Arthur Wallace and Richard E. Terry
Environmental Chemistry of Selenium, edited by William T. Frankenberger, Jr., and Richard A. Engberg
Principles of Soil Chemistry: Third Edition, Revised and Expanded, Kim H. Tan
Sulfur in the Environment, edited by Douglas G. Maynard
Soil–Machine Interactions: A Finite Element Perspective, edited by Jie Shen and Radhey Lal Kushwaha
Mycotoxins in Agriculture and Food Safety, edited by Kaushal K. Sinha and Deepak Bhatnagar
Plant Amino Acids: Biochemistry and Biotechnology, edited by Bijay K. Singh
Handbook of Functional Plant Ecology, edited by Francisco I. Pugnaire and Fernando Valladares
Handbook of Plant and Crop Stress: Second Edition, Revised and Expanded, edited by Mohammad Pessarakli
Plant Responses to Environmental Stresses: From Phytohormones to Genome Reorganization, edited by H. R. Lerner
Handbook of Pest Management, edited by John R. Ruberson
Environmental Soil Science: Second Edition, Revised and Expanded, Kim H. Tan
Microbial Endophytes, edited by Charles W. Bacon and James F. White, Jr.
Plant–Environment Interactions: Second Edition, edited by Robert E. Wilkinson
Microbial Pest Control, Sushil K. Khetan

Soil and Environmental Analysis: Physical Methods, Second Edition, Revised and Expanded, edited by Keith A. Smith and Chris E. Mullins

The Rhizosphere: Biochemistry and Organic Substances at the Soil–Plant Interface, Roberto Pinton, Zeno Varanini, and Paolo Nannipieri

Woody Plants and Woody Plant Management: Ecology, Safety, and Environmental Impact, Rodney W. Bovey

Metals in the Environment: Analysis by Biodiversity, M. N. V. Prasad

Plant Pathogen Detection and Disease Diagnosis: Second Edition, Revised and Expanded, P. Narayanasamy

Handbook of Plant and Crop Physiology: Second Edition, Revised and Expanded, edited by Mohammad Pessarakli

Environmental Chemistry of Arsenic, edited by William T. Frankenberger, Jr.

Enzymes in the Environment: Activity, Ecology, and Applications, edited by Richard G. Burns and Richard P. Dick

Plant Roots: The Hidden Half, Third Edition, Revised and Expanded, edited by Yoav Waisel, Amram Eshel, and Uzi Kafkafi

Handbook of Plant Growth: pH as the Master Variable, edited by Zdenko Rengel

Biological Control of Crop Diseases, edited by Samuel S. Gnanamanickam

Pesticides in Agriculture and the Environment, edited by Willis B. Wheeler

Mathematical Models of Crop Growth and Yield, Allen R. Overman and Richard V. Scholtz III

Plant Biotechnology and Transgenic Plants, edited by Kirsi-Marja Oksman-Caldentey and Wolfgang H. Barz

Handbook of Postharvest Technology: Cereals, Fruits, Vegetables, Tea, and Spices, edited by Amalendu Chakraverty, Arun S. Mujumdar, G. S. Vijaya Raghavan, and Hosahalli S. Ramaswamy

Handbook of Soil Acidity, edited by Zdenko Rengel

Humic Matter in Soil and the Environment: Principles and Controversies, Kim H. Tan

Additional Volumes in Preparation

Molecular Host Resistance to Pests, S. Sadasivam and B. Thayumanavan

Soil and Environmental Analysis: Modern Instrumental Techniques, Third Edition, edited by Keith A. Smith and Malcolm S. Cresser

Humic Matter in Soil and the Environment
Principles and Controversies

Kim H. Tan
*University of Georgia
Athens, Georgia, U.S.A.*

MARCEL DEKKER, INC.   NEW YORK · BASEL

Library of Congress Cataloging-in-Publication Data
A catalog record for this book is available from the Library of Congress.

ISBN: 0-8247-4272-9

This book is printed on acid-free paper.

Headquarters
Marcel Dekker, Inc.
270 Madison Avenue, New York, NY 10016
tel: 212-696-9000; fax: 212-685-4540

Eastern Hemisphere Distribution
Marcel Dekker AG
Hutgasse 4, Postfach 812, CH-4001 Basel, Switzerland
tel: 41-61-260-6300; fax: 41-61-260-6333

World Wide Web
http://www.dekker.com

The publisher offers discounts on this book when ordered in bulk quantities. For more information, write to Special Sales/Professional Marketing at the headquarters address above.

Copyright © 2003 by Marcel Dekker, Inc. All Rights Reserved.

Neither this book nor any part may be reproduced or transmitted in any form or by any means, electronic or mechanical, including photocopying, microfilming, and recording, or by any information storage and retrieval system, without permission in writing from the publisher.

Current printing (last digit):
10 9 8 7 6 5 4 3 2 1

PRINTED IN THE UNITED STATES OF AMERICA

PREFACE

A large amount of information has accumulated on humic acids and related substances, which warrants the creation of an independent science of humic compounds. Two different concepts have emerged from the maze of data, one claiming humic compounds to be operational or fake compounds, produced by the analytical extraction procedures, and the other considering them to be natural compounds occurring in soils, rivers, lakes, oceans and their sediments. Apparently the two opposing opinions have caused considerable confusion among scientists, students, and professionals alike about exactly what humic acid is, or what the difference is between soil organic matter, humus, and humic acid. Several of the books and especially the symposium proceedings on humus and soil organic matter are guilty of making the chaos worse, by using different terms and concepts interchangeably and by only covering "specialty" topics. The need for a book providing comprehensive coverage, on definitions, concepts, genesis, extraction, properties, and the impact of humic matter on agriculture, industry, and environment, is apparent.

This book tries to address the problem of complete coverage as highlighted above. In addition to its value as a textbook, it can be used equally well as a reference book by all interested in humic matter. The issues and controversies associated with humic acids are analyzed from the two different viewpoints mentioned above. The advances of the past century, and the prospects for advancing humic acid science in the new millennium, are explored from both viewpoints. The text also carries a message for increasing awareness of the appearance of more and more data, emphasizing the ubiquitous presence of humic compounds in nature and their impact on the environment, soils, and agriculture. The intensified application of humic substances in industrial and pharmaceutical operations is discussed, underscoring the significance of humic acids as highly important organic substances in nature. The production and use of therapeutic chemicals from humic acids and the manufacture of commercial humates for use in soils, which has grown lately into a multimillion dollar business, are addressed. These are issues of considerable interest to people studying, practicing, and producing medicines and fertilizers, and therefore enlarges the audience for this book beyond the scope of soil, agricultural, and chemical science.

The book starts by examining the concepts of humus and humic matter from the two different standpoints. Definitions are given in Chapter 1 to delineate soil organic matter, humus, and humified substances. The term *"humic matter"* is defined and adopted in this book as the humified fraction of humus, and the controversy of whether it is present as an artifact or as a true compound in nature is addressed. Questions are raised on the significance of studying fake compounds, especially in institutions where *"publish or perish"* prevails.

Chapter 2 discusses the nature and distribution of humic matter in soils, wetlands and peat, in aquatic environments, and in geologic deposits. A classification of the different types of humic matter based on origin is provided. The chapter contains a discussion of *anthropogenic humic matter*, developed from agricultural waste, polluting the environment. The notorious deposits from the so-called CAFOS, confined animal feeding operations, located on top of the recharge zones of aquifers in Texas, are explained as being too close for comfort. The topic of domestic waste, fouling drainage ditches and canals, is included to cover humic matter produced by these rotten pollutants.

Extraction, isolation, and fractionation of humic substances are featured in Chapter 3, starting with the search for the 'best' inorganic and organic reagents. Detailed analytical procedures are given according to the International Humic Substances Society, the Soil Science Society of America, and the methods presented by Stevenson and Tan. The extraction of aquatic humic matter is discussed separately and the use of XAD resins evaluated. A descriptive analysis

Preface

is provided at the end on fulvic acids, humic acids, and humin, highlighting their definitions, properties, and significance in soils.

Chapter 4 is on the genesis of humic matter. The components from which humic matter is formed are defined here as *precursors*, and distinguished into (1) major precursors, e.g., lignin, phenols, quinones, protein, amino acids, and carbohydrates, and (2) miscellaneous humic precursors, e.g., lipids, sterols, and nucleic acids. Growth promoting substances such as auxin, gibberellin, and vitamins are included in the latter group, and their biotic origin and decomposition are examined in relation to claims that humic acids display hormone-like actions. The section above is then followed by a probing discussion of the processes of formation of humic matter, defined here as *humification*. The three major theories, ligno-protein, phenol-protein, and sugar-amine condensation theory, are addressed in relation to the biopolymer degradation and/or polymerization or condensation concept. As a final topic, a detailed analysis is given of the significance of statistical modeling of humification, including the use of stability coefficients, humification indexes, and models.

Chapter 5 discusses the chemical composition of humic matter, which is distinguished into an elemental and a group composition. The significance of using weight and atomic percentages is studied, underscoring the importance of C/N ratios, atomic ratios, and functional group contents in the formation of formula composition. Molecular structures of humic acids are created by applying simple basic reactions, and the structural models obtained by the newest advances in computer modeling are a major challenge to the idea that humic substances lack formulas and structures.

Chapter 6 is about characterization of humic substances by molecular weight and spectroscopic analysis. The types and ranges of molecular weight values are described, and their effect on size and shape of humic molecules is evaluated, including the importance of frictional ratios, f/f_o. The usefulness of spectrophotometric color ratios and infrared group frequency and fingerprint regions in the identification of fulvic and humic acids is studied. Characterization by electron spin resonance (ESR), nuclear magnetic resonance (NMR), and electron microscopy is addressed in detail. Characteristic visible light, infrared, ESR, and NMR spectra and electron micrographs are provided, with detailed descriptions for each of the humic compounds. They are valuable for classroom teaching, and/or for use as standard reference in research and other scientific or industrial analysis; hence they are assets that make this book stand out over any other book published on the subject.

Chapter 7 discusses electrochemical properties of humic matter. Negative and positive charges are examined and their magnitude is explained using the Henderson-Hasselbalch equation, pK_a, and pK_b values. The issue of COOH and phenolic-OH group contents affecting negative charges and total acidity of humic matter is addressed. Definitions and formulation of surface charge density are studied and the electric double layer theories amended to include a new concept called *fused double layer*. The proper definitions are given for adsorption, cation exchange, complex, chelation, and bridging reactions, and deviations from the concepts are questioned as aberrations. The importance of these interactions in soils, agriculture, and the environment are addressed, and the role of pK_a and stability constants in the reactions evaluated.

The agronomic importance of humic matter is featured in Chapter 8, highlighting its effect on soil physical, chemical, and biological properties. The significance of humic matter for terrestrial and aquatic life is explained, and the role of humic matter in the carbon and nitrogen cycles underscored. The action of humic acids as a redox agent is analyzed in the overall soil's redox system. The direct and indirect effect of humic matter on plant growth and crop production are discussed in detail.

Chapter 9 covers the environmental and industrial importance of humic matter. The outstanding role humic matter plays in preservation of soil organic matter, mobilization and immobilization of elements, and biological detoxification is presented by underscoring the issue of degradation of the soil ecosystem. The use of humic matter in industry is discussed, stressing the production of agrochemicals, e.g., biofertilizers and biopesticides, and the salient features of humic acids considered for use as drilling fluids, paint, ink, tanning, ceramics, and silicones. An assessment is also made of the increased importance of commercial humates, and their production, types, and controversies over their use as fertilizers are addressed. The significance of humic

Preface

matter as a source for the production of pharmaceuticals is examined, and claims of humic acid derived medicines for antiviral, anticancer, and eye disease treatments is discussed.

Finally, I would like to acknowledge the scientists and publishers who offered their generous support. Special thanks are due to Dr. Hans-Rolf Schulten, Professor, Institute for Soil Science, Rostock University, Rostock, Germany, for his generosity in supplying from his personal files the 3D-structural models of humic acids. Thanks are also conveyed to Dr. Patrick G. Hatcher, Professor and Co-Director, Ohio State University EMSI, Department of Chemistry, Newman and Wolfram Laboratory, Ohio State University, Columbus, Ohio, for his permission to use his 2D-structural model of humic acid. Appreciation is extended to the American Chemical Society, Washington, D.C., and to Elsevier Publishers, UK and Amsterdam, for their approval to quote or reproduce figures or photographs. Last but not least, I wish to thank my wife for her loyal assistance and encouragement, enabling me to devote my time and efforts to producing this book.

Kim H. Tan

CONTENTS

Preface	iii
Chapter 1 THE ISSUE OF HUMIC MATTER	1
1.1 Concept of Humus	1
1.1.1 The Early Concept of Humus	1
1.1.2 Concept of Humus in the Third Millennium	3
1.2 Concept of Humic Matter	5
1.3 The Issue of Artifacts	6
1.4 The Issue of Real Compounds	8
1.5 The Issue of Chemical Composition	11
Chapter 2 THE NATURE AND DISTRIBUTION OF HUMIC MATTER	14
2.1 Concepts and Historical Background	14
2.1.1 Historical Concepts	14
2.1.2 Concepts in the Early Twentieth Century	15
2.1.3 The Dawn of Modern Concepts	17
2.2 Distribution of Humic Matter	19
2.2.1 Humic Matter in Soils	20
2.2.2 Humic Matter in Soils of the Wetlands	22
2.2.3 Humic Matter in Aquatic Environments	23
2.2.4 Humic Matter in Geologic Deposits	26

Contents

	2.2.5 Humic Matter in Agricultural, Industrial and Municipal Waste	27
2.3	Classification of Humic Matter	29
	2.3.1 Terrestrial or Terrigenous Humic Matter	30
	2.3.2 Aquatic Humic Matter	31
	2.3.3 Wetland or Peat Humic Matter	32
	2.3.4 Geologic Humic Matter	32
	2.3.5 Anthropogenic Humic Matter	33

Chapter 3 EXTRACTION AND FRACTIONATION OF HUMIC SUBSTANCES 34

3.1	The Search for Extractants	34
	3.1.1 Inorganic Reagents	37
	3.1.2 Organic Reagents	39
	3.1.3 Reagents for Collecting Aquatic Humic Substances	42
3.2	Terrestrial Humic Matter	45
	3.2.1 Extraction Methods	45
3.3	Fractionation of Humic Substances	50
	3.3.1 Fractionation of Humic Acid	50
	3.3.2 Fractionation of Fulvic Acid	57
3.4	Aquatic Humic Matter	58
	3.4.1 Extraction Methods	58
	3.4.2 Fractionation of Aquatic Humic Matter	61
3.5	Types of Humic Substances	63
	3.5.1 Fulvic Acid	65
	3.5.2 Humic Acid	68
	3.5.3 Humin	71

Chapter 4 GENESIS OF HUMIC MATTER 75

4.1 Major Pathways of Humification	75
4.2 Precursors of Humic Matter	76
4.2.1 Lignin	77

	4.2.2 Phenols and Polyphenols	84
	4.2.3 Quinones	89
	4.2.4 Protein and Amino Acids	90
	4.2.5 Carbohydrates	94
	4.2.6 Miscellaneous Humic Precursors	99
4.3	Theories of Humification	111
	4.3.1 The Ligno-Protein Theory	112
	4.3.2 The Phenol-Protein Theory	116
	4.3.3 The Sugar-Amine Condensation Theory	119
4.4	Statistical Modeling of Humification	121
	4.4.1 Humification Indexes	122
	4.4.2 Stability Coefficient of Humus	123
	4.4.3 Humification Model	124

Chapter 5 CHEMICAL COMPOSITION OF HUMIC MATTER   127

5.1	Elemental Composition	127
	5.1.1 Weight Percentage	127
	5.1.2 The C/N Ratio	131
	5.1.3 Atomic Percentage	133
	5.1.4 Internal Oxidation of Humic Substances, ω	135
	5.1.5 Atomic Ratios	138
	5.1.6 Group Composition	142
	5.1.7 Calculation of Formula Weights	152
5.2	Molecular Structures	155
	5.2.1 Structures Based on the Ligno-Protein Concept	157
	5.2.2 Structures Based on the Phenol-Protein Concept	159
	5.2.3 Structures Based on the Sugar-Amine Condensation Concept	162
5.3	Computer Modeling of Humic Acid Structures	162

Contents

Chapter 6 CHARACTERIZATION OF HUMIC SUBSTANCES — 169

6.1 Chemical Characterization — 169
6.2 Molecular Weights — 169
    6.2.1 Number-Average Molecular Weight, M_n — 170
    6.2.2 Weight Average Molecular Weight, M_w — 171
    6.2.3 Z-Average Molecular Weight, M_z — 171
    6.2.4 Characterization by Molecular Weight — 171
    6.2.5 Relationship between Molecular Weight and Size or Shape — 173
6.3 Ultraviolet and Visible Light Spectrophotometry — 176
6.4 Infrared Spectroscopy — 180
    6.4.1 Infrared Spectra of Humic Matter — 181
    6.4.2 Classification of Infrared Spectra — 185
6.5 Nuclear Magnetic Resonance Spectroscopy — 189
    6.5.1 Electron Paramagnetic Resonance — 189
    6.5.2 Carbon-13 Nuclear Magnetic Resonance — 192
    6.5.3 Nitrogen-15 Nuclear Magnetic Resonance — 197
    6.5.4 Phosphorus-31 Nuclear Magnetic Resonance — 201
6.6 Electron Microscopy of Humic Matter — 203
    6.6.1 Transmission Electron Microscopy — 203
    6.6.2 Scanning Electron Microscopy — 204

Chapter 7 ELECTROCHEMICAL PROPERTIES OF HUMIC MATTER — 210

7.1 Origin and Types of Electric Charges — 210
    7.1.1 Negative Charges — 210
    7.1.2 Positive Charges — 217
7.2 Surface Charge Density — 221
7.3 Electric Double Layer — 223
    7.3.1 Fused Double Layer — 224
7.4 Chemical Reactions and Interactions — 225
    7.4.1 Adsorption — 226
    7.4.2 Cation Exchange Capacity — 235

7.5	Complex Reaction and Chelation	238
	7.5.1 The Significance of COOH groups	241
	7.5.2 The Significance of pK_a	243
	7.5.3 Stability Constants of Chelates	243
	7.5.4 Effect on Soil Genesis	247
	7.5.5 Statistical Modeling	249
7.6	Bridging Mechanism	250

Chapter 8 AGRONOMIC IMPORTANCE OF HUMIC MATTER 254

8.1	Importance in Soils	254
	8.1.1 Effect on Soil Physical Properties	254
	8.1.2 Effect on Soil Chemical Properties	257
	8.1.3 Effect on the Soil Redox System	259
	8.1.4 Effect on Soil Biological Properties	267
8.2	Importance in Plant Growth	280
	8.2.1 Effect on Plant Nutrition	280
	8.2.2 Effect on Plant Physiology	286

Chapter 9 ENVIRONMENTAL AND INDUSTRIAL IMPORTANCE OF HUMIC MATTER 292

9.1	Importance in the Environment	292
	9.1.1 Preservation of Soil Organic Matter	293
	9.1.2 Mobilization and Immobilization of Elements	295
	9.1.3 Biological Detoxification	297
9.2	Degradation of the Soil Ecosystem	299
9.3	Importance in Industry	301
	9.3.1 Production of Agrochemicals	302
	9.3.2 Production of Commercial Humates	304
9.4	Importance as Pharmaceuticals	309

Appendix A   Greek Alphabet   313

Contents

Appendix B Atomic Weights of Major Elements in Soils	314
References and Additional Readings	317
Index	356

Humic Matter in Soil and the Environment

CHAPTER 1

THE ISSUE OF HUMIC MATTER

1.1 CONCEPT OF HUMUS

1.1.1 The Early Concept of Humus

Soil organic matter is derived from the soil biomass, and strictly speaking it consists of both living and dead organic matter. It is the most important fraction in soils and has attracted considerable attention since the early days of agriculture because of its pronounced effect on the physical, chemical, and biological condition of soils (Russell and Russell, 1950; Tan, 2000). The growth and yield of crops have always been noted to be better when plants are grown in soils rich in organic matter. Lack of scientific data in the past is perhaps one of the reasons for the presence of some reservations as to its beneficial effect on soils and plants (Stevenson, 1994).

The term soil organic matter is frequently used to indicate the dead organic fraction only, and the live fraction, though of equal importance, is usually ignored. The dead fraction is formed by chemical and biological decomposition of organic residues and can be distinguished into (1) organic matter at various degrees of decomposition, in which the morphology of plant material is still visible, and (2) completely decomposed materials with no traces of the anatomical structure of the material from which they have been formed. The first group mentioned above, containing most of the undecomposed material, has a prominent influence on soil physical properties. It also finds practical application in Soil Taxonomy, where it is distinguished into a *fibric* and *hemic fraction*, a distinction based on the relative degree of decomposition. The fibric (Latin fibra = fiber) fraction is the least decomposed, whereas the hemic (Greek hemi = half) fraction is the partly decomposed fraction (Soil Survey Staff, 1990). A third fraction is recognized pertaining to the completely decomposed part and this will be explained below in the second group. From the standpoint of soil chemistry and humic matter the nondecomposed group is of minor importance though it is the source for formation of the decomposed fraction. The term *litter* is often used for this type of organic matter when it lies on the soil surface. In forest and grassland soils, litter is particularly important in the process of nutrient cycling.

The second group of dead organic matter is called *humus* (Latin for soil or vegetation), a name that seemed to be accepted for this dead residue by many scientists from the early days till late in the twentieth century. This decomposed dark colored organic matter in soils, referred to above as humus, is extractable by alkali (Russell and Russell, 1950). Though Achard (1786) was reported as the first scientist to extract peat in his study of humus, it was De Saussure (1804) who was given credit for introducing the name humus in soil science. Since then humus has been the subject of a lot of research attention, which during the early years resulted in the discovery of several different types of humus. Based on origin, Sprengel (1826) recognizes (1) *acid humus*, formed from peat and other type of 'acidic' vegetation, and (2) *mild humus*, which has been formed from deciduous hardwood vegetation. As noted by Sprengel, the acid type of humus is generally more stable to decomposition than the mild humus. In forestry, it is common to recognize *mor* and *mull* humus, names introduced by Müller (1878) in Denmark for differentiating the types of humus formed under different forest vegetation. It is generally noted that mor humus occurs under a heath vegetation or coniferous forest, hence it is comparable with Sprengel's acid humus. The concept of mor humus includes a non-

The Issue of Humic Matter 3

decomposed layer, lying on top of a F-layer (F = fömultningskit, German for fermentation), underlain by some structureless dark layer, called H-layer (H=humus). The undecomposed top layer is comparable to the fraction referred earlier as litter. On the other hand, mull humus is formed more under a deciduous hardwood forest, and hence, contains more bases, especially Ca, than mor. It is perhaps comparable to Sprengel's mild humus and generally believed to be affected by soil organisms, especially earthworms, in its formation. In mor humus, the fungi are credited for the decomposition of the raw material (Russell and Russell, 1950). Attempts have also been noticed to distinguish humus by its alleged function relative to soil and plant growth, and names such as *dauerhumus* (resistant humus) and *nährhumus* (nutrient humus) have been suggested by Scheffer and Ulrich (1960).

1.1.2 Concept of Humus in the Third Millennium

The concept of humus today has not changed drastically from the early theory. It still pertains to the decomposed part of the dead organic fraction in which, as defined in the preceding section, the structure of the original material has disappeared completely. In Soil Taxonomy, it is referred to as *sapric material*, from the Greek term *sapros*, meaning rotten and hence is the most highly decomposed organic fraction. Sapric materials are commonly dark gray to black in color and will change very little physically and chemically with time in comparison to the fibric and hemic fractions discussed above (Soil Survey Staff, 1990). A small though significant change in the definition of humus can be noticed perhaps as a result of the following ideas promoted by several scientists. The use of the term humus is suggested by Page (1930) to be dropped and to be replaced by the names *nonhumic matter* and *humic matter*. As opposed to the dark colored humic matter, fulvic acid and colorless decomposition products of organic matter are grouped by Page under the name nonhumic matter. In contrast, Waksman (1938) prefers to delete all the terms, and proposes the use of the name humus only for referring to the humic substances. Apparently both Waksman's proposal in retaining the name humus and Page's idea of distinguishing nonhumic matter and humic matter have been melded together to develop today's concept of humus. By current standards humus is distinguished into a nonhumified and humified fraction (Stevenson, 1994; Tan, 1998). The nonhumified fraction is an extended version of Page's definition and is now defined to include all substances released by decomposition of residues of plants and other organisms. It is composed of substances with definite characteristics, e.g., carbohydrates, amino acids, protein, lipids, waxes, nucleic acids, lignin, and many other organic substances. This part of humus is believed to contain in general almost all of the biochemical compounds synthesized by plants and other soil organisms. These substances are usually subject to further degradation and decomposition reactions, and are the main sources for the synthesis or formation of the humified fraction by a process called *humification*. Often, they are adsorbed by the inorganic soil components, such as clay, and they may also occur under anaerobic conditions. Under these conditions, the compounds above will be relatively protected from further decomposition reactions, enabling their accumulation in soils.

The discussion in the preceding sections indicates that humus is part of, but not equal to, the total organic matter content in soils. Considering it synonymous with SOM, defined as the total soil organic matter content (Schnitzer, 2000), will not only bring confusion in the concept of humus, but will also ignore or erase the achievements attained in the development of the many theories on humus. By using the term SOM indiscriminately, one wouldn't know whether it is referring to the total soil organic matter content or to just that part of the soil organic matter called humus. Ignoring the live fraction, soil organic matter or SOM includes the dead organic fractions at various degrees of decomposition as defined in section 1.1.1. It is very unfortunate of Schnitzer citing that Stevenson (1998) also considers humus synonymous with SOM. Special efforts are made, in fact, by Stevenson (1998) to distinguish humus from SOM in his section on 'modern-day concepts' of humus, though his references to and usage of SOM in the text are indeed confusing and misleading.

Aside from the issue above, the suggestion is made here to use names such as SOM only when it is necessary, since it opens a precedent for other scientists to use TOM for total organic matter, OM for organic matter and the like. The ensuing proliferation of acronymical names, such as SOM, TOM, DOM (dissolved organic matter), OM, TOC (total organic carbon), OC, DOC, is not only undesirable but also somewhat ridiculous. To be called a soil scientist is acceptable, and to be named a TOM scientist is very amusing, but who wants to be called a DOM scientist (Dutch dom = stupid, dumb).

1.2 CONCEPT OF HUMIC MATTER

The concept of humic matter is very confusing, since many people are using interchangeably the terms soil organic matter (SOM), humic substances, humic material, or the black decomposed organic substances, and the like. Since the beginning an exact definition of humic matter has been missing and only lately have efforts been made by a limited number of scientists to come up with a more precise definition of humic matter. Nevertheless, many authorities in humic acid science tend to stick to the name SOM, though according to the exact meaning of the term soil organic matter, the fraction called litter and/or the nondecomposed organic fraction are included.

As explained above, humus is defined today as a mixture of nonhumified and humified organic material. The humified fraction is identified by Christman and Gjessing (1983) as *humic material*. It is called here *humic matter*, a name used earlier by Page (1930) for the dark colored high molecular weight organic colloids as discussed earlier, but applied by Tan (1998) in analogy to the term organic 'matter.' Like soil organic matter being a mixture of nondecomposed and decomposed organic components, or clay being a collection of an assortment of clay minerals, so is humic matter composed of a variety of *humic substances*, e.g., humic acid, hymatomelanic acid, fulvic acid, and humin. In the German and Russian literature, humic matter is called humus acid or humussäure (Döbereiner, 1822; Scharpenseel, 1966; Orlov, 1985). This is the fraction that was assumed in the past to be soil humus (Russell and Russell, 1950; Waksman, 1938). Considering humus as equivalent to humic matter was very common in those years, and even now Flaig (1975) and Haider (1994), both prominent authorities on the subject, use the term humus and humic matter or humic substances interchangeably. Kumada (1987) adds to the confusion by using in his book the terms SOM (soil organic matter) and humus synonymously, whereas Schnitzer's (2000) statement that he personally, as a SOM scientist, prefers the use of the term SOM for humic substances, makes the issue worse.

The humic substances make up the bulk of humus and the nonhumified fraction is usually present in a relatively smaller amount (Tan, 1998; Stevenson, 1994). These humic compounds are the most chemically active compounds in soils, with electrical charges and exchange capacities exceeding those of the clay minerals. They are essentially new products in soils synthesized from the nonhumified compounds released during the decomposition of the plant and animal residue without or with the assistance of microorganisms. Like clay, humic matter is a building constituent of soils, and the process of its formation is called earlier humification.

Summarizing the above, it is perhaps correct to state that humic matter as defined today is a mixture of amorphous, polydispersed substances with yellow to brown-black color. The humic substances are hydrophilic, acidic, and high in molecular weight, ranging from several hundreds to thousands of atomic units or daltons. They originate from the decomposed organic fraction by *'new formation'* called humification, and are usually obtained from soils by extraction, fractionation and isolation procedures with basic and acidic solutions.

1.3 THE ISSUE OF ARTIFACTS

The issue of artifacts appears to have started early during the birth of the science of humic substances, and has increased since then in seriousness to become a major concept today. Criticisms were launched in the early days at the reality of crenic and apocrenic acids – old names for fulvic acids – isolated by Berzelius (1839) and Mulder (1862). Opponents argued that they were oxidized products of humic substances, and disputed the correctness of the chemical composition (Hermann, 1845; Kononova, 1966). With the accumulation of data

The Issue of Humic Matter 7

during the years, more and more people seem to resist considering laboratory products similar to natural compounds. In the beginning of the twentieth century the concept that humic substances are not definite chemical compounds at all, but are merely substances formed by the extraction procedures, has been firmly adopted (Waksman,1938; Bremner, 1950). This concept implies that they are in essence fake compounds or substances not present in nature. They are assumed to be heterogeneous in composition (Felbeck, 1965) and refractory in nature, hence they cannot be placed into a definite class of compounds as commonly executed in the classification of chemical compounds into categories, such as polysaccharides, proteins, amino acids, lignin, etc. (Gaffney et al., 1996). Because of this presumption, the notion is that humic substances do not exhibit clear molecular structures. As summarized by Clapp et al. (1997), they are gross mixtures of macromolecules, and two humic molecules in any batch will likely be dissimilar in nature to each other. Aiken et al. (1985) and Hayes et al. (1989) also indicate that these organics have not been formed biologically for performing specific biochemical functions, hence cannot be translated into specific functional terms.

Since to a large number of people the concept of humic acids has also been vague and confusing, especially during the period of 1970-1980, the problem seems to be exacerbated by an apparent identity crisis on what exactly humic matter is. It attracted at that time relatively little interest, and the few soil scientists and organic chemists struggling with the problem face opposition from fellow scientists who consider humic substances as 'dirt' hardly worth studying. Such resentment becomes so non-productive for advancing humic acid science that Schnitzer (1982) felt compelled to contest it in the International Congress of Soil Science in New Delhi. The recent founding of an International Humic Substances Society has evidently contributed to providing more exposure to humic substances in the quest for some basic legitimacy for their presence. The major scientists responsible for the establishment of this scientific society are R. L. Malcolm and coworkers from the US Geologic Survey, Water Division, and M. H. B. Hayes of the Chemistry Department, University of Birmingham, England. In its first meeting, several of the society scientists opted for the use of the term *operational compounds* for these substances, which reflects their formation due to specific analytical operations (Aiken et al., 1985). This concept has not changed with the dawn of the new century, since the latest published Proceedings of the Symposium of the International Humic Substances Society (Clapp et al., 2001) seems to use and enforce the idea of humic substances being operational compounds. Although this approach only sidesteps the issue of artifacts, the implication is that it provides some justification for further continued effort in the study of humic substances. Such a concept seems to be shared by other scientists, since it is used as a working hypothesis by members of the American Chemical Society in their humic acid research (Gaffney et al., 1996).

All the above raise questions why such great importance should be given to fake substances, or whether the current technology is incapable of identifying them properly. The statement given by Hayes and Malcolm (2001) in their most recent paper that it can be justified on the basis of scientific curiosity and the advancement of scientific awareness raises more questions than providing answers. Research efforts solely for scientific curiosity are seldom career oriented and are nonproductive for scientists at research institutions and universities, especially where the motto prevails of *'publish or perish.'* It is still a mystery why the abundance of humic substances and the role they play in soils and water environments, though firmly recognized, are not reasons enough for considering them real compounds in nature.

1.4 THE ISSUE OF REAL COMPOUNDS

With the advances in humic acid chemistry during the last two decades, another concept of humic substances, completely opposite to the one discussed above, seems to surface. Some are vocal about it whereas others remain more discreet. Though many of the proponents have not stated it specifically, their efforts to present especially a molecular structure show their convictions about humic substances being real compounds with definite compositions. As indicated by Hayes et al. (1989) a molecular structure as defined by a discrete elemental composition can be constructed only for a real compound.

The Issue of Humic Matter

The concept, considering humic substances as real compounds in nature, is not new in fact and finds support apparently from a greater number of people than expected as noticed from the following discussion. It originated perhaps with Achard (1786) and Berzelius (1839) when they started to study humic substances. In addition to these pioneers, Mulder (1840, 1862) believes that humic substances are definite chemical compounds characterized by a specific composition. Though the initial idea of assessing humic substances as real compounds was challenged several times, as discussed in the previous section above, the concept seemed to continue to simmer over the years, especially in Russia and Europe, the birthplace of humus and humic acids. It flares up again at the beginning of the twentieth century when new evidence presented by Oden (1914, 1919) and others provided the right fuel. Kononova (1966), who wrote a skeptical review about the early fights, was not exactly refuting the idea but was more concerned about and in disagreement with the elemental composition. In Kononova's opinion humic substances must contain N, as can be noticed in the hypothetical structure advanced by the author for humic acid. Orlov (1985) is more straightforward, indicating that the reality of humus acids must be accepted for many reasons. They are present in soils as well as in aquatic environments and in marine and lacustrine sediments. He indicates that they can be isolated from natural waters without using the usual procedures employing alkaline and acidic reagents. In his opinion, spectral characteristics are not only reproducible, but the spectral features are identical between extracted and nonextracted humus acids in whole soils. Among the German scientists, it is perhaps Flaig (1975) whose studies are based on the concept of formation of humic substances as really being chemical compounds in nature, characterized by specific molecular structures. He notes that the changing nature of their chemical composition is due to the dynamic or transient nature of humus – Flaig's term for *humic substances* – because of the never-ending decomposition and new-formation in nature. However, far more noteworthy in this respect is Ziechmann (1994), who states that 'humic substances are natural compounds existing in reality.' In his opinion humic substances exhibit specific characteristics that can be measured and that distinguish them clearly from other soil organic compounds, e.g., acetic acid.

Consequently, they are a separate group of compounds with a composition of their own. He believes that by accepting the division of humic substances into humic acid and fulvic acid the presence of a specific chemical composition is implied. An alternative concept is proposed by Ziechmann in dealing with the debate on chemical composition, which will be discussed as a separate section below. For the reasons discussed above and in view of the increased application of humic substances in other fields, e.g., technology, industry, pharmacy and medicine, Ziechmann questions the merits in defending any further the operational definition. In the United States, it is Stevenson (1994) who views humic substances as distinctive soil components, different from the biopolymers of plants and microorganism, although his statement that the term humic substances should be regarded as a 'generic name' is confusing the issue. However, Stevenson's opinion, that humic substances can be placed into a chemical category different from carbohydrates, protein, amino acids and the rest of the nonhumified substances, infers the humic substances to possess a definite chemical composition. Stevenson's dimer concept, revealing a molecular structure of humic substances to be composed of two monomers, is an extension of such perception and supports the concept of humic substances to be real compounds. This perception reinforces the opinion of Ziechmann (1994) as discussed above. Another prominent scientist in the Western Hemisphere indicating discreetly that humic substances are natural organic products on the earth surface is Schnitzer (1975; 1972). His efforts on constructing molecular structures show his convictions about humic substances being discrete natural compounds (Schnitzer, 1994). Together with Schulten, Schnitzer has presented several concepts on the molecular structure of humic acid assembled from products found in their degradation analyses (Schulten and Schnitzer, 1993;1995). Such an assessment of Schnitzer's view is in agreement with the opinion of Hayes et al. (1989) on molecular structures as the property of real compounds, as discussed earlier. However, it is perhaps the work of Schulten (2002; 2001) that has provided a mortal blow to the *'artifact concept.'* Schulten's brilliant presentation of molecular structures of humic substances appears to indicate that structure is no longer an issue. The presence of structures seems to be undisputable, and strongly suggests

The Issue of Humic Matter 11

the presence of real compounds. Therefore, considering humic substances as structureless artificial compounds created by the analytical extraction procedures may in the long run become very difficult to defend.

1.5 THE ISSUE OF CHEMICAL COMPOSITION

The term chemical composition can refer to the elemental composition, the functional group composition, and the many compounds considered 'building blocks' that make up the molecule of the humic substance, though according to Hayes and Malcolm (2001), humic substances do not contain peptides, sugars, nucleic acid and the like. The following discussion will be confined mostly to elemental composition, and the other issues on composition will be discussed in separate sections later in the book.

In the aforementioned sections, a controversy was presented on the problem of a characteristic composition, with one group claiming there is none, whereas the other side insisted that humic substances exhibit a definite chemical composition that distinguishes them from other organic compounds. The first group maintains that the refractory nature allows for the isolation of humic fractions with a variety of compositions. However, none of the chemical compositions are characteristic for the humic substances, since any composition can be obtained, depending on the isolation and fractionation procedures. Among the staunchest proponents in this group are Hayes et al. (1989), who believe that humic substances cannot have unique molecular weights. The elemental composition, considered by these authors the most fundamental property, can only be exhibited by a discrete chemical compound. It is their firm belief that humic substances, being composed of complex nonstoichiometric mixtures, cannot be represented by empirical formulas, since their elemental compositions are only the average values of large clusters of molecules. Such an idea is in Stevenson's (1994) opinion very unfortunate for teaching and learning purposes. The use of a simple structural unit would be of advantage in explaining the properties and reactions of humic compounds to students. Ziechmann (1994) declares it as a 'dead-end street,' which leaves investigators with no alternatives. He suggests abandoning further searches on chemical composition when carried out with the preconception of the absence of a molecular structure. This is part of the reason why no progress has been achieved or even will be obtained in the future.

On the other hand, the group that believes in the presence of a chemical composition has yet to determine the composition characterizing the humic substances, which up till now has allegedly eluded all efforts, though chemical formulas for humic substances have been presented as early as in 1839. Formulas such as $C_{24}H_6O_{-2}$ and $C_{24}H_{12}O_{16}$ were presented by Berzelius (1839) for his crenic and apocrenic acids, the fulvic acids of today. More recently, a chemical formula of $C_{308}H_{328}O_{90}N_5$ has been reported by Schnitzer (1994) for a humic acid molecule. Two years later Schulten (1996) claimed to have determined by computer modeling monomers of humic acid with a composition of $C_{308}H_{335}O_{90}N_5$, which is almost the same as Schnitzer's formula. Schulten's formula shows the humic acid monomer to correspond with a molecular weight of 5478 and a nitrogen content of 1.28%. That such a composition may apply to artificially prepared materials is without doubt, but whether it characterizes a really humic acid molecule is still subject to argument. This will be discussed in more detail in Chapter 5.

According to several scientists the refractory nature of humic substances, showing an apparently endless variation in composition, is probably more the result of their natural formation in soils. As mentioned earlier, they are transient organic components in soils (Flaig, 1975; Ziechmann, 1994), forming a dynamic system that decomposes and is formed again continuously, changing their composition in the process. The problem becomes even more complex because of the many reactions and interactions, adding and losing organic constituents during the humification process. Consequently, the nonhumified compounds and humic substances in all phases of formation and degradation make up the humus mixture in the soil, which is sampled and extracted for humic substances. This accounts for the extreme difficulties in the identification of a characteristic monomer of a humic substance, e.g., humic acid. Which of the substances isolated should then be considered to exhibit the characteristic composition of humic acid? They are all real organics formed in nature and the extracted preparations, labeled operational compounds or artifacts, are nevertheless noted to be reproducible in behavior and properties (Orlov, 1975). The issue is perhaps created only by a disagreement among scientists on what to call humic substances. A good analogy can perhaps be given with protein, which can be extracted from soils and plants. Everybody agrees that they are discrete substances, built up from a variety of amino acids and other organic compounds. Carbohydrates as well as nucleic acids may participate in their formation as noted by the presence of gluco- and nucleoproteins in nature. Therefore, they exhibit a multitude of formulas with different compositions, since their formulation depends on the various types of amino acids and other components, and their configuration in the protein molecule. This is also true of humic acid, since the data presented so far underscore the humic molecule to contain phenolic or benzene, amino acid and carbohydrate components (Flaig, 1988; Schnitzer, 1994, 1986; Christman et al., 1989). The controversy on chemical composition can perhaps be eliminated by accepting now the various compositions noted in the so-called operational humic substances as the result of a multitude of interactions and arrangements of the different types of components much in the same way as protein molecules are formed. The different phases caused by the humification and decomposition processes, though very important, may pose only minor problems, since most of the components are known today, with each exhibiting a discrete empirical formula. An example of how to apply an analytical chemical composition of a humic substance has been presented by Steelink (1985), who uses atomic ratios, as suggested earlier by Van Krevelen (1961) and Visser (1983), in the determination of a formula. Another system is presented by Orlov (1985), who claims that the molecular weight of humic substances can be calculated from their mean elemental composition. In this concept, any humic fragment or phase, containing a residue of amino acid, is assumed to have a *minimum molecular weight*, which is obtained by conversion of the total nitrogen content to one atom of N in the formula. More details about Steelink's and Orlov's concepts will be provided in Chapter 5.

CHAPTER 2

THE NATURE AND DISTRIBUTION OF HUMIC MATTER

2.1 CONCEPTS AND HISTORICAL BACKGROUND

2.1.1 Historical Concepts

Humic matter as defined earlier is composed of a variety of substances that can be obtained by fractionation on the basis of their solubility in alkaline and acidic solutions. The first humic substances isolated were compounds soluble in bases, acid solution and water. They were extracted in Sweden by Berzelius (1839), who assigned them the names of crenic and apocrenic acids, the fulvic acids of today. This investigation was continued by Mulder (1840), a former student of Berzelius, who in the following years isolated additional humic fractions. On the basis of color and solubility, Mulder (1862) classified them into (1) crenic and apocrenic acid, the yellowish to brown fractions soluble in water, (2) ulmic acid and humic acid, the brown and black fraction, respectively, soluble in alkali but insoluble in acid, and (3) ulmin and humin, the fractions insoluble in alkali, acid and water. Mulder's accomplishment was followed twenty-seven years later by the discovery of hymatomelanic acid, isolated by Hoppe-Syeler (1889) as the ethanol-soluble fraction of humic acid. Since then no further achievements of major importance can be noticed in the study and isolation of humic substances until Oden's (1914; 1919) concept surfaces at the start of the twentieth century. Considered by many scientists as based on a more solid scientific foundation, it is in essence a revision of Mulder's classification. Oden recognizes (1) fulvic acid, which replaces the use of the terms crenic and apocrenic acid, (2) humic acid for the fraction soluble in alkali and insoluble in acid, (3) hymatomelanic acid, a name used earlier by Hoppe-Syeler (1889) for the humic acid fraction soluble in ethanol, and (4) humus coal, for replacing humin and ulmin.

2.1.2 Concepts in the Early Twentieth Century

Oden's theory, though challenged many times during the years, has set the stage for the development of the concepts and types of humic substances followed today. It triggered in the beginning a flurry of investigations and the *ligno-protein* theory was introduced during this period by a number of people (Fuchs, 1930a, b, c; 1931; Hobson and Page, 1932a, b, c; Waksman, 1938), a concept that has dominated humic acid chemistry and formation till today. This theory assumes humic matter to be the product of reactions mainly between lignin and protein, two important components of plant tissue. Although it is a very viable theory that is used by Flaig (1975) as the foundation of his humic acid concept, other theories were presented when more became known about humic acid chemistry toward the end of the century. This will be discussed in more detail in Chapter 4 on biochemistry and formation of humic matter.

The new burst of research activity has also resulted in the development of new names, often amounting only to noise in the nomenclature of humic substances. Names, such as *rotteprodukte* (rotten or decomposed products) and *echte huminsäuren* (real humic acid), have been proposed by Simon and Speichermann (1938). Springer (1938) added *humoligninsäuren* and *lignohuminsäuren* (lignin- or lignohumic acid) to the confusion. Attempts, especially of fractioning humic acid further into several subtypes, have come up with more names. By manipulation of the fractionation procedures, it has been reported that humic acid can be subdivided into α- and β-*humic acid*. (Russell and Russell, 1950). The name α-humic acid is used for the initial or original humic acid fraction before ethanol extraction for isolation of hymatomelanic acid, whereas the humic acid residue, remaining after ethanol extraction, is called simply humic acid. The name β-humic acid has been reserved by Waksman (1936) for a precipitate produced after adjusting the pH of a fulvic acid solution to 4.8 with NaOH. According to Stevenson (1994) this β-fraction is an Al-humate with properties similar to an organic substance obtained by Hobson and Page (1932a, b, c) named *neutralization fraction*. Not only is such a nomenclature very confusing, but the identification of a fulvic acid fraction as a humate is questionable. It is also difficult to call it an *Al-fulvate,* since it is insoluble in acid condition. Hence the only alternative is the possibility that the separation of humic acid from fulvic acid in the soil extract has been conducted improperly or a proper fractionation procedure is indeed unavailable. Consequently, the analysis is fraught with many errors or uncertainties, justifying claims for the production of artifacts. Although several people agree with α-humic acid as defined above, it makes more sense if this name is assigned to the insoluble part of humic acid remaining after ethanol extraction. This residue has changed in composition and it seems more reasonable to assign it the symbol α and retain the name humic acid for the original substrate before ethanol treatment. Hence, humic acid can be fractionated by ethanol into an α-fraction and hymatomelanic acid. In analogy to the above, reference can be made to the division of humic acid into a *brown* and *gray fraction* that has attracted considerable attention in Germany. By using neutral salt solutions, Springer (1938) succeeded in separating humic acid into a (1) brown fraction (Braunhuminsäure), soluble in NaCl, and (2) gray fraction (Grauhuminsäure), insoluble in NaCl. The brown humic acid is said to be highly dispersible, contains a lower carbon content, and according to Stevenson (1994) has characteristics of humic matter in peat and in brown coal. However, Kononova (1966) is highly critical of humic matter originated from coal or peat, since these two materials are formed in anaerobic conditions completely opposite to the aerobic system present in soils responsible for formation of soil humic matter. Springer's gray humic acid has a

The Nature and Distribution of Humic Matter 17

low degree of dispersion, and is easily coagulated. In Stevenson's opinion, it looks similar to the humic acid in mollisols.

2.1.3 The Dawn of Modern Concepts

With research interest declining among soil scientists facing resentment due to an apparent identity crisis on humic acids late in the twentieth century as discussed in Chapter 1, scientists from a wide variety of other disciplines have taken the lead in humic acid research. Chemists, geochemists, hydrologists and environmentalists have become fascinated by the ubiquitous presence of humic matter in the ecosphere. In contrast to most soil scientists, they recognize its profound role in environmental issues, and its effect on migration and immobilization of industrial and nuclear waste and other pollutants. In industry, medicine and pharmacy, humic substances are recognized as potential sources for production of valuable chemicals. They are considered commercially viable to be used as surfactants and as drilling fluids in oil exploration, as well as in medicines for human health. Research in humic matter took a sharp turn away from soil science to make rivers, lakes and oceans the centers for explorations and investigations of humic substances. The result is that new humic compounds, assigned exotic names, have been discovered, enlarging our concept of humic matter. The name *copropel* is presented for a humic substance, labeled as humus by Swain (1963) and Stevenson (1994). It has been formed from the decomposition of microscopic plants in eutrophic lakes and marshes. A black mass of humified material located at deeper hypolimnetic areas of lakes and bays is called *sapropel*, whereas a pondweed type of sapropel, believed to originate from cellulose-rich plants, is called *förna* by the authors above. A deposit in dystrophic lakes consisting of an allochthonous precipitate of humic acid and detritus is referred to as *dy*. Marine slime resulting from settled decomposed plankton detritus is called *pelogoea* and an amorphous, gummy accumulation of humic substances beneath or within peat bogs is *dopplerite*. Recently, less exotic names have been used for humic matter present in the water medium. It is known by geochemists under the collective name of *aquatic humic materials* as opposed to *terrestrial humic materials* for humic compounds in soils (Christman and Gjessing, 1983).

The few soil scientists, who have continued with humic acid research, have also yielded some results. An alleged new humic acid fraction, identified by the name *green humic acid*, has been isolated by Kumada and Sato (1962). Chromatography using a cellulose column separated a humic acid extract of a spodosol into a green and brown fraction. The green fraction, called green humic acid, is believed to be derived from a fungal metabolite (Kumada and Hurst, 1967) and has attracted considerable attention, especially in Japan and New Zealand. In Japan, much value has also been placed in the use of visible light spectrophotometry in the identification of humic substances. The nature of absorption spectra and values of $\Delta \log K$ (= $K_{400} - K_{600}$ in which K = extinction) are applied to distinguish humic acids into four major types, e.g., types A, B, R_p, and P (Kumada, 1965; 1987; Kumada and Miyara, 1973; Yoshida et al., 1978). The P-type of humic acid produces, after separation by gel filtration and column chromatography with cellulose powder or sephadex, a P_b (brown) and P_g (green) fraction. The P_g humic fraction corresponds to the green humic acid discussed above. However, the existence of green humic acid is later rescinded by Kumada (1987), who considers the name as incorrect since humic acid is by definition brown to black in color. In his opinion the green fraction is an impurity commonly co-extracted with the brown (P_b) fraction.

In the United States, with many of the prominent authorities adhering to the concept of humic substances being operational compounds, no new discoveries have been noticed in recent years. Though not really a major breakthrough, Stevenson's (1994) suggestion for distinguishing *generic humic substances* should perhaps be mentioned in all fairness, though this term had been recognized earlier by Kononova (1966) and Kumada (1987). The name of generic fulvic acid is used by Stevenson for fulvic acid purified by the XAD-resin procedure. Accordingly, he believes that fulvic acid can be distinguished into a (1) generic or true fulvic acid, obtained by purification with amberlite-XAD resins, and (2) fulvic acid, obtained after purification using conventional ion exchangers and dialysis procedures. However, the analogy presented by Stevenson in reference to the

The Nature and Distribution of Humic Matter

subdivision of humic acid by MacCarthy et al. (1979) makes the generic concept a very confusing issue. Peat humic acid is in fact separated into two subfractions by MacCarthy and coworkers using a pH gradient elution technique, whereas Stevenson's generic fulvic acid is just a XAD-resin purified version of fulvic acid, involving no separation into subfractions at all. Consequently, the correctness of using the term 'generic' is still open for questions, whereas the subdivision of humic acid as discussed above is also suspect. More convincing research data are needed differentiating unequivocally the generic from the conventional type of fulvic acid. No supporting data have been presented confirming Stevenson's contention that generic fulvic acid is lower in carbohydrate and peptide contents than conventional fulvic acid.

In summary it can be stated that several new concepts on humic matter and a variety of new humic substances have been presented or discovered toward the modern era. This has no doubt broadened the concept of humic matter. Some of them have been used occasionally today, though several scientists tend to consider them only of academic importance. However, Waksman's (1938) proposal to delete all the names and replace them by humus has not found wide acceptance. On the other hand, Oden's concept on humic acid, fulvic acid and hymatomelanic acid and Mulder's idea of humin seem to have weathered all criticisms. The trend can be noticed that they are widely used today by the majority of scientists, though reluctantly by some. Names such as green humic acid, gray and brown humic acid have been used sometimes, depending on research purpose and interest.

2.2 DISTRIBUTION OF HUMIC MATTER

The distribution of humic matter is not limited to the soils ecosystem and to climatic conditions. Thought at first to be present only in soils, humic matter is currently assumed to be the most widely distributed organic carbon containing material on the earth's surface. It is present in soils, in water of streams, lakes and the oceans, and in their foam and sediments, from the tropics to the arctic regions. Its presence as a major constituent of the huge deposits of peat, lignite or leonardite, coal and oilshale adds to the dimension of its wide occurrence in the world. Geochemists are even of the opinion that the greatest storehouses of humic matter are the oilshales (Swain, 1975). Though most of the humic matter is a natural product synthesized in the environment, some are now assumed to be *anthropogenic* in origin, such as in polluted waterways, drainage ditches, and sewage ponds or sewage lagoons. Especially in the 'Old World,' such as in Europe, anthropogenic humic matter, identified as humus of harbor and city agglomeration sediments has started to become recognized (Ciéslewicz et al., 1996). No doubt, such types of humic matter are also abound in other parts of the world, where stable civilizations have had the opportunity to accumulate a lot of organic waste during the centuries.

2.2.1 Humic Matter in Soils

Humic matter occurs in all kinds of soils since it is the major fraction of soil humus. Though variations as to its content can be noticed due to differences in climate and drainage, humic matter can be found in soils from the lowlands to high in the mountains of warm tropical to frigid arctic climates. It is also present in humid region to arid region soils. Famous for their high organic matter contents are mollisols, soils under grass vegetation of the semihumid regions. Contents have been reported as high as 5 to 6% in terms of organic carbon (Stevenson, 1994). This is equivalent to approximately 9-10% organic matter, and half of this is estimated to be humic matter. The latter has a composition, characterized by a humic acid content slightly dominating that of fulvic acid, as noticed by its fulvic acid/humic acid ratio ranging from 0.9 to 0.6 (Tan, 1978). Another soil with a similar high humic matter content is the andosol, a soil occurring in the humid tropics to the arctic regions (Arnalds et al., 1995; Tan, 1984; Theng, 1980). However, the composition and type of its humic matter differ markedly from those of the mollisols. The fulvic acid content makes up more than half of the humic matter in andosols, which is in sharp contrast with that in mollisols. The fulvic acid /humic acid ratio is often noticed to range between 4.0 -1.0 in andosols of the humid tropics (Tan, 1965; 1964). Another marked difference is that the humic matter in andosols is present in close association with Al and allophane, whereas the humic matter in mollisols is more likely present as Ca-humate and Ca-fulvate. Large amounts of humic matter are also noticed in spodosols, where they are concentrated in the B_h or spodic horizons. This is in contrast to mollisols and andosols where the A horizons contain most of the humic matter. The B_h horizon deposit of humic matter is often so thick that it becomes a valuable source for commercial humate production as is the case in Florida (Lobartini et al., 1992; Burdick, 1965). Spodosol humic matter has been a favored material for investigations in Canada, where the results are taken to apply also for humic matter from other soils (Schnitzer and Khan, 1972; Schnitzer, 1972; 1976). Judging from its formation due to leaching from the A and E horizons, the general opinion is that spodosol humic matter is composed of large amounts of fulvic acids. However, recent studies with a spodosol, on the border of Georgia and Florida, reveal its humic matter to have a composition characterized by a FA/HA ratio = 0.13, suggesting a humic acid concentration 10 times higher than that of fulvic acid (Lobartini et al., 1991). This finding supports an earlier report for the Unicamp Company in Florida showing the Florida source of its humate products to contain 91.3% humic acid and 8.7% fulvic acid (Tan et al., 1988). Since spodosols are very acidic soils, conditions favoring dissolution of large amounts of Fe and Al, the humic substances are mostly in the form of Fe- and Al-humates and fulvates.

Other groups of soils containing humic matter are the ultisols and oxisols, soils generally low in organic matter content due to a rapid rate of decomposition. The oxisols of the humid tropics are notorious for their low organic matter contents, with contents often reported to be as low as 1% organic carbon. The humic matter in these soils is often noticed to contain more fulvic acid than humic acid and these substances are assumed to be present as Fe- and Al-fulvates or humates (Tan, 1978). However, the lowest organic matter contents are detected in aridisols, soils of the arid regions and the sandy soils in the deserts. Because of deficiency of water, biomass formation and decomposition reactions in these dry regions are very limited.

2.2.2 Humic Matter in Soils of the Wetlands

The soils discussed above are generally well-drained, and the aerobic conditions encourage the occurrence of rapid decomposition processes. However, under the influence of poor drainage, such as occurring in marshes and swamps or in general in soils of the wetlands, anaerobic decomposition prevails (Tan, 2000). Due to lack of oxygen, the decomposition process is very slow, if not inhibited, and incomplete, and hence contributes to accumulation of huge amounts of organic matter. Many of the wetlands and some of the lake areas are eutrophic, encouraging an excessive growth of aquatic weeds and other plants. The latter provide an overabundance of dead organic residue filling gradually the inundated or wet areas. The partly decomposed organic matter will eventually develop into bogs, peat, and muck, which are often believed to be precursors for formation of coal and ultimately fossil fuel (Hatcher et al., 1985). When conditions are favorable, sapropel, copropel and the like may also develop into peat and bogs. The only conditions required are anaerobic environments for the accumulation and development into peat and its eventual conversion into coal.

Peat deposits are also not limited to climatic conditions and can be found all over the world where large amounts of biomass are available and where decomposition of organic residue is inhibited. They are distributed from the tropical Amazon basin of Brazil and the coastal regions of Sumatra, Indonesia, to the Baltic coast in Europe and tundras in Alaska and other arctic regions. In addition to excess water, the frigid temperature in the tundras is another reason for inhibiting decomposition of organic residue. In his opening address at the 1972 International Meeting of Humic Substances at Nieuwersluis, The Netherlands, Golterman (1975) underscored the importance of peat as the producer of humic substances affecting the living environment of the Dutch people. Called sometimes peatlands or mires in Europe and Canada and known as histosols in the United States, these organic deposits are believed to cover an area of 500 million hectares worldwide, representing an organic carbon reserve of 10^{12} metric tons (Mathur and Farnham, 1985), and only the organic carbon reserve in oilshale is believed to exceed this amount. Swain (1975) presented data showing the earth's crust to possess a total organic carbon reserve of 19×10^{15} metric tons, and most of it (18×10^{15} metric tons) is stored in oilshales.

In contrast to a mineral soil system, where the organic matter content makes up only a small fraction compared to the mineral fraction, the organic matter content is an integral and substantial part of peats or histosols. By definition organic soils contain >80% organic matter and <20% mineral matter (Brady and Weil, 1996). A more complex definition, used in Soil Taxonomy (Soil Survey Staff, 1990), indicates that organic soils must have an organic carbon content >18% in the presence of > 60% clay, or > 12% in the absence of clay. Most of the information indicates that peats contain large amounts of humic acids (Zelazny and Carlisle, 1974; Kononova, 1966), though occasionally it is reported that humic matter in peat is composed mostly of fulvic acids (Schnitzer, 1967). The humic acid content appears to increase from peat to muck, with the more humified muck noticed to contain its humic matter mostly in the form of humic acid (Preston et al., 1981). The elemental composition, spectral characteristics and other chemical properties of peat humic acids are believed to be similar to those of humic acids in mineral soils (Mathur and Farnham, 1985). However, as indicated before, Kononova (1966) was highly critical of humic matter originated from peat, since the material has been formed in anaerobic conditions, completely opposite to the aerobic system present in soils.

2.2.3 Humic Matter in Aquatic Environments

It is now an established fact that the distribution of humic matter is not limited to soils, but it has also been detected in streams, lakes, and oceans and in their sediments. These humic substances may influence ground water properties and are considered to play an important role in the geochemical cycle of organic carbon in aquatic systems. They are distributed in what they call dissolved organic matter (DOM) or dissolved organic carbon (DOC), which according to Aiken (1985) can be distinguished into two big groups, *hydrophobic* and *hydrophilic groups*. Each of the two groups can be subdivided again in acidic, basic, and neutral subgroups. Humic substances form the bulk of the *hydrophobic acidic fraction* of the DOC, with concentrations reported to amount to 20 µg/L in ground water and to ≥30 µg/L in surface water (Thurman and Malcolm, 1981). The DOC in water from different lakes in the United States is reported by Steinberg and Muenster (1985) to contain 80% humic substances. The concentration is noticed to be 10 to 20 times greater in natural aquatic foam of rivers, lakes and sea than in the water itself, and 90% (by weight) of the DOC in foam is humic matter (Mills et al., 1996). The presence of DOC or DOM often lends to make the water yellow or darker in color. Large amounts of dark brown, organic rich water can be seen all over the world flowing from swamps and poorly drained areas into creeks and rivers, especially after a rainfall. Named *black water* by Tan et al. (1990), it is noticed flowing in the tributaries of the Amazon river in Brazil, where the Rio Negro is one of the most striking examples, and in the coastal streams of Sumatra and Papua running between the vast expanses of peat deposits, to the rivers, lakes and marshes in central Africa, Scandinavia and the tundras in the arctic region. Many of the coastal plain streams and swamps of the Southeastern United States carry also dark colored water, attributed to the presence of organic substances in solution.

Identified at first as *gelbstoff* (German for yellow material) by Kalle (1938) or yellow organic acids by Shapiro (1975) and Dawson et al. (1981), the dissolved organic matter has attracted gradually a lot of research attention. It has become the subject of extensive investigations by many chemists and hydrologists, especially due to the fear of a health hazard from the chlorination of drinking water containing colored substances (Bellar et al., 1974; Aiken et al., 1985). As a result of analysis of samples from streams in Alaska, the Georgia-Florida border, Washington and California, Lamar (1968) believed the yellow organic substances to be complex polymeric hydroxy carboxylic acids. According to Beck et al. (1974), the spectral characteristics of these complex acids resemble closely those of fulvic acids. More recent investigations by Tan et al. (1990) confirm the presence of humic matter as the reason for the dark color. As a result of further investigations using black water samples of the Okefenokee swamp, Satilla river and Ohoopee river in the Georgia coastal plain of the United States, Tan et al. (1991) show the composition of the humic matter to be characterized by a fulvic acid/humic acid ratio = 2.6. This means that fulvic acid is the dominant fraction, making up more than twice the concentration of humic acid in aquatic humic matter.

Mixed opinions exist on the nature of these aquatic humic substances. A number of authors believe that stream humic matter is similar in nature to its counterpart in soils (Stevenson, 1985; Beck et al, 1974; Shapiro, 1975; Black and Christman, 1963). Their chemical properties are reported to be comparable to those of soil humic substances (Mayer, 1985; Steinberg and Muenster, 1985). However, other scientists are of the opinion that stream humic matter is different from soil humic matter (Malcolm, 1985). Another possibility is that some of the stream humic acid is of terrestrial origin, but after transfer into the streams and lakes has been subjected to further distinct changes in the anaerobic environment (Jackson, 1975). This issue has resulted in the distinction of *autochthonous* and *allochthonous* varieties of aquatic humic matter. The autochthonous types are formed from indigenous aquatic organisms, whereas the allochthonous variety originates from the soil. Such a division is especially important for humic matter in lagoons, bays, and other lake or marine environments where substantial amounts of plankton, algae and kelp can grow. It is believed that in big lakes and oceans most of the humic matter is autochthonous and only on the beaches or coastal zones and in the lagoons or estuaries is there a mixture with soil humic matter. In the ocean remote from any terrestrial influence the gelbstoff mentioned earlier is assumed to be formed from dead phytoplankton (Kalle, 1966). Ishiwatari (1985) indicates that humic matter derived from phytoplankton cells will undergo diagenetic changes when deposited at lake bottoms. The deeper it is buried in the lake bottom sediments with time, a gradual decrease tends to occur in humic acid and fulvic acid content, accompanied by a corresponding increase in the amount of humin. Differences in aquatic systems have also been cited recently to yield different types of aquatic humic matter. Humic substances from dystrophic environments are believed to be more aromatic than those from eutrophic and mesotrophic systems (Klavins and Apsite, 1997).

2.2.4 Humic Matter in Geologic Deposits

Humic matter occurs also in the vast geologic deposits, such as lignite, coal, oilshale, and fossil fuel. Extensive deposits of lignite in the United States are located in Utah, North Dakota, Idaho, New Mexico and Texas. The lignite in North Dakota is often called leonardite. High quality grade lignite or leonardite may contain 80% to 90% humic matter. Though commercial grade lignite is on the average composed of 60-70% humic matter, contents of 30% to 60% have been reported occasionally (Stevenson, 1986).

As discussed in the aforementioned sections, these geologic deposits rich in humic matter have been formed in time from bog, peat, and muck. The transformation process is called *diagenesis* by geochemists (Hatcher et al., 1985), but in soil science such a transformation is called *metamorphism*, since high pressures and temperatures are required to induce the needed compaction and drastic chemical changes (Miller and Gardiner, 1998; Tan, 2000). During the conversion of peat into coal, it is believed that humic acid is undergoing drastic changes. The theory presented by Van Krevelen (1963) and Stach (1975) assumes that humic acid is converted into humin by a *condensation* process during the transformation of peat into lignite and subbituminous coal. In soil science the reaction is called *polymerization* instead of condensation. The amount of humin is considered to increase during further *coalification* into bituminous coals, anthracite, and finally graphite. Breger (1963) and Hatcher et al. (1985) recognize two types of coal derived from peat and peat bogs: (1) humic coal formed from peat containing humic substances derived from vascular plants, and (2) sapropelic coal, formed from peat derived from sapropel composed of algae remains.

The geochemical condensation and coalification theory seems to suggest that lignite and coal deposits are rich in humin. Though Hatcher et al. (1985) admitted that more data are required to confirm it, these authors indicate that with increasing coalification, the material is more likely to become less extractable with dilute alkali solutions. They expected less humic acid and fulvic acid to be extracted with increasing rank from peat, lignite to coal. However, more recent investigations are unable to confirm the contention above. As a result

The Nature and Distribution of Humic Matter

of analysis with commercial sources of humates, Lobartini et al. (1992) show a commercial grade lignite from North Dakota to be composed of substantial amounts of humic acids. The humic acid content is 99.0% in contrast to that of fulvic acid, which amounts to 1.0% only. Earlier investigations yielded similar results. Lignite or leonardite samples from eight different locations in Utah and North Dakota were also noted on the average to be composed of 99.5% humic acid and 0.5% fulvic acid (Tan and Rema, 1992; 1993). At the present stage of knowledge, the reason for such low contents of fulvic acids is not known yet. While awaiting more information, it can perhaps be speculated that most of the fulvic acids, if not all, have been polymerized into humic acids, whereas the remainder was lost by leaching during the time of deposition.

For completeness the theory embraced by geochemists that humic matter is present in substantial amounts in a variety of prehistoric sedimentary rocks should perhaps also be mentioned. Precambrian (approximately 4,700 millions years ago), lower and upper Paleozoic and Mesozoic to Cenozoic (recent life, today) rocks are considered to contain humic acids (Swain, 1975). Swain believes that organic soils may have become widespread by the appearance of large terrestrial plants with extensive root systems in those prehistoric days. The increased occurrences of petroleum and coal deposits during the upper Paleozoic period have been used by the author above as markers or tracers for organic matter production and deposition in the past. This type of humic matter can be called geologic humic matter as suggested by Stevenson (1994) or better *paleontologic humic matter*.

2.2.5 Humic Matter in Agricultural, Industrial and Municipal Waste

Agricultural, industrial, and other operations in our modern society produce large amounts of waste, some inorganic and some organic. The soil is traditionally the site for disposal of all these wastes. People have been discarding waste since prehistoric times, but there was little concern in the old days about pollution and degradation of the environment because the human population was still small and there was plenty of space on earth for the amount of waste produced. However, with population growth and the revolutions in industry and agriculture, huge amounts of waste and a variety of new types of pollutants have been produced. Because of concern about contamination and especially the degradation in environmental quality starting in the 1970's, a number of waste disposal methods were explored. Today, waste is buried in landfills or dumped in the sea, whereas the organic part is incinerated or used again on farmlands. When waste is used again as soil amendment or for any other useful purpose, it is no longer considered waste but rather a valuable resource material (Tan, 2000). This is the case today with poultry litter, sewage sludge, and compost. The poultry industry in the southeastern United States is estimated to produce 11 million tons of poultry litter each year (Tan et al., 1975). In the state of Georgia alone, the official 1975 assessment amounts to a production of 720,000 metric tons of manure at the time of removal from the point of production of broilers. To this amount should be added 450,000 metric tons of commercial hens' manure, 79,200 tons of breeders manure, and 100,000 tons of turkey manure (Muller, 1975). The ever-expanding poultry industry, resulting in more increases in manure, may result in sanitary disposal problems, though part has found application as soil amendments (Campbell, 1973). Aside from being an organic fertilizer, poultry manure is a valuable source of humic matter. The water soluble fraction has been identified to carry properties similar to those of fulvic acids (Tan et al., 1975; 1971). Due to the huge amounts of poultry manure available each year, its potential as a humic acid source is comparable to that of lignite or peat deposits. Not much information is available on this matter today.

Another type of waste containing humic matter is sewage sludge, produced in amounts even more enormous than poultry manure. The figures reported by Larson and Schuman (1977) of 5,680 m^3 of sewage sludge barged daily from Chicago, IL, to its dumping site, and 5,450 m^3 of raw sludge transported daily by gravity through open ditches to landfills in Melbourne, Australia, demonstrate the colossal dimension of production. Sludge, as a reservoir of humic substances, attracted considerable research attention during the period of 1970-

1980. The NaOH extract yields substances with properties similar to those of humic acid and fulvic acids (Tan et al., 1971). The fulvic acid fraction was the subject of extensive investigations for its metal chelating power by Sposito et al. (1978; 1981). Senesi and Sposito (1984) regard sludge derived fulvic acid equivalent in its complexing capacity to soil fulvic acid.

Other examples of human-made wastes containing humic matter are compost and deposits of cattle manure in feedlots. The so-called CAFOS, for confined-animal feeding operations, in Texas are notorious for animal-waste production, creating annually an estimated 120 million metric tons of manure (Ivins, 2001). These huge amounts of animal wastes, not confined only to cattle manure, but also including hog and chicken manure, are perhaps sources for fulvic acids that being soluble are reported fouling the water supply in the Lake Waco area, the Playa lakes in the high plains, and east Texas. Sitting right on top of the recharge zones of the Paluxy, Trinity and Ogallala aquifers, these deposits are too close for comfort. Waste from food processing plants is an additional source of humic substances. Especially beer breweries yield large amounts of residues that after fermentation and decomposition become potential sources of humic and fulvic acids. Wastes from canning industries are not important as a source of humic substances, since most of them are too valuable and can be recycled into animal feed or into fertilizers. Finally, the thick dirty black material burdening heavily polluted drainage ditches, rivers, and lakes, which can be noticed in developing as well as in developed countries, should be mentioned. This often foul smelling black mass is an often forgotten source of humic matter. Unfortunately, not much is known yet on the humic matter contained in man-made wastes.

2.3 CLASSIFICATION OF HUMIC MATTER

From the preceding sections it is perhaps possible to recognize five general groups of humic matter, e.g., soil humic matter, aquatic humic matter, humic matter from wetlands, geologic humic matter, and anthropogenic humic matter.

2.3.1 Terrestrial or Terrigenous Humic Matter

This was explained earlier as the group of humic matter in soils, composed of substantial amounts of humic acid and fulvic acids. According to the ligno-protein theory the humic substances are mainly lignoprotein complexes. Lignin, considered as the fundamental constituent of these humic substances, can be distinguished into three major types on the basis of its monomers: (1) softwood lignin, characterized by coniferyl alcohol, derived from softwood or coniferous plants (gymnosperm), (2) hardwood lignin, characterized by sinapyl alcohol, common in hardwood (dicotyledonous angiosperm) vegetation, and (3) grass or bamboo lignin, characterized by coumaryl alcohol, common in grasses and bamboo (monocotyledonous angiosperm). Consequently, the suggestion is presented for a possible division of soil humic matter into three subgroups (Tan, 1998):

Softwood Soil Humic Matter. - This type of humic matter has been formed from softwood lignin monomers, and is structurally characterized by coniferyl alcohol.

Hardwood Soil Humic Matter. - This is humic matter made up mostly of hardwood lignin monomers, and structurally characterized by sinapyl alcohol.

Grass or Bamboo Soil Humic Matter. This is the humic matter formed from grass or bamboo lignin, and structurally characterized by coumaryl alcohol.

It should perhaps be realized that a sharp distinction among the three subgroups above can only be established in soil humic matter affected by monocultural environments. In nature, it is common to have a hardwood forest mixed with a conifer stand and an underbrush vegetation of grasses and bamboos. In such a complex ecosystem, the humic matter will likely be more of a mixture of the three groups above. The idea is presented here to promote more research on the

The Nature and Distribution of Humic Matter 31

structure of humic substances. The identification of the type of aromatic nucleus of the so-called humic acid monomers may shed light on the viability of the proposed division of humic matter into the three groups as stated above.

2.3.2 Aquatic Humic Matter

This is the humic matter in streams, lakes, and oceans, and their sediments. The term kerogen is sometimes used for humic matter in aquatic sediments (Swain, 1975). However, Mayer (1985) considers it to be a condensed form of aquatic humic acids, produced by diagenesis of the latter (Hatcher et al., 1985).

In the discussions above, it is indicated that fulvic acid is the dominant substance in the humic matter carried by water, whereas humic acid is only a minor constituent. However, humic acid may be present in appreciable amounts in the humic matter deposited at lake or sea bottoms. On the basis of origin, two groups of aquatic humic matter have been recognized:

Allochthonous Aquatic Humic Matter. This is humic matter brought from the outside into the aquatic environment (water). The humic matter is formed in soils and after formation leached or eroded into rivers, lakes, and oceans. Although physical and chemical changes may have been induced by the aquatic system, the nature of the humic matter is still related to soil (terrestrial) humic matter.

Autochthonous Aquatic Humic Matter. – This is humic matter formed in the aquatic environment from cellular constituents of indigenous aquatic organisms. In marine sediments, this kind of humic matter consists of carbohydrate-protein complexes (Degens and Mopper, 1975; Jackson, 1975). Sapropel, copropel and the like can perhaps be grouped in this category.

2.3.3 Wetland or Peat Humic Matter

This is humic matter derived from material formed in poorly drained ecosystems accumulating as thick deposits of bog, peat, and mucks. It is composed of fulvic acid and humic acid, with the humic acid content reported to increase from peat to muck. Humin was also mentioned earlier as a possible fraction present in substantial amounts. At the present stage of knowledge, it is not known whether the properties of the humic substances may be different in the different kinds of peat, since peat from sphagnum is chemically different from peat of heath vegetation or peat formed from woody trees. In peat deposits, formed in systems resembling the soil ecosystem except for the poor drainage, the humic acid is reported to exhibit properties closely related to those of terrestrial humic acid. Peat humic acid is believed to have properties similar to brown humic acid, though some are skeptical about this. With future developments in technique and instruments, differences can perhaps be detected between humic matter formed under well-drained and poorly drained systems. The properties of humic acid in sapropelic peat are assumed to be far remote from those of its soil counterpart.

2.3.4 Geologic Humic Matter

This is the humic matter in lignite or leonardite and the various types of coal. It is composed mostly of humic acids, though many believe that it also contains a lot of humin. Because of the aging process, most of the fulvic acids have been squeezed and polymerized into humic acids by diagenesis reactions. Environmental processes, such as leaching, may have assisted by decreasing the fulvic acid content further. A subdivision into geologic and paleontologic humic matter can perhaps be made on the basis of the geologic ages of the deposits.

2.3.5 Anthropogenic Humic Matter

This is the humic matter derived from agricultural, industrial, and domestic wastes, and from material in polluted waterways. From the scant data available, it is expected that this type of humic matter is composed of fulvic acid and humic acid. In the polluted drainage canals and ditches, the water is often yellowish to brownish in color suggesting perhaps the presence of large amounts of fulvic acids. Not much is known yet about these anthropogenic humic substances, though at one time poultry manure and sludge derived fulvic acids are considered comparable to soil fulvic acid.

CHAPTER 3

EXTRACTION AND FRACTIONATION OF HUMIC SUBSTANCES

3.1 THE SEARCH FOR EXTRACTANTS

A proper analytical procedure is necessary for achieving the purpose of extracting the real humic substances. Contrary to the general opinion, isolation and purification are considered here as integral parts of the extraction procedure. The aim is to obtain the real humic compounds isolated from each other, and free from co-extracted materials or contaminants. A mixture of humic acid, fulvic acid, protein and other compounds cannot be called humic acid, before the latter is separated from fulvic acid and cleaned by a purification procedure. In the chemical industry a mixture of formic acid and other organic acids is also not called formic acid, unless this acid is separated from the other acids and adequately purified.

The success of the extraction procedure depends on the use of the most correct extraction reagent. The early methods have used NaOH, which is accepted as the ideal extractant for humic substances in the old days. However, with the advancement of humic acid chemistry in the twentieth century, its suitability has been challenged

Extraction and Fractionation of Humic Substances

by a number of scientists (Chaminade, 1946; Bremner, 1950; Dubach et al., 1963; Flaig et al., 1975). Concerns of creating artifacts by NaOH extraction have started a worldwide search for new, perhaps more suitable reagents for the extraction of humic substances. Hayes (1985) has provided a detailed account on the theory and chemical characteristics that a proper extracting reagent should have for an effective extraction of humic substances and quoted four chemical properties for a reagent to be accepted as the ideal solvent. He considers water to also be an excellent reagent for extraction of humic matter. In contrast, a different concept for setting criteria has been used by Stevenson (1994), who believes that an ideal extraction procedure should meet the following criteria:

1. The method should lead to the isolation of unaltered material

2. The extracted humic materials must be free of inorganic contaminants, such as clay and polyvalent cations

3. Extraction is complete, or nearly so, thereby insuring representation of fractions from the entire molecular weight range

4. The method is universally applicable to all soils.

Criterion 2 is perhaps unrealistic and nonpractical. By using a soil sample, it is unavoidable to co-extract inorganic and organic contaminants. However, this is a matter of purification and not an issue of extraction. The remaining three criteria have in essence been covered earlier, and have been circulating for quite some time in a shorter version as presented below (Flaig et al., 1975; Bremner, 1954):

1. The reagent should have no effect on changing the physical and chemical nature of the humic substances extracted, and

2. The reagent should be able to quantitatively remove or isolate the

humic substances from soils.

Over the years a number of inorganic solvents have been tested (Schnitzer and Khan, 1972). In addition, several complexing agents and a variety of organic solvents have also been examined. In Hayes' (1985) opinion these reagents must be capable in dealing with the polyelectrolyte behavior of humic substances and in overcoming secondary forces in the solvent and macromolecular systems, whatever they are, if a proper dissolution is desired. Several methods have since been developed and are currently available for the extraction and isolation of humic substances from soils. To name a few, analytical procedures have been proposed by the International Humic Acid Society (Hayes et al., 1997), Soil Science Society of America (Schnitzer, 1982), Stevenson (1994), Tan (1996), and many more. Ignoring the variations present in the different methods for manipulation of the analyses, the key ingredient in all the methods above is still NaOH, hence the uncertainty facing investigators of extracting really humic compounds from soils is still alive. However, some consolation can be found from several studies indicating the methods using NaOH to be reliable within certain limits and suitable for use with the intended purpose (Tan et al., 1994). This observation is supported by Orlov (1985), who, as indicated before, is a firm believer that NaOH extraction yields humic products that are reproducible from the quantitative as well as from the qualitative aspects.

While all the above applies only to extraction of humic matter from soils, peat, and other deposits, extraction and isolation of aquatic humic matter require a somewhat different approach. The humic substances are already present in solution, or at least in fine suspension form. However, because of their low concentrations in the aqueous medium, a preparative method for concentrating the humic matter is necessary before sufficient amounts can be isolated for analytical and other purposes. Several methods are available to meet this objective, and one of the most modern methods is the resin adsorption technique, which will be discussed separately in a section below.

3.1.1 Inorganic Reagents

Over the years many inorganic solvents have in fact been evaluated for their effectiveness in extracting humic compounds (Stevenson, 1965; Schnitzer et al., 1959), usually with mixed results in meeting the two conditions for an ideal extractant as indicated above. Some of the reagents, e.g., dilute bases, can meet the condition set for the quantitative removal of humic fractions. However, all of these bases are suspected to have some effect on modifying the physical and chemical properties of the extracted humic substances (Flaig et al., 1975). The possibility of creating artifacts is still confronting the investigator today. Some of the inorganic reagents used in extraction are listed in Table 3.1. Among the reagents listed, especially NaOH is believed to approach the two points above for an ideal extraction reagent more closely than the other listed chemicals. The use of NaOH can be traced back to the first attempts to extract humic matter by Achard, Doebereiner, Berzelius, Mulder, and other pioneers in humic matter science (see Chapters 1 and 2). Since Oden's rediscovery of this reagent in 1919, the method of extraction with NaOH seems to have been the most widely accepted procedure for some time. Today, the basics remain the same, but some modifications and refinements to Oden's original extraction procedure have been instituted corresponding to new modern standards. NaOH is believed to be the most effective reagent in quantitatively isolating humic substances from soils. Its easy removal during the purification process is an additional benefit. However, the use of this reagent is suspected to induce autooxidation of humic substances, and humic acids extracted by NaOH are reported to differ in C, N, and O contents from those extracted by other reagents (Hayes, 1985). To alleviate this problem, it is usually recommended to conduct the extraction under a N_2-gas atmosphere (Hayes, 1985; Schnitzer, 1982; Choudri and Stevenson, 1957). However, indications are presented from other studies that the differences in amounts extracted and the properties of humic substances, attributed to NaOH extraction under air and N_2-gas, are very small. No definite trend is noted for extraction under a N_2-gas atmosphere to be superior to the conventional method with plain air (Bremner, 1950; Tan et al., 1991; 1994). The problem lies perhaps not

Table 3.1 Inorganic Reagents Used for Extraction of Humic Substances

Acids	Bases and Salts
0.1 N HCl	0.1 N NaOH
0.025 N HF	0.5 N NaOH
1% H_3BO_3	0.1 M Na_2CO_3
	0.5 M Na_2CO_3, pH 10.5
	0.2 M Na citrate, pH 7.0
	0.1 M NaF
	0.1 M $Na_4P_2O_7$, pH 7.0
	0.1 M $Na_4P_2O_7$, pH 9-10
	0.2 M Na_2-EDTA
	0.1 M $Na_2B_4O_7$
	Urea

so much in the type of extracting reagent, but more in the standard for comparison of humic compounds. No accurate standard humic acid is available, or better scientists cannot agree about a humic acid compound that accurately represents the real compound in nature. Efforts made by the International Humic Substances Society (IHSS) for standardization of extraction procedures are helpful, but its method has not been thoroughly tested and still requires a whole lot of comparison with other methods.

In using the NaOH method, the use of a 0.1 M NaOH solution is suggested, because of its milder nature for extraction than 0.5 M NaOH (Tan, 1995; Pierce and Felbeck, 1975), decreasing in this way the chances for harmful alterations. For qualitative investigations, a very weak NaOH solution, of 0.001 to 0.01 M, is even better than 0.1 M NaOH. By rule it is known, that the stronger the NaOH solution the more will be extracted, but the greater will be the chances for chemical changes to occur in the extracted humic matter.

Sodium pyrophosphate, $Na_4P_2O_7$, although not as effective as NaOH, is used frequently for the extraction of humic matter from soils

Extraction and Fractionation of Humic Substances

high in sesquioxide contents. Chelation by the phosphate of Al, Fe and other metal sesquioxidic ions is believed to increase the solubility of the humic substances, hence enhances their extraction (Kononova, 1961). Other chelating agents, e.g., EDTA, have also been used for a similar purpose in the extraction of humic matter. To increase the effectiveness of the pyrophosphate, a solution with a pH of 9 to 10 is recommended. Nevertheless, the amount extracted is usually considerably less than that obtained with NaOH, but Stevenson (1994) believes that the chances for alterations are less if $Na_4P_2O_7$ is used, especially at pH 7.0. Although reports to the contrary are present, the use of pyrophosphate often eliminates the need of decalcifying calcareous soils prior to extraction, a pretreatment required with the NaOH procedure. The disadvantage of the $Na_4P_2O_7$ method is that it is very difficult to purify the extracted humic matter. Phosphate is known to be chelated strongly by humic substances. Though in a few cases, it has been noticed that humic fractions, isolated with $Na_4P_2O_7$, yield infrared spectra with better resolutions (Tan, 1978), the infrared features are often not characteristic for humic compounds. A comparative study on the effectiveness of NaOH and $Na_4P_2O_7$ extraction of humic acids by Orioli and Curvetto (1980) provides additional information showing that the pyrophosphate method has apparently not extracted three high molecular weight fractions that were identified in the NaOH extracts.

Extraction with acids, e.g., HCl, as conducted by Schnitzer et al. (1959), technically yields only fulvic acids, since only these humic fractions are soluble in acidic solutions. By definition, humic acids are insoluble in acidic condition, and the 'humic acids' identified by Schnitzer and coworkers above are most probably degradation products due to hydrolysis by the acidic reagent.

3.1.2 Organic Reagents

The selection for using organic reagents in the extraction and isolation of humic matter from soils originates from the desire to avoid producing chemical changes in the extracted product. A variety of or-

Table 3.2 Organic Reagents Used for Extraction of Humic Substances

Acids	Non-acids
Formic acid	Acetonitrile
Oxalic acid	Acetyl-acetone
	Benzene
	Chloroform
	Dichloromethane
	Dimethylformamide (DMF)
	Dimethylsulfoxide (DMSO)
	Dioxane
	Dodecylsulfate
	Ethanol
	Ether
	Ethylenediamine
	Formamide
	Hexamethylenetetramine
	Methylisobutyl ketone
	Phenol
	Pyridine
	Tetrahydrofuran

Sources: Schnitzer and Khan (1972); Hayes (1985); Ziechman (1994); Stevenson (1994); Hayes and Malcolm (2002).

ganic solvents have been tested and some are listed in Table 3.2. These substances have been used alone, as single reagents, or as mixtures at varying concentrations. Thus far, none of them has been satisfactory (Schnitzer and Khan, 1972), and the use of organic solvents appears to create more problems than with the inorganic solvents. Not only have they proven to be weaker extractants, but in contrast with the inorganic reagents, e.g., NaOH which is easy to remove, the organic reagents are more difficult to remove in the purification process. Their possible interaction with humic substances is assumed to make it more difficult to purify the extracted humic fractions. In addition, the chances are enhanced not only for producing chemical changes, but also for creating artifacts, due to incorporation of the organic reagents in the molecular structure of the humic molecule. Indications that the organic reagents are extracting lower amounts of humic matter than the inorganic solvents have been presented by Hayes (1985), whose analysis shows ethylenediamine (EDA) to be a poor solvent, and only by repeated extractions with EDA was the combined yield of humic matter comparable to that extracted by a 0.5 M NaOH solution. The data of the author above with methylsulfoxide and dimethylformamide even show considerably less humic substances extracted than that by ethylenediamine. Indications for the possibility of chemical changes in the extracted humic matter are provided by the higher N contents noticed in humic substances extracted by Hayes (1985) with ethylenediamine, a compound containing N. To this fact can be added reports claiming that humic acids isolated by 0.5 M and 0.1 M hydroxymethyl amine have carboxyl contents substantially different from those extracted with NaOH (Orioli and Curette, 1980). On the other hand, Alberts and Filip (1994) show that the chelation capacity of humic acids extracted by hexane and methylene chloride solutions is not much different from that of humic acids isolated with a mixture of 0.1 N NaP_2O_4 and NaOH solutions. The latter finding indicates that further investigations are perhaps needed to bring more light on the issue of chemical changes in humic matter caused by organic solvents.

Organic acids, similarly as the inorganic acids, are poor choices as extracting reagents, because their acidic reactions are incompatible with the concept of humic acids. Nevertheless, a lot of work has been conducted on extraction of humic substances with formic acid. This reagent is reported to extract 50% of the humic matter content in soils only in the presence of LiF, LiBr, or HBF_4, added to the formic acid solution. The added chemicals are assumed to break the hydrogen- or metal-complex bonds, enhancing in this way the dissolution of the humic substances (Stevenson, 1984).

Organic chelating agents have also proven to be less suitable for extraction of soil humic matter. The organic reagent acetyl-acetone (Table 3.2) has been used for isolation of humic substances in the B_h

horizons of spodosols. It is reported to function by chelation of Al and Fe, enhancing the dissolution and extraction of the humic substances from the spodosols. However, the reagent is believed to be ineffective for extraction of humic matter in other soils (Stevenson, 1994).

According to Hayes and Malcolm (2001), good organic solvents have an electrostatic factor (whatever this is) > 140, and a $pK_{HB} > 2$, as exhibited by DMF and DMSO. The pK_{HB} is explained as the capacity of the solvent as a H-acceptor, hence the organic reagents DMF and DMSO are Brønsted-Lowry bases. The authors add that to be considered the best, the solvents must exhibit also δ_p (dispersion force), δ_H (force of H-bonding), and δ_b (force as proton acceptor) greater than 6.5 and 5, respectively. Apparently, the authors forgot a third value, since three kinds of δs were stated. They concluded that water satisfied all the criteria, but unfortunately it is a very poor solvent and extractant for humic substances. It is also puzzling why so much emphasis is placed on the parameters above, since at the end, it is concluded that the use of DMF and DMSO also present problems in extraction and isolation of humic substances.

3.1.3 Reagents for Collecting Aquatic Humic Substances

As indicated earlier, a different approach must be taken in the extraction and isolation of aquatic humic matter due to the fact that the humic substances are already in finely dispersed forms, or in 'solution' as considered by some authors. They are in fact in suspension as macro-molecular colloids in the 0.45 µm or smaller size fraction (Gaffney et al., 1996), a size below the clay- size limit of 2 µm. Because of their low concentrations in water, it is necessary to concentrate them first in order to be able to isolate them properly, and at the same time obtain sufficient amounts of samples for analysis and investigations. A great number of methods have been developed for this purpose, which perhaps can be distinguished into *physical* and *chemical methods*. Aiken (1985) has given a detailed account on the merits of several of these techniques.

The major physical methods for concentrating the finely dispersed humic substances are vacuum distillation, freeze drying, freeze concentration, superspeed centrifugation, and ultra-filtration. Since they do not require the use of chemical reagents, the proponents of these methods believe that formation of physical and chemical changes in the humic substances is avoided or minimized.

Chemical methods, on the other hand, employ either chemical agents or adsorbents. The chemical reagents are used in the precipitation method by which the humic substances are precipitated either with metals or by adjusting the pH to 2.0. The methods using adsorbents apply the basics of *gel chromatography* and appear to have attracted the most attention. A variety of adsorbents have been tested, e.g., anion resins, polysaccharides, polystyrene, polyamides, aluminum oxides, cellulose, and agar. Cationic resins are unsuitable, since aquatic humic matter is also negative in charge. The gel materials include cross-linked polymers of polysaccharides, polystyrene, polyamides, and the like, distributed under names such as biogel, cellogel, sephagel, sephadex. However, these resins do not only isolate and concentrate, but also fractionate the humic substance, which in this case is not the purpose of the analysis. XAD amberlite-resins have been suggested by Thurman and Malcolm (1981) as the most suitable sorbents for the preparative isolation of 'dissolved' humic substances. In contrast to sephadex, sephagel and the like, the XAD resins do not fractionate but function mainly in isolating and concentrating the humic substances from water. This XAD-resin technique is reported to be superior to the physical methods, and capable of concentrating humic matter from solutions with concentrations as low as < 50 µg/L DOC, such as in groundwater, by repeated cycles of adsorption and desorption. A variety of amberlite XAD resins is available for this purpose, e.g., XAD-1 to XAD-12. They are nonionic copolymers exhibiting macropores and large amounts of surface area. The sorption capacity of the resins is assumed to be affected by solution pH, hydrophobic behavior and solute's solubility in water (Aiken, 1988; 1985;), which Thurman et al. (1978) explain as follows. At low pH, the resin will adsorb humic substances, since the latter are noncharged (neutral or protonated) in the acidic medium. On the other hand, at high pH, the humic matter is negatively charged, and charged particles will not be adsorbed by the resin. Of the types in the XAD series, the resins in the form of acrylic-esters (XAD-7 and XAD-8) are preferred to those composed of styrene-divinylbenzene (XAD-1, XAD-2, and XAD-4). Recovery by elution of adsorbed material is more efficient and rapid with the two acrylic esters, which are hydrophilic in nature, than with the three other resins, which are more hydrophobic in character. Aiken (1985) claims that due to the hydrophilic character, the acrylic ester resins, XAD-7 and XAD-8, are capable of adsorbing more water and becoming wet more easily. These are essentially factors favoring development of water bridging enhancing adsorption of noncharged humic matter by relatively weak bonding. On the other hand, the difficulties in elution from the styrene-divinyl benzene resins are explained by the authors as effected by strong bonding resulting from hydrophobic and chelation interactions between the resins and the humic material. Between XAD-7 and XAD-8, Aiken (1985) suggests the use of XAD-8, because of the serious problems obtained with XAD-7 and NaOH, known as *bleeding*. However, Chen (1977) is of the opinion that XAD-12 is a better adsorbent than XAD-8, especially for the isolation of humic acids. Its stronger hydrophilic property than that of XAD-8 is assumed to result in attraction of more water molecules, producing more bridges for the sorption of humic acids as indicated above.

Though many believe that the XAD resins are the best choices, other resins are also available for the purpose of isolation and concentrating humic substances from dilute solutions. For example, anion exchange resins have the potential for use in sorption and isolating humic substances from aquatic environments. The negatively charged humic molecules will be adsorbed readily by the positively charged resins, and the recovery of the adsorbed material is expected to be achieved easily by elution with a NaOH or by a neutral salt solution. Aiken (1988) recommends Duolite A-7, which he assumes to have a high adsorption capacity for anionic organic solutes that can be eluted rapidly when the amount adsorbed is limited from 1/3 to 1/2 of the resin's loading capacity. However, today a variety of more modern types of anion exchange resins are in fact available for use in a rapid and simple isolation of aquatic humic matter. For example, Dowex SAR, a styrene-DVB copolymer containing trimethyl benzyl ammonium active groups, Dowex SBR with dimethyl benzyl ammonium active groups, and the Ionac NA-38, one of the Ionac series,

Extraction and Fractionation of Humic Substances 45

an alkyl quaternary ammonium polystyrene resin, are possible reagents for use with the same purpose as Duolite A-7. This list should be realized as a scientific statement only and is not an endorsement of a specific product at all. Unfortunately most of them have not attracted too much research attention and are still awaiting testing for their suitability in sorption of dispersed humic substances in natural waters. To be effective in sorption of humic matter, Aiken (1988) quoted Abrams (1975) that the anion exchange reagent should exhibit a macroporous structure, contain weak-base functional groups, and have a hydrophilic matrix which is highly negatively charged at pH 10. However, the latter is perhaps in error and is subject to a lot of argument. As a rule anion exchange resins carry positive charges that are attracting negatively charged ions, called anions, and this is the reason why they are called anion exchange resins. A resin carrying positive and negative charges is not an anion exchange resin but an amphoteric compound. The anion exchange resins listed above are all strong bases, meaning that at pH 10 they are positively charged.

In contrast with the XAD resins where nonionic (neutral) humic substances are adsorbed, with the anion exchange method, only the ionic form or negatively charged humic substance will be adsorbed. The force for attraction is electrostatic attraction, which provides for a stronger bonding than the water bridge bond with the XAD resins. Nevertheless, it is not expected to result in difficulties for recovery by elution of the adsorbed material. The disadvantage is that other negatively charged organic compounds are co-adsorbed making purification of the eluted humic substances somewhat more complex and tedious.

3.2 TERRESTRIAL HUMIC MATTER

3.2.1 Extraction Methods

As mentioned earlier the several methods available for extraction of humic substances from soils, peat, and geologic deposits are in essence not much different from the basic procedure using NaOH. They only vary in pretreatments of the sample, fractionation, and in purification of the extracted materials. Hence the basic procedure widely used by the majority of scientists will be outlined below, and the variation in pretreatments, fractionation, and purification according to the International Humic Substances Society, Soil Science Society of America, and other authors will be provided in separate sections.

Pretreatments

The International Humic Substances Society (IHSS) recommends acidifying the sample prior to extraction (Hayes, 1985). The soil is equilibrated first with 0.1 M HCl for two to three hours, after which it is washed thoroughly with water followed by adjusting the soil pH to 7.0 with NaOH. The Soil Science Society of America (SSSA) procedure follows a similar procedure (Schnitzer, 1982), but Tan (1996; 1994) suggests using such a pretreatment only for calcareous soil samples. Removal of the high calcium content seems to enhance extraction of humic matter. On the other hand, acidification of acidic soils, e.g., andosols, ultisols, and oxisols, is unnecessary and may even prove to create problems in the extraction of the humic substances.

Extraction Procedure

The basics of a procedure most people adhere to are to mix a weighed amount of soil with a 0.1 M solution of NaOH at a soil: solution ratio = 1:5, and shake it continuously overnight. A soil: solution ratio = 1:10 is used by the International Humic Substances Society. In general the ratio to be used depends on the organic content of soils. The higher ratio is preferred for use with soil samples rich in organic matter, hence a ratio of 1:10 is recommended for extraction of peat. In the IHSS and SSSA procedures the mixture is placed and shaken in a nitrogen atmosphere. Choudri and Stevenson (1957) even

Extraction and Fractionation of Humic Substances 47

suggest adding stannous chloride as antioxidant to prevent autooxidation of the humic substances. However, Stevenson (1984) deleted the use of both the nitrogen atmosphere and $SnCl_2$. This makes the extraction procedure faster and easier, and as indicated earlier it has no effect at all in the quantity and quality of the extracted product.

An example of the basic extraction procedure is outlined in Figure 3.1, and will be discussed in detail as follows. Weigh 10 grams of soil (sieved to pass a 2-mm sieve) in a propylene centrifuge tube. If

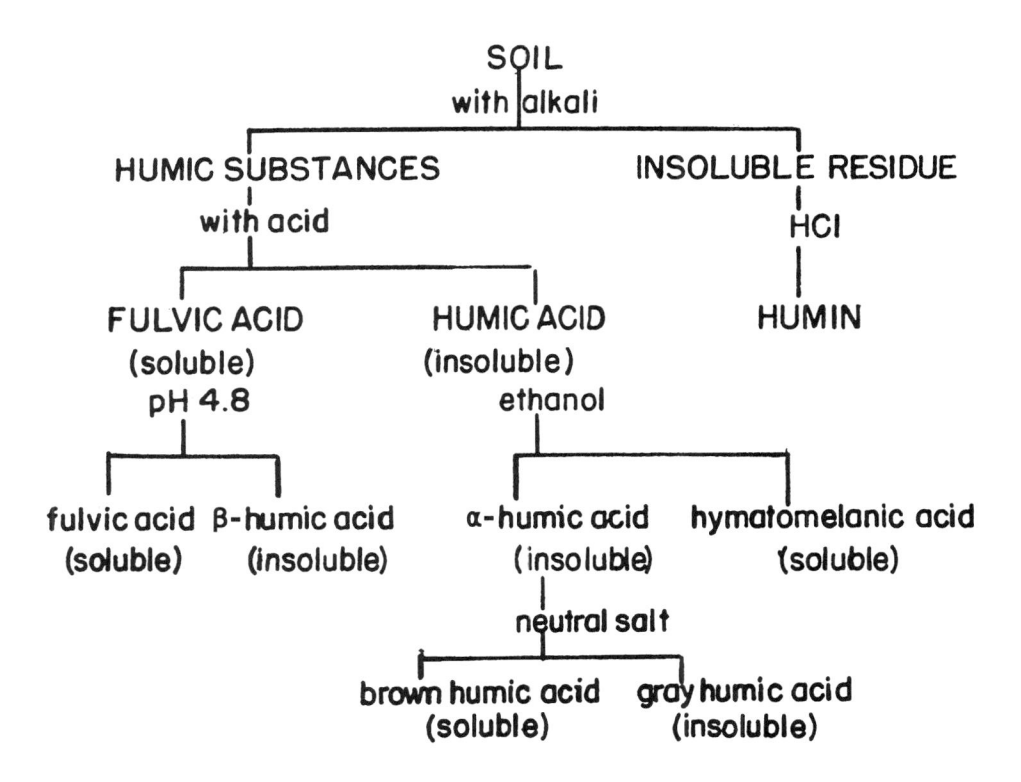

```
                          SOIL
                       with alkali
         ┌─────────────────┴─────────────────┐
   HUMIC SUBSTANCES                  INSOLUBLE RESIDUE
       with acid                            HCl
   ┌────────┴────────┐                       │
FULVIC ACID      HUMIC ACID                HUMIN
 (soluble)       (insoluble)
  pH 4.8           ethanol
 ┌───┴───┐       ┌─────┴─────┐
fulvic  β-humic  α-humic    hymatomelanic
 acid    acid    acid        acid
(soluble)(insoluble)(insoluble)(soluble)
                    │
                neutral salt
                ┌───┴───┐
           brown humic   gray humic
              acid         acid
           (soluble)   (insoluble)
```

Figure 3.1 Flow sheet for extraction of humic substances from soils, peat and other terrestrial deposits.

a N_2-atmosphere is desired, place the soil in a vacuum Erlenmeyer flask. Add 50 mL 0.1 M NaOH and bubble N_2 gas into the flask until all the air is replaced by nitrogen. Stopper the flask airtight and shake the mixture continuously overnight. The dark colored supernatant is separated from the soil the next day by centrifugation at 10,000 rpm for 15 min. After decanting the supernatant in a separate flask, the soil residue is washed with 50 mL distilled water and the colored water is centrifuged, decanted and combined with the previous extract. If desired for quantitative purposes, the extraction can be repeated once or several times by adding again 50 mL NaOH, etc.

The combined supernatants, containing the humic substances, are centrifuged again at 15,000 rpm for 15 min. to ensure complete removal of fine colloidal clays. Discard the precipitate and collect the humic solution, which should be as clear as a pair of dark sunglasses. It is then acidified to pH 2.0 by adding drops of HCl in order to precipitate the humic acid fraction. Separate the supernatant, containing fulvic acid, from the precipitate (=humic acid) by centrifugation at 10,000 rpm for 5 - 10 min. Both the humic acid and the fulvic acid fractions are collected for further purification and fractionation as discussed below. Obtaining humic and fulvic acid is considered here not as a fractioning process of a humic substance. It is an extraction and isolation procedure necessary to obtain both substances present in the mixture.

Purification of Humic Acid

The crude humic acid precipitate obtained above is redissolved with 0.1 M NaOH, and centrifuged at 10,000 rpm. The undissolved fraction is discarded, and the dissolved fraction acidified again to pH 2.0. Separate the precipitate after acidification from the solution by centrifugation at 10,000 rpm for 5 min., and discard the supernatant. Shake the HA precipitate at the bottom of the flask loose with 50 mL of a very dilute HCl+HF mixture in order to reduce the ash and silica content. Centrifuge after shaking and discard the HCl+HF extract, but save the HA precipitate. Wash this precipitate thoroughly with distilled water, centrifuge and discard the wash-water. For a final purification step, Tan (1996) suggests redissolving the HA precipitate with 50 mL 0.1 M NaOH and allowing the solution to pass through a H-saturated cation exchange (Dowex 50-X8) column, a procedure adapted from Lakatos et al.(1977). The eluted HA solution has a pH between 2.0 and 3.0, because the humic acid is highly protonated. The HA remains in 'solution,' though apparently as an unstable solution. Such a behavior of HA corresponds to that of aquatic HA, which stays in solution in the Okefenokee swamp black water with a pH of 3.8 (Tan, 1993). This purified HA can be used as such for analysis or it can be freeze dried and stored in an amber-colored flask for later use.

An alternative purification procedure without the use of a cation exchanger is to transfer the humic solution after HCl+HF treatment into dialysis tubings, and dialyze it against distilled water for three to four nights. Care must be taken to refresh the water every 6 to 12 hours. This procedure is more time consuming but the results are the same as with the cation exchange method.

Purification of Fulvic Acid

The colored supernatant containing the fulvic acid (FA) fraction is purified by passing through an amberlite XAD-8 column, as recommended by the International Humic Substances Society. For the preparation of an XAD-8 column see the section on extraction of aquatic humic matter. Fulvic acid retained by the XAD resin is washed twice by elution with distilled water to remove carbohydrates and extraneous co-adsorbed compounds. The washed FA is then eluted from the column with a 0.1 M NaOH solution, after which it is allowed to flow through a H-saturated cation exchange (Dowex 50-X8) column for a final purification. The purified FA can be used as such for analysis or it can be freeze-dried, weighed and stored in an amber-colored flask for later use. This XAD-purified fulvic acid is called generic fulvic acid by Stevenson (1994).

3.3 FRACTIONATION OF HUMIC SUBSTANCES

3.3.1 Fractionation of Humic Acid

As stated above, the procedure of obtaining humic acid and fulvic acid is in fact an isolation process, and purification is an integral part of the extraction method. Extraction and isolation are not fractionation procedures as many authors believe them to be. The following analogy is used to emphasize this contention. In a mixture of soil organic matter composed of clay, silt, sand, leaves, twigs, flowers, and roots, the extraction and isolation of the leaves and twigs from the soil mixture can hardly be called a fractionation process. The term fractionation can be used only when after isolation of the leaves from the twigs, the leaves are cut into leaf-blades and stems. This applies also to the fractionation of humic and fulvic acids, as embodied in the definition presented by Swift (1985) stating that fractionation is the subdividing of humic substances according to some property. This is supported by a statement of Leenheer (1985) implying that fractionation procedures must be clearly distinguished from extraction and isolation procedures. Hence, what both Swift and Leenheer want to imply is that after the humic substances have been extracted and humic acid properly isolated from fulvic acid, each of them can then be broken down into subfractions. Three reasons have been cited by Swift (1985) for the necessity of fractionation of humic substances: (1) to facilitate determination of the physical and chemical properties, (2) to determine the range of variation in properties, and (3) for use in characterization or in fingerprinting. These reasons can only be realized when the humic substances, humic and fulvic acid, have been isolated as single, individual entities, ignoring the issue of their being real or fake entities.

A large number of chemical and physical methods are now available for fractionation of humic substances. The chemical methods are generally based on the application of differences in some physico-chemical properties, such as differences in solubilities in a chemical reagent, charge characteristics, differences in adsorption, density, particle size, and molecular weight. The physical methods of importance are filtration and centrifugation. Swift (1985) has presented an excellent account of the significance and merits of several of these methods. The methods of fractioning discussed below do not apply only to subdividing humic acids; by changing and selecting the proper parameters, they can be adapted for use in fractioning fulvic acids.

Differences in solubilities of humic acid in chemical reagents, such as in ethanol and in neutral salt solutions, are the basis for fractioning humic acid into different subtypes or subfractions. Fractioning on the basis of charge characteristics is conducted by the electrophoretic technique and ion exchange method. The procedures of fractioning using differences in adsorption vary widely from methods using anion exchange resins to gel chromatography, which applies other types of adsorption resins. The selection of the resins has been discussed earlier, and the purpose for use of a specific resin also varies from one to another resin. For example, anion exchange resins are chosen for attracting negatively charged humic compounds, as indicated before. Other adsorption resins are applied for attracting noncharged humic molecules as in gel chromatography, and fractioning them into different molecular sizes or into different molecular weight size measurements.

As indicated above, fractionation of a humic substance can also be achieved by physical means. One of the most frequently applied methods is ultrafiltration, using filters with varying pore sizes. Filters generally separate on the basis of linear sizes or dimensions and not necessarily by molecular sizes, and many people use size measurements as identical to molecular weight sizes, which is scientifically incorrect. In pure colloid chemistry, colloidal sizes can be expressed in terms of linear dimensions (µm or nm) or in terms of mass using molecular weights or daltons. The choice depends on the purpose of the study and no direct statistical conversion of linear sizes into mass units is available. However, a few daltons are usually considered equivalent to a diameter of 1 nm. Generally, the size of colloids represents a continuum, ranging from 0.001 to 1.0 µm in diameter (Ranville and Schiermund, 1998), hence defining upper and lower size limits of colloidal particles is arbitrary. Though a size limit of 0.45 µm is most widely accepted, Ranville and Schiermund (1998) believe that the range of 2-5 µm would better describe the hydrodynamic behavior of large colloids, which coincides with the usual clay-silt boundary (2 μm) in soil science. The most common filters used for concentrating of humic matter have pore sizes of 0.45 μm based on the assumption that humic colloids are at the size range of < 0.45 μm (Gaffney et al., 1996). Fractioning can now be carried out by a filtering process using a series of filters varying in pore sizes. This will be discussed below in more detail in fractioning by filtration.

Fractioning by Dissolution

The fractioning is based on the capacity of a chemical reagent to break down the humic molecule into several subfractions with different solubilities. A variety of reagents have been used for this purpose, but in general most of them function through a dissolution process producing soluble and insoluble humic fractions. For example, humic acid can be fractionated into two fractions by shaking in ethanol. The dissolved fraction is called hymatomelanic acid, whereas the undissolved part is called α-humic acid.

As indicated in Chapter 2, by shaking in neutral salt solutions, e.g., 1% NaCl or KCl, humic acid can also be divided into a brown (soluble) fraction and a gray (insoluble) fraction. This process is called by Swift (1985) *salting-out*. In soil chemistry, salting-out is used for suppression of the double layers, composed of counterions surrounding the charged surfaces of colloidal particles in suspension. The presence of thin double layers causes precipitation of the colloidal particles and no fractionation of the colloids has in fact taken place.

Fractioning by Gel Chromatography

This is perhaps a simple and relatively effective method in achieving molecular fractionation of fulvic and humic acids. The variety of resins available for use has been discussed earlier, and some of them are suitable for fractioning, e.g., sephadex, sephagel and the like, whereas others are not, e.g., XAD resins. As an example, the method using *sephadex*, synthetic cross-linked polydextranes

Extraction and Fractionation of Humic Substances 53

(Pharmacia Fine Chemicals, Uppsala, Sweden), will be explained below A column of swollen sephadex gel beads is prepared, and a determined amount of a humic acid solution is placed on top of the column. As soon as it has drained into the column, it is eluted with distilled water at a controlled flow rate of 40 mL/hour. The subfractions collected are purified and freeze dried. This gel filtration procedure can be repeated several times until sufficient subfractions are collected for use in analysis and research (Tan, 1977; Tan and Giddens, 1972).

The filtration is diagrammatically illustrated in Figure 3.2. The pores between and within the sephadex gel beads act as a chromatographic medium. The large open circles in the figure are the sephadex

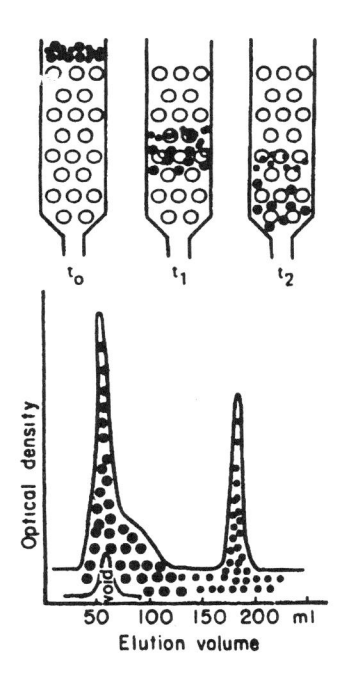

Figure 3.2 Top: schematic diagrams of sephadex gel filtration of humic acid into a high molecular weight (●) and low molecular weight fraction (•); t_0= time zero; t_1=time interval 1; t_2= time interval 2. Bottom: elution curve characterized by two peaks, a peak representing the high molecular weight and a second peak representing the low molecular weight fraction.

beads. The small and large black dots represent a mixture of small and large molecules of humic acids. As the mixture of humic molecules passes through the column, the larger molecules are eluted first and the smaller molecules are eluted last. Consequently, the elution curve may be characterized by two peaks: the first peak representing the higher molecular weight, and the second peak representing the low molecular weight fraction. Different types of sephadex resins with different molecular weight cutoffs can be used separately or in sequence (tandem), e.g., Sephadex G-50, G-20, and G-10, with molecular weight cutoffs of 30,000, 5,000 and 1,000, respectively. Sephadex resins with lower molecular weight cutoffs are now available. For more details reference is made to Tan (1998) and the Handbook for Sephadex-Gel Filtration in Theory and Practice, Pharmacia Fine Chemicals, Uppsala, Sweden (1969).

Fractioning with Anion Exchange Resins

The method of fractioning applying anion exchange resins is in fact closely related to gel chromatography discussed above. Anion exchange resins have been tried several times in subdividing humic substances (Wright and Schnitzer, 1960; Roulet et al., 1963; Barker et al., 1968). The adsorbed humic substance is eluted with a buffer or salt gradients, applied at gradually increasing or decreasing concentrations. However, what exactly the subdivisions are is still not clear. A subdivision by molecular weight size was supposed to be the objective, but anion exchange functions through attraction by electrical charges. What then is the salient feature of subdividing a humic molecule on electrical charge properties. Swift (1985) shows a fulvic acid eluted by a salt (tris+NaCl) gradient yielding an elution curve with two peaks. The first peak is produced by elution with the tris reagent, and the second peak is the result of elution with NaCl. However, Swift fails to explain whether the peaks represent fractions with different sizes or the like, or are they two subfractions of fulvic acid different in electrical charges? How are differences in molecular weights related to differences in electrical charges? These questions have to be answered before one can conclude that ion exchange media

Extraction and Fractionation of Humic Substances

lend themselves very readily to fractionation of humic substances.

Fractioning by Filtration

Two general types of filters are available for this purpose, e.g., membrane filters and depth filters. The membrane filters, such as cellulose-acetate, cellulose-nitrate, and silver membrane filters, function by sieve action. All humic particles larger than the filter's pore size will be retained. Depth filters, e.g., glass fiber filters, are also functioning as sieves, but the sieve action is caused by a matrix of fibers forming a labyrinth of meandering flow channels with variable pore sizes. These fiber filters are sometimes noticed to pass particles larger than their pore size. The main disadvantage with all filters is the problem of clogging when large amounts of humic material are collected, requiring the use of adequate suction. For the merits of the various filters with respect to pore size, chemical composition, and flow characteristics reference is made to Aiken (1985). Fractionation of a humic substance is achieved by using a series of filters of different pore sizes, either in ascending or decreasing order of size limits. By using only one filter size, the process will only concentrate the fraction of the humic substance above the pore size cutoff limit.

Although the filtration technique is by design to be used for separating the humic substances according to linear size limits, in the 1990s the method has been expanded to sort out humic substances into different molecular weight fractions by using several filters. Membrane filters have been made available with pore sizes from several micrometers (µm) to several nanometers (nm). Filters with the larger pore sizes are used for sieving relatively larger size particles and the process is called *microfiltration*. On the other hand, filters with the smallest pore sizes can be applied for filtering molecules in solution, and this is called *ultrafiltration*.

Though linear sizes have been discussed before as not being equivalent to molecular weight sizes, and a few daltons are taken for convenience to equal 1 nm. Tan and McCreery (1975) note that degree of polymerization of humic substances affects their linear sizes as well as their molecular weights. The data of the authors summarized in

Table 3.3 demonstrate the relation between size of a molecule and its molecular weight. By assuming that the humic molecules are spherical, the larger the size of the sphere, the larger will be the numerical value of the molecular weight of humic acid. It is true that other factors, e.g., density, will spoil this relationship, while many scientists may also disagree that humic substances are spherical, but until a better concept can be advanced, the hypothesis of the authors above seems to solve the issue satisfactorily for the time being. Most of the scientists are using linear size limits interchangeably with molecular weight limits anyway, as can be noticed from the discussion below.

Membrane filters are now available with ultrafine pores, assumed equivalent to molecular weight cutoff limits between 50 and one million. Such an ultrafiltration technique was introduced by Lobartini et al. (1997) using an Amicon cell model 8050 adapted for use with a continuous flow. The method, employing a series of cellulose membranes, starts with a membrane with a 10,000-dalton exclusion limit. A humic solution is placed in the cell and sufficient amount of distilled water is passed with the aid of a peristaltic pump until the solution flowing from the cell is totally colorless. The fraction passing through the filter is the fraction with a molecular weight of < 10,000 daltons. The following fractions are then obtained by treating the remaining HA (m.w. >10,000) in the Amicon cell with the same proce-

Table 3.3 Molecular Weights and Sizes (in Å and nm) of Humic Acid Fractions Obtained by Sephadex Gel Filtration (Tan, 2000; Tan and McCreery, 1975).

Molecular weight	Molecular volume Å	Radius Å	nm
30,000	23,622	17.8	1.78
5,000	3,937	9.8	0.98
1,500	1,181	6.6	0.66
1,000	787	5.7	0.57

Extraction and Fractionation of Humic Substances

dure but by using at each step membranes of different exclusions limits in order to yield fractions with molecular weights between 10,000 - 30,000; 30,000 - 50,000; 50,000 -100,000; and >100,000 daltons, respectively. Each fraction is purified and concentrated by passing through a column of XAD-8 resin.

3.3.2 Fractionation of Fulvic Acid

Due to its small size or molecular weight size, fractionation of fulvic acid has attracted less research attention than that of humic acid. Only a limited number of attempts have been reported in the past for separating fulvic acid further into subfractions. The methods at our disposal for this purpose are in essence similar to the chemical and physical methods used for fractionating humic acid.

Sephadex gel filtration has been reported to yield two fulvic acid subfractions, a high molecular weight and a low molecular weight fraction (Tan, 1998; Khan and Schnitzer, 1971). However, Forsyth (1947) claims to have subdivided fulvic acid into four fractions, A, B, C, and D, with activated charcoal. By using a different procedure, Stevenson (1994) seems to support the separation of fulvic acid into two fractions only. The acid supernatant following separation of humic acid by centrifugation is transferred by the author above into a dialysis bag and dialyzed against distilled water. After equilibration for three to four nights both the dialysate (colored water) and the nondialyzable fraction within the bag are collected and separately purified and concentrated with the XAD-8 resin technique. The purified nondialyzable fulvic fraction is called by Stevenson (1994) the high molecular weight generic fulvic acid, whereas the purified dialysate is called by the author above the low molecular weight generic fulvic acid.

Waksman (1936) also claimed in the early days to have separated fulvic acid into two fractions. By adjusting the pH to 4.8 of the fulvic acid solution, after the humic acid fraction was removed by acidification and centrifugation, a precipitate is produced that he called β-humic acid. As discussed before in Chapter 2, the correctness of such fractionation is open for discussion.

Today fractionation of fulvic acid into several subfractions is made possible by the availability of filters with pore sizes ranging from 1 to 450 nm (0.001 to 0.45 µm). The process, applying several of these *ultrafilters* with very small pore size limits in ascending or descending order, earlier called ultrafiltration, has not been exploited with fulvic acid.

3.4 AQUATIC HUMIC MATTER

3.4.1 Extraction Methods

As discussed above, a different approach has to be taken in extraction of aquatic humic matter. The process has to take into account concentrating the humic colloids in solution that are generally present not only in small concentrations, but also with a composition dominated by fulvic acids, the smallest humic substances by size or molecular weight. A variety of methods, including physical and chemical methods have been discussed previously, and according to Aiken (1985), it is recommended to employ a variety of methods, rather than a single method alone, if a high-quality product is to be produced. The author above cited four steps for obtaining the highest quality material: (1) filtration through a micropore filter with a pore size of ≤ 0.45 µm, (2) concentration by adsorption on XAD or Duolite A-7 resin, (3) isolation from inorganic and organic contaminants, and (4) preservation by freeze drying. Step 3 is, however, stated in the wrong context, since removal of inorganic and organic contaminants is a purification and not an isolation issue.

Extraction by XAD-8 Gel Filtration

This is a chemical method recommended by the International Humic Substances Society (IHSS). The method was first presented by Thurman and Malcolm (1981) and later adopted as the procedure of the IHSS (Aiken, 1985). It is reported to be capable of extracting aquatic humic matter from a number of surface and ground water samples with a DOC as low as 0.7 mg C/L. Tan et al. (1991) have used it for extraction of humic substances in black water of streams and swamps in southeast Georgia, USA, with an average DOC content of 10.0 mg/L.

The IHSS method requires the construction of a properly prepared XAD resin column, which is prepared as follows. Amberlite XAD-8 resin (40-60 mesh; Rohm and Haas) is cleaned by shaking it successively in 0.1 M NaOH and ethanol. The resin is rinsed several times with distilled water before it is loaded into a polyethylene chromatographic column. After leaching alternately with 0.1 M NaOH and 0.1 M HCl as specified by Thurman and Malcolm (1981), the column is ready for use.

Five to 10 L of black water are then filtered through a micropore filter with a pore size of 0.45 μm to remove suspended inorganic matter. The filtered water is acidified with HCl to pH 2.0 and allowed to flow by gravity through the prepared XAD-8 column at a flow rate of 200 mL/hour. The humic substances retained by the column are washed thoroughly with distilled water, and eluted with 0.2 M NaOH. The dark colored solution is collected and acidified with HCl to pH 1.0, to precipitate the humic acid from the solution. The humic acid precipitate is separated from the fulvic acid that remains in solution by centrifugation at 10,000 rpm for 15 minutes. After collection, the humic acid is then purified by twice redissolving, reprecipitation, and centrifugation. The final precipitate is then washed once with a very dilute HCl+HF mixture, dialyzed against distilled water, or subjected to a final purification with a hydrogen-saturated cation exchange procedure as discussed earlier.

The fulvic acid (FA) in solution is concentrated by filtering again through the XAD-8 column. The adsorbed FA is washed twice with distilled water and eluted with 0.1 M NaOH after which it is allowed to flow through a H-saturated cation exchange (Dowex 50-X8) column for a final purification. The purified FA can be used for further analysis as such, or it can be freeze dried, weighed and stored in an amber-colored flask.

Extraction by Freeze Drying

This is a very simple physical method for the collection of aquatic humic matter. It has been used by Tan et al. (1990; 1991) in the extraction and isolation of humic substances in black water of streams and swamps from southeast Georgia, USA. By this method a determined amount of black water is filtered by microfiltration (0.45 µm) and freeze dried to concentrate and collect the humic matter. The humic and fulvic acid is then extracted and isolated from the freeze-dried sample by the 0.1 M NaOH procedure as described above. Purification, considered an integral part of the extraction procedure, is conducted as described earlier. As can be noticed from the data in Table 3.4, this method yields the same amounts of humic matter as the IHSS-recommended XAD-8 resin technique. No significant differences can be noticed in the total (combined) concentrations of humic acids and fulvic acids extracted between the two methods, indicating that both methods are equally suitable for extraction and isolation of aquatic humic matter. However, a considerable difference can be noticed in the proportional composition of fulvic acid and humic acid extracted from the aqueous medium. The humic acid content obtained by freeze drying appears to be invariably larger than that extracted by the XAD-resin method. This is clearly reflected by the FA/HA ratios.

Table 3.4 Humic (HA) and Fulvic Acid (FA) Content in Black Water of Streams in Southeast Georgia, USA, Extracted by the XAD-8 Resin and Freeze-Dry Methods (Tan et al., 1991; 1990).

Black Water (Georgia, USA)	XAD-8 resin			0.1 M NaOH		
	FA	HA	FA/HA	FA	HA	FA/HA
	---- g/kg ----			----g/kg----		
Okefenokee swamp	720	280	2.6	675	375	1.8
Satilla river	726	274	2.6	600	400	1.5
Ohoopee river	725	275	2.6	696	304	2.3

3.4.2 Fractionation of Aquatic Humic Matter

The opinion exists that fractionation techniques for aquatic humic substances have not been developed to the same extent as their concentration and isolation techniques (Leenheer, 1985). Such a statement is out of context, since a great variety of fractionation techniques can be adopted from terrestrial humic acids. The issue with aquatic humic substances is that they are present in very low concentration, hence fractionation is only possible after the aquatic humic substances have been isolated and collected properly in adequate amounts. Concentration and isolation techniques are the first steps for subdividing aquatic humic matter. Now that efficient procedures are available for concentration and isolation of aquatic humic substances, fractioning of these compounds is expected to be a relatively simple matter. Contradicting his statement above, a variety of chemical and physical fractionation methods are nevertheless cited by Leenheer (1985), such as precipitation methods, solvent extraction, adsorption chromatography, electrophoresis, ultrafiltration and ultracentrifugation. This list is identical to, though shorter than the fractionation procedures discussed earlier for terrestrial humic and fulvic acids. However, the author's discussion on each of the methods leaves us wondering what the purpose is for listing the methods at all, if they are considered unsuitable. For example, the precipitation method, by adding metals or inorganic and organic acids, is according to the author above a very crude method, and is more suitable for isolation and concentration purposes. Leenheer (1985) also believes that partitioning of aquatic humic substances by solvent extraction is not possible, because of their amphiphatic character that will form dark films at the liquid-liquid interfaces. The author seems not to realize that humic acid can be subdivided into hymatomelanic acid and α-humic acid by dissolution in ethanol. At present a number of effective fractioning procedures have been developed for terrestrial humic matter. The present author can see no reason why both humic acid and fulvic acid from the aquatic environment cannot be fractionated by the same physical and chemical procedures created for terrestrial humic matter.

Fractionation by Ultrafiltration

As indicated in preceding sections, the development of filters with pore sizes < 0.45 µm makes it possible to extract, isolate and fractionate the humic substances in soils and water. An example of fractionation of humic acids by ultrafiltration using flatbed membrane filters has been provided in an earlier section above.

For fractionation of aquatic fulvic acid, a series of filters with very fine pore sizes is recommended. Hence, methods using filters with pore size limits from 0.001 µm to 0.0025 µm should be able to produce fulvic acid fractions partitioned by molecular weights from 1,000 to 30,000 daltons. Additional research has to be conducted to verify this contention. From sephadex G-50 (fractionation range = 1,500 – 30,000 daltons) gel filtration, indications have been presented that fulvic acid can be subdivided into two fulvic acid fractions (Tan, 1977: Khan and Schnitzer, 1971).

More recently the use of hollow-fiber ultrafilters has been suggested for fractionation of aquatic humic and fulvic acids by a couple of scientists from the American Chemical Society (Gaffney et al., 1996). Hollow-fiber ultrafilters are now available, capable of size fractioning humic compounds from 1 to 450 nm (0.001 to 0.45 µm). These filters are considered to differ from the flatbed filters in that they prevent *polarization* of the filter when large volumes have to be processed. What the authors mean by polarization is piling-up of the (filtered) trapped humic molecules on a flatbed filter during analysis of large volumes of water. The trapped molecules are believed to act as filters themselves, and smaller substances can be trapped in the pores of the layered larger molecules. This polarization effect is said to be prevented by using a hollow-fiber filter, because water is pumped from the inside to the outside of the filter during filtration, causing a flow of water parallel to the hollow-fiber filter. By first using hollow-fiber filters for fractioning fulvic acids to molecular size fractions down to 1 nm (equivalent to molecular size limits of 1000), these humic compounds can be fractionated further by using flatbed filters down to molecular weight sizes of 500.

From the discussions above and those in earlier sections, the question can be raised whether fractionation by ultrafiltration will yield legitimate fractions differentiated by molecular sizes, since it appears that the method would yield any molecular weight fraction as imposed by any exclusion limits used in the procedure. Unfortunately, no definite answer can be given at the moment, and more research has to be conducted for solving this issue. However, for consolation it can perhaps be stated that the presence of a seemingly endless variety of molecular fractions concurs with the condition in a natural ecosystem where decomposition at all stages and new synthesis of humic substances are taking place in a never-ending process.

3.5 TYPES OF HUMIC SUBSTANCES

Most scientists consider humic acid and humic acid fraction (or fulvic acid and fulvic acid fraction) as identical. Through the whole literature these terms are used interchangeably, although types and fractions usually carry different connotations. The type is the original humic compound and this can be broken down into several fractions. Hence, the names humic acid and fulvic acid are in essence referring to the types of the humic substances. These two compounds, regardless of their being real, artifacts or operational compounds, are neither obtained by fractioning of the humic substances nor are they fractions of humic acid or fulvic acid, respectively. As indicated before, humic matter is composed of a mixture of humic substances, and humic acid and fulvic acids are two of the humic substances making up the mixture. This problem can be illustrated again perhaps by using the analogy with humus, which is a mixture of carbohydrates, protein, lignin, nucleic acids, enzymes, humic matter, etc. The carbohydrate, protein or lignin is the type of component making up the humus, and it is rather confusing to realize that a carbohydrate fraction or protein fraction infers to the respective component itself. The analogy can be illustrated further by using clay as an example. The soil's clay complex is composed of a variety of clay minerals, e.g., kaolinite, halloysite, smectite, vermiculite, and chlorite. Kaolinite is one of the types of clay and is not considered a kaolinite fraction of the clay complex. It is true that in both cases, the use of the term *humus fraction* or *clay fraction* is often used and admissible, but its application then carries a different meaning.

With this in mind and in view of the discussion in the preceding sections, three major types of humic substances can be recognized, which is unfortunately not much different from the categories in the old theories. Each of the three types can be fractionated into several fractions. The concepts of these three types of humic substances are apparently valid for terrestrial, aquatic, peat, geologic, and anthropogenic humic matter. At the present time no information is available for the presence of other types of humic compounds due to differences in origin or method of extraction. Speculations on the presence of differences between terrestrial and aquatic humic acids have off and on been presented.

Recently a humic substance called kerogen has been brought to attention as a very specific type of humic acid present only in aquatic environments. As stated earlier, it is considered by some a condensed form of aquatic humic acid with diagenesis as the mechanism for its formation. A classification scheme for kerogen has even been proposed distinguishing it into type I, II, and III (Vanderbroucke et al., 1985). Type I kerogen is found in lacustrine deposits whereas type II kerogen is typical for marine shales. Type III shows the influence of the continents. However, at the present state of knowledge very little is known about kerogen and no information has been presented about its extraction, isolation, fractionation, and characterization.

In line with the divisions of kerogen, perhaps terrestrial humic acid can also be distinguished on the basis of the monomeric type of lignin, e.g., softwood humic acid, hardwood humic acid, and grass or bamboo humic acid. However, all the above needs to be investigated more in detail and some kind of definite evidence has yet to be provided.

Substances labeled copropel, sapropel, förna, dy, pelogoea, and dopplerite have at times also been considered as major types of aquatic and marine humic acids (Swain, 1963; Stevenson, 1994), though some geochemists tend to use them for subdivisions of humic substances at lower categories. Subclasses of copropelic humic acid, sapropelic humin and the like have been suggested by Hatcher et al. (1985) and Breger (1976). The potential is present for recognition of these aquatic or

Extraction and Fractionation of Humic Substances

marine types of humic acids, but it appears that more research data are required showing similarities and differences in properties and behavior in order to establish their qualifications as members of the types of humic substances.

As summarized in Table 3.5, the present author will focus on the three established types of humic substances, differentiated by their solubilities in basic and acidic solution. Each will be discussed in more detail below, followed by their fractions. The copropelic and sapropelic substances and the other aquatic and marine counterparts will also be covered but with fewer details.

3.5.1 Fulvic Acid

As noticed from Table 3.5, fulvic acid is the type of humic compound that is soluble in alkali, acid, and water. It is assumed to be a colloidal polydispersed, amorphous humic substance with yellow to brown-black color. Fulvic acid exhibits hydrophilic and highly acidic properties, and of the three major types of humic substances it is relatively the smallest in molecular size, ranging from a few hundreds to a couple thousands of atomic units or daltons. The number-average molecular weight is reported to range from 175 to 3,570 (Schnitzer and Skinner, 1968). Fulvic acid is believed to be the agent of major importance in the mobilization of sesquioxides in the podzolization

Table 3.5 The Three Major Types of Humic Substances Differentiated by Differences in Solubility in Acids, Alkalis, and Water

Type of Humic Substance	Alkali	Acid	Water
Fulvic acid	soluble	soluble	soluble
Humic acid	soluble	insoluble	insoluble
Humin	insoluble	insoluble	insoluble

process (Forsyth, 1946). Freeze-dry fulvic acid is sometimes difficult to dissolve in water, because of polymerization of the compound during freeze drying. The freeze-dried material is often electrostatically charged, since the fine fulvic acid particles are attracting themselves to a metal or plastic spatula. Fulvic acid will decompose upon heating without burning, but through a gradual charring process. It is reported to start burning at 190 - 200°C if an oxidizing agent is present (Orlov, 1985; Hoffman and Schnitzer, 1968). Because of this and since irreversible changes are noticed to occur at 50°C, drying fulvic acid is suggested to be carried out at 35 - 40°C in vacuum over P_2O_5.

Fulvic acid is often ignored and has attracted research attention like a roller coaster. It was perhaps the first humic compound that was extracted in the early days but research interest dropped considerably in the 1960's-1970's. In Stevenson's (1994) opinion, this is due to difficulties in recovering fulvic acid from the acidified soil extract following separation of humic acid. The remaining acidic soil extract contains substantial amounts of inorganic contaminants and NaCl salt as a result of acidification of the NaOH extract with HCl. The author above believes that removal of these contaminants always results in considerable losses of fulvic acid. However, the attention to fulvic acid skyrocketed again with the discovery of aquatic humic matter. In aquatic environments fulvic acid is the dominant type of humic matter. Depending on whom to believe, fulvic acid is considered either a precursor for formation of humic acid or a degradation product of humic acid. This will be discussed in more detail below in the section on bioformation of humic substances.

Fulvic Acid Fractions

High Molecular Weight Fraction. - This is the fulvic acid fraction separated by sephadex G-50 (Tan, 1977). It is the fraction, characterized by a molecular weight > 1,500, which is eluted first from the sephadex column. It makes up approximately 20 - 25% of the total fulvic acid compound.

By applying sephadex G-10 with a smaller fractionation range (molecular weight cutoff = 700), a fulvic acid fraction is obtained with a molecular weight >700, called fraction D_2 by Khan and Schnitzer (1971). It makes up 16.5% of the fulvic acid. Fraction D_2 may also include the fulvic acid fraction with a molecular weight range of >1,500 as obtained above with sephadex G-50.

Generic High Molecular Weight Fraction. - This is Stevenson's (1994) version of a high molecular weight fraction of fulvic acid, obtained by dialysis of fulvic acid in a membrane tubing. The fraction retained within the membrane tubing is, after purification by the amberlite XAD-8 resin technique, called generic fulvic acid. The name high molecular weight is apparently used in a relative sense with respect to the dialysate, a fraction that has passed the membrane tubing and is assumed to be lower in molecular weight. No mention is made of any limit specifying the molecular weight cutoff of the membrane tubing.

Low Molecular Weight Fraction. - This is the fulvic acid fraction that is eluted last from the sephadex G-50 column. It has a molecular weight in the range of 1,500 or smaller. It is the dominant fraction, making up approximately 75 - 80% of the total fulvic acid compound.

By using sephadex G-10, Khan and Schnitzer (1971) have obtained a fulvic acid fraction with a molecular weight ≤ 700, called fraction D_1. It is also the dominant fraction and its amount of 83.5% does not differ much from that obtained by sephadex G-50 filtration. Apparently, fulvic acid fraction obtained by sephadex G-50 contained fraction D_1.

Generic Low Molecular Weight Fraction. - As indicated above, this is Stevenson's (1994) version of a fulvic acid fraction that has passed a dialysis membrane tubing, and has been purified by the XAD-8 resin method.

Fractions A, B, C, and D. - These are the fulvic acid fractions adsorbed by activated charcoal and separated by successive elution with different reagents (Forsyth, 1947).

Fraction A, containing substances eluted by water and $0.1N$ HCl, includes sugars and amino acids. It is a colorless or slightly yellow filtrate. Fraction B is the acetone soluble eluate, and is the most deeply colored fraction. It is believed to be composed of tannins or phenolic glycosides. The remaining residue in the charcoal column yields fraction C upon elution with distilled water. It is a colorless eluate which produces a white precipitate upon addition of alcohol. This fraction is believed to contain polyuronides and sugars. Fraction D, the last fraction in the column, is eluted with 0.5 N NaOH. The eluate has a deep wine-red color, and is assumed to be rich in nitrogen, organic phosphates and sugars. According to Tan (1977), only fractions A, B, and D are of importance, since fraction C seems to be small in amount and is mostly composed of tailings from fraction B. Results of infrared analysis by the author above indicate that the tannins identified by Forsyth are most likely to be fulvic acid fractions. The sugars originate perhaps from polysaccharides, since the latter are found to be important components of fulvic acids (Tan, 1968; Clark and Tan, 1969).

3.5.2 Humic Acid

This is the type of a humic substance that is soluble in basic solvents, but insoluble in acidic conditions and in water (Table 3.5). Humic acid is now believed to be present in soils as well as in peat and aquatic environments. A condensed form of aquatic humic acid is called earlier kerogen. At the present state of knowledge, most of the research data suggest that humic acid from aerobic soils is similar to that from anaerobic environments (Hatcher et al, 1981).

Humic acid is generally characterized by a dark brown to black color. It is an odorless, colloidal polydispersed substance, and assumed to be hydrophilic and acidic in nature. In the moist state, humic acid is said to have a slightly bitter and acid taste (Orlov, 1985). Its molecular weight is very high, ranging from several hundreds to thousands of daltons. Though humic acid is considered to be amorphous, after freeze drying most of the humic acid particles are rhombic or rhombohedral in shapes like the granular crystals of sugar. They seem to possess high electrostatic charges, since the particles are instantly attracted to a metal spatula, which is especially true with the finer particles.

Extraction and Fractionation of Humic Substances

The humic acid compound is usually low in ash content when the purification process includes treatment with a dilute mixture of HCl + HF, as indicated earlier. Humic acid, obtained by extraction with pyrophosphate solutions, usually contains a higher ash content than that extracted by the NaOH method. According to Stevenson (1994) this is due to higher metal contents in the pyrophosphate extract, and a subsequent treatment for reducing the ash content often results in a significant loss in humic acid.

As with the case of fulvic acid, humic acid will not melt upon heating. However, it will decompose upon heating in an air or nitrogen atmosphere or even in vacuum. During heating in vacuum it is reported to form a resin-like substance accompanied by the release of water (Orlov, 1985). Similarly as with fulvic acid, the decomposition of humic acid takes place without burning, but through a gradual process of charring. However, it may start to burn at 150 -240°C in the presence of oxidizing agents (Hoffman and Schnitzer, 1968). Irreversible changes start to appear clearly on heating at 105°C, and Orlov recommends avoiding the use of elevated temperatures in the study of humic acids. Drying of humic acid is suggested as a better method to be conducted at 50°C in vacuum over P_2O_5.

Humic Acid Fractions

Hymatomelanic Acid. - This is the humic acid fraction soluble in ethanol. Of all the other fractions, it is perhaps the best-known fraction of humic acid. It is believed to contain polysaccharide components in ester linkages (Tan and Clark, 1969; Tan, 1975).

Alpha (α) Humic Acid. - This is the remaining humic acid fraction after separation of the hymatomelanic fraction, hence it is the fraction insoluble in ethanol. In the past, this ethanol insoluble fraction was called humic acid, and the original humic acid prior to ethanol treatment was given the name α-humic acid (Russell and Russell, 1950). However, as explained in a preceding section above, it is more logical to assign the name α-humic acid to the remaining portion of the humic substance after separation from the hymatomelanic part and retain the name humic acid for the original substance.

Beta (β) Humic Acid. – This is a fraction isolated by Waksman (1936) from a fulvic acid solution. As indicated earlier, by adjusting the pH of the acidic solution of fulvic acid to 4.8, a precipitate is produced which is assigned the name *β – humic acid*. Though many seem to agree with this, the question was raised earlier of how fulvic acid can yield a humic acid fraction. To some scientists, the substance is comparable to Al-humate. However, the present author is of the opinion that the name Al-fulvate is perhaps more suitable. Though fulvic acid is by definition soluble in acidic solutions, in the form of an Al-fulvate it can precipitate at pH = 4.8.

Brown Humic Acid. – This is the humic acid fraction soluble in neutral salt (0.1 M NaCl or KCl) solutions. It is highly dispersable, and acidic in reaction. The brown color is reported to fade away after long standing. This fraction is believed to be prevalent in the humic acids of peat, brown coal, and alfisols (Stevenson, 1994).

Gray Humic Acid. – This is the humic acid fraction insoluble in neutral salt (0.1 M NaCl or KCl) solutions. It is less acidic ('milder') than the brown humic acid fraction, and the gray color is relatively stable upon long standing. Though this gray humic fraction is reported to disperse in water, it is also noticed that it can be easily coagulated. Stevenson (1994) believes that gray humic acid is a typical fraction of humic acids in chernozems and rendzinas, soils classified today as mollisols.

High Molecular Weight Fraction (m.w. > 30,000). – This is the humic acid fraction that is eluted first as a result of separation by sephadex G-50 gel filtration. It makes up approximately 50% of the total humic acid compound (Tan and Giddens, 1972).

Low Molecular Weight Fraction (m.w. = 30,000 – 15,000). – This is the humic acid fraction that is eluted last from a sephadex G-50 column. It amounts to approximately 50% of the total humic acid compound.

Extraction and Fractionation of Humic Substances

Fractions A, B, R_p, and P. - Kumada (1987) claims to be able to distinguish four types of humic acids, types A, B, R_p and P. From his descriptions, types A and B are in fact humic acid fractions, fractionated using 1 N $MgSO_4$ according to the old method of Simon and Speicherman (1938). The fraction insoluble in $MgSO_4$ is called type A, whereas the soluble fraction is called type B. This is then the reason why these 'types' of humic acids are placed in this section about fractions. Whether the remaining R_p and P humic acids can also be considered as types is not clear. Humic acid types R_p and P are distinguished by Kumada by differences in their absorption spectra and Δ log k values. Humic acid type P is reported to originate from humic acids extracted from spodosols, but can be divided further by gel chromatography into a P_b (brown) and P_g (green) fraction. As discussed earlier, the existence of a green humic acid fraction was later invalidated by Kumada.

3.5.3 Humin

This is a type of humic substance that is insoluble in alkaline and acidic solvents. It is the substance that remains behind after extraction of the soil with dilute alkali for humic and fulvic acids. Because of its insolubility and consequent difficulties in extraction and isolation, humin is the compound that has attracted the least research attention, though its presence has been known since the early days of humic acid science. The opinion exists that humin is a mixture of highly condensed humic acids strongly bonded to clay minerals, fungal melanins and paraffinic compounds (Stevenson, 1994; Kononova, 1966; Somani and Saxena, 1982). Hatcher et al. (1985) disagree about humin being a clay-humic acid complex, whereas others believe that humin has a composition close to that of humic acids, though perhaps slightly less aromatic than the latter (Schnitzer and Khan, 1972). This is assumed to be related to aliphatic polysaccharide components in the humin molecule, which cause humin to be insoluble in alkaline solvents. The higher polysaccharide content in humin is sometimes used to distinguish it from humic acid, though emphasis has recently been placed on the paraffinic constituents for differentiating it from humic acid (Hatcher et al., 1985). The paraffinic substances are believed to be derived from nonvascular plants contributing to the formation of peat.

Extraction Procedures

Humin may be extracted by boiling the soil residue, following extraction of humic and fulvic acids, in a 0.5 N NaOH solution. Repeated treatments with a mixture of HCl and HF of the residue are suggested prior to alkaline extraction. Since humin is believed to exist as a humic-clay complex, the treatment above is considered required to break the bond with the clay and silica materials, enhancing extraction of the compound (Stevenson, 1994). A more rigorous method is refluxing the soil residue with 6 N HCl (Passer, 1957), whereas Hatcher et al. (1985) recommend pretreatment of the soil residue successively with benzene/ethanol to remove lipids and 1 N HCl, prior to extraction of humin. The use of a concentrated HCl + HF mixture (1:1 v/v) is even suggested for marine sediments in order to remove by hydrolysis the strongly bonded polysaccharide and protein components. The authors above believe that such a drastic HCl + HF treatment has no effect on the extracted humin, but others are of the opinion that even the use of HCl alone may alter the extracted humin. The acidified sample is then extracted for humin by washing with 0.1 N HCl. This procedure is supposed to yield a hydrogen-saturated humin.

Recently a new method was introduced for isolation of humin by partitioning a 0.5 M NaOH soil (or peat) extract acidified to pH 1.0 between methyl isobutyl ketone (MIBK) and water (Rice and MacCarthy, 1989). The material suspended in the MIBK phase is according to the authors above humin.

Types of Humin

Hatcher et al. (1985) make a distinction between *humin in aerobic* and *humin in anaerobic (peat) soils*. Both types or groups of

Extraction and Fractionation of Humic Substances 73

humin appear to be similar in composition, and differ only in degree and rate of decomposition of the polysaccharide and paraffinic constituents in the humin molecule. These components are better preserved in anaerobic than in aerobic environments. The authors above assume that in the formation of humin some kind of selective preservation of the organic components occurs. Polysaccharides, or carbohydrates in general, are known to decompose first from soils. In addition, Hatcher and coworkers believe that humic acids are the degradation products of humin rather than precursors or are compounds genetically unrelated to humin. They consider humic acids in ancient sediments nothing more than degradation products of humin. In addition, the authors assume that with geologic age, humin can be transformed under anaerobic conditions into coal or kerogen without passing first into a humic acid form.

Aerobically Derived Humin. -- This is the type of humin present in terrestrial soil where aerobic conditions prevail. It is believed to be different from humic acid, because of its higher aliphatic character due to the presence of high amounts of polysaccharide and paraffinic constituents in its molecule. In the older literature, humin is reported to contain 40% carbohydrates (Russell and Russell, 1950).

Anaerobically Derived Humin. -- This is the type of humin in peat and other aquatic environments. A distinction is made by Hatcher et al. (1985) between humin from peat and marine deposits. The authors believe that *peat-humin* is composed of lignin, polysaccharides, and paraffinic substances. Some are even claimed to be a two-component system, composed of 60% paraffinic and 40% lignoid substances, because of selective degradation processes. Carbohydrates are considered to be the major constituents in humin at the peat surface, but decrease in concentration with depth due to breakdown by decomposition. Paraffinic substances are said to be selectively preserved in the humin with depth in the peat deposit. In contrast to peat-humin, the authors above claim that *marine-humin*, from algal-sapropel deposits, is composed almost entirely of polysaccharides and paraffinic substances. However, a similar selective decomposition process is reported to occur with this type of humin. The authors above note that the polysaccharide content in the humin also decreases with depth in the sapropel deposit, whereas the paraffinic components increase due to selective preservation processes.

Humin Fractions

Inherited Humin. – This is a humin fraction separated by ultrasonic treatment (Almendros and Gonzalez-Vile, 1987). It is assumed to be a mixture of degradation products of lignin polymers and subcellular components of plants.

Humin Fractions of Rice and MacCarthy (1989). – The humin in the MIBK phase above is, in the authors' term, *disaggregated* by adding water. After vigorous shaking three fractions are produced: (1) a lipid fraction in the MIBK phase, referred to as bitumen by Stevenson (1994), (2) a humic acid-like fraction in the alkaline solution phase, and (3) an insoluble non-humic fraction, usually settling out at the bottom of the flask.

Not much is known yet on the physico-chemical and biological characteristics of humin. The electrochemical properties including cation exchange and chelation capacity remain a mystery, providing new frontiers for detailed investigations. The significance and impact of the results of such research on the soil ecosystem and the environment will no doubt also be very prominent, because humin is an important part of the organic cycle. The results of limited investigations showing total acidity values of humin, ranging from 326 to 636 cmol/kg (Rice and MacCarthy, 1989), are signs for the presence of enormous chemical activities, though less than those of fulvic and humic acid. However, more research is needed to advance or support this knowledge. What is found most important today is the contribution of humin to the carbon cycle and its possible conversion with geologic age into fossil fuel.

CHAPTER 4

GENESIS OF HUMIC MATTER

4.1 MAJOR PATHWAYS OF HUMIFICATION

The process by which humic matter is formed has been called *humification*, which involves a number of biochemical reactions. It is closely connected to the organic and nitrogen cycles in the environment. Though some people are of the opinion that the mechanisms for synthesis are not clear, a number of hypotheses have in fact been presented on how humic matter is formed. In general, these theories differ in the way the sources of original or raw materials are utilized in the synthesis of humic substances. Whereas one group of theories is based on depolymerization of biopolymers causing their direct transformation into humic substances, the other group envisages polymerization of small molecules, liberated by complete decomposition of the biopolymers, in the formation of humic matter. All agree that the materials for formation originate mostly from plant material, though in practice animal residue can also be transformed into humic matter. The depolymerization theory, called *biopolymer degradation* by Hedges (1988), assumes that the biopolymers in plants are gradually transformed into humin, which eventually will be degraded successively into humic acids and fulvic acids. The lignin theory of Waksman (1932) and its modern version are considered examples of the biopolymer degradation theory. In contrast, the polymerization theory claims that the plant biopolymers are decomposed first into their monomers or smaller organic components. Humic substances are then formed by interaction reactions between these small components. This theory assumes fulvic acid to be formed first, which by polymerization or condensation can be transformed into humic acids. The polyphenol or phenol, quinone, and sugar-amine condensation theories belong to the category of the polymerization theory. This second pathway of humification has recently also been called the *abiotic condensation process* (Hayes and Malcolm, 2001). The lignoprotein theory of Flaig et al. (1975; 1988), focusing on the breakdown of lignin and further oxidation of the degradation units into quinone derivatives, is an excellent example of the polymerization or abiotic condensation theory. Hayes and Malcolm (2001) believe that the rate of depolymerization depends on the oxygen content, and humification will be retarded in anaerobic conditions. It is true that a lot of oxygen is required for oxidation reactions, but the issue can be raised whether a lack of oxygen will severely inhibit the humification process. As discussed in Chapter 2, huge deposits of peat and bogs, rich in humic matter, are instead formed in wetlands, where anaerobic conditions prevail.

Another important question is whether biopolymer degradation is really a humification process. Is humification a decomposition or a polymerization process? The present author would like to refrain from assessing judgment now and let the readers draw their own conclusion after reading the sections below on humic precursors and several theories on humification processes.

4.2 PRECURSORS OF HUMIC MATTER

The plant biopolymers of importance in humic matter synthesis are for convenience called precursors of humic substances. The major components of higher plants, important as sources for formation of humic matter, are lignin, cellulose and hemicellulose, called poly-

Genesis of Humic Matter

saccharides, and proteins. Phenols and amino sugars synthesized by microorganisms have recently been added as important raw materials for the synthesis of humic substances. Since degradation of lignin can also produce phenols, two sources of phenolic compounds can be distinguished in soils. All these compounds, present originally in the form of large molecules in the plant tissue and soils, will be discussed in more detail below in order to give a better picture of their characteristics and reactions related to the formation of humic substances. Moreover, many people are often confused about what the biopolymers are, what aromatics are and what the difference is between phenol and quinone. Even some hard-core scientists wonder about terms such as phenolic-OH and the like. It sounds like basic organic biochemistry, but it is not, though some of the basic definitions are needed to explain the chemical behavior of the compounds, which is necessary in understanding their interaction reactions in humic matter formation.

4.2.1 Lignin

Lignin is a system of thermoplastic, highly aromatic polymers of the phenylpropane group. The name is derived from the Latin term *lignum* = wood. It is one of the three major components of wood, with the other two being cellulose and hemicellulose (Schubert, 1965). The bulk of lignin occurs in the secondary cell walls where it is associated with cellulose and hemicellulose. It is noted to coexist with the cellulosic plant components in such an intimate association that its isolation requires drastic chemical treatments that often alter the structure of the lignin itself. The latter raises questions about the assumption held by most biochemists that the lignin is associated physically, rather than chemically, with the polysaccharides. The nature of the lignin-polysaccharide complex has still to be resolved and more definite data need to be presented refuting one or the other or supporting the presence of both physical and chemical interactions.

The quantity of lignin increases with plant age and stem content. It is not only an important constituent of the woody tissue, but it contains the major portion of the methoxyl content of the wood. A large amount of lignin is also detected in the vascular bundles of plant tissue. The purpose is perhaps to strengthen and make the xylem vessels more water resistant. By virtue of the presence of larger amounts of vascular bundles, the lignin content of tropical grasses is considerably larger than that of temperate region grasses (Tan, 2000; Minson and Wilson, 1980). Consequently, soils under tropical grasses are expected to have higher lignin contents than soils under temperate region grasses. These differences may produce differences in the nature of humic substances formed.

Lignin Monomers

The building stones of lignin are monomeric lignin possessing a basic *phenylpropane* carbon structure. Three types of lignin monomers can be distinguished on the basis of the type of wood or plant species, e.g., coniferyl, sinapyl, and ρ-coumaryl monomers (Figure 4.1). The coniferyl type characterizes lignin in softwood or coniferous plants, and the sinapyl type represents lignin in hardwood, whereas the coumaryl type is typical of lignin in grasses and bamboos. Several of these monomers are linked together to form the total lignin polymer. The process, called polymerization, forms a very complex and long series of a lignin polymer structure (see Tan, 2000).

Aromatization

The ultimate source for formation of lignin is carbohydrates or intermediate products of photosynthesis related to carbohydrates. The process of conversion of the nonaromatic carbohydrates into substances containing phenolic groups characteristic of lignin is called *aromatization*. Enzymatic reactions are required to effect such a drastic transformation of nonaromatic carbohydrates into aromatic precursors of lignin. Several theories have been advanced on the aromatization process, e.g., aromatization of carbohydrates through a *dehydration* process and the *shikimic acid pathway.*

Genesis of Humic Matter

SOFTWOOD	HARDWOOD	GRASS-BAMBOO
Gymnosperm	Dicot. angiosperm	Monocotyledons

Coniferyl alcohol — Sinapyl alcohol — p-Coumaryl alcohol

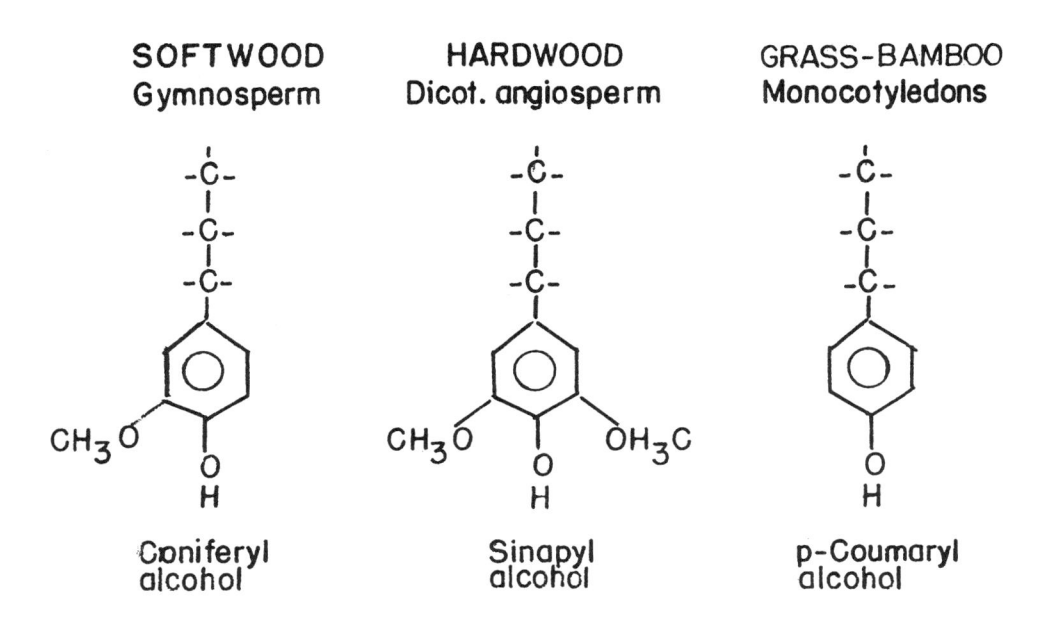

Figure 4.1 Lignin monomers from softwood, hardwood, and grass or bamboo.

In dehydration theory, carbohydrates, such as fructose, are releasing three water molecules, and with the assistance of enzymatic reactions, three possible aromatic end products are produced, e.g., pyrogallol, hydroxyhydroquinone, phloroglucinol, or a combination thereof (Figure 4.2).

The shikimic acid pathway has been adopted from the theory for the biosynthesis of aromatic amino acids from carbohydrate precursors with the help of enzymes originating from *Escherichia coli* bacteria (Schubert, 1965). The end products, phenylpyruvic acid and p-hydroxyphenylperuvic acid, yield by transamination reactions phenyl-

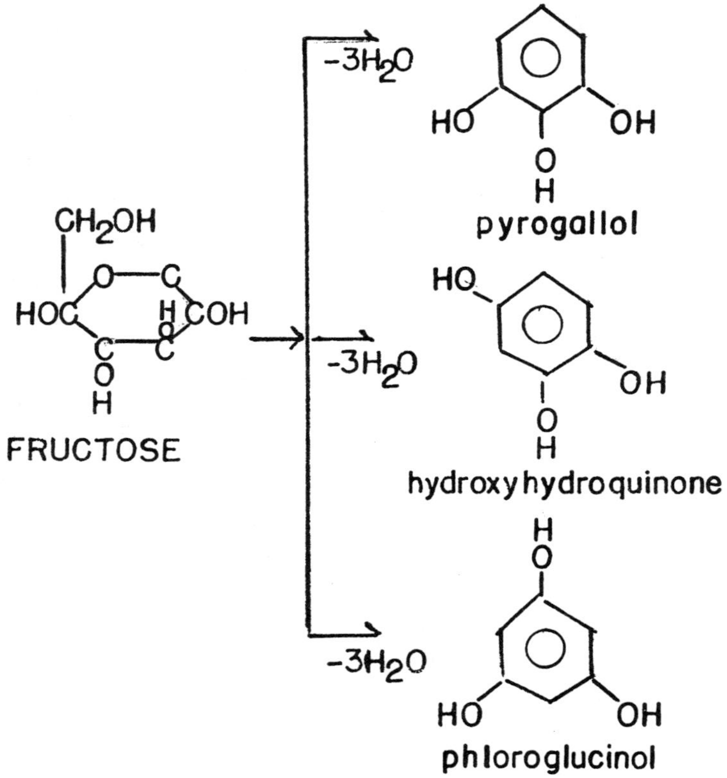

Figure 4.2 Aromatization of fructose through a dehydration process.

alanine and tyrosine, respectively. As illustrated in Figure 4.3, the chemical structures of these compounds show close similarities to those of the monomeric units of lignin. In particular, the structure of p-hydroxyperuvic acid is almost the same as that of p-coumaryl lignin, leading to the assumption that lignin monomers may have been formed through similar processes. In addition, the structures of phenylalanine and tyrosine are also very similar to those of ligno-protein compounds, the humic substances according to the ligno-protein theory. These find-

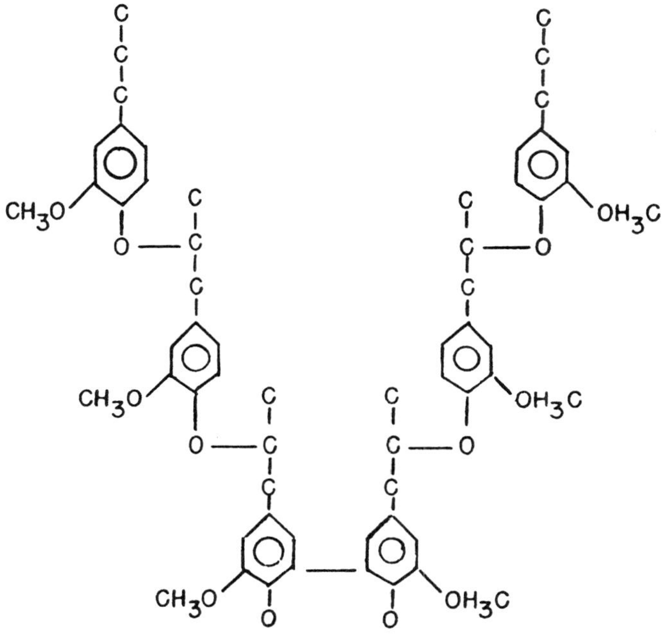

Genesis of Humic Matter

Figure 4.4 A hypothesis for a softwood lignin structure by a systematic linkage of coniferyl alcohol monomers.

cetes (Schubert, 1965; Paul and Clark, 1989). Several forms of these so-called *lignolitic fungi* have been reported as the major organisms responsible for the partial decomposition of lignin, e.g., *white-rot*, *brown-rot*, and *soft-rot fungi*. In well-aerated soils, the white-rot fungi are reported to decompose wood containing lignin into CO_2 and H_2O. Patches of a white substance are often formed in the residue, hence the name white-rot. These white patches have been identified as pure forms of cellulose. According to Paul and Clark (1989), the brown-rot fungi are useful for the removal of the methoxyl, $-OCH_3$, group from lignin, leaving the hydroxyphenols behind, which upon oxidation in the air produce brown colors. However, Schubert (1965) believes that the cellulose and other associated carbohydrates are attacked preferentially, leaving the lignin behind, which turns the residue brown in color. The soft-rot fungi are most active in wet soils and are specifically adapted to decomposing hardwood lignin.

The hydroxyphenol units resulting from demethylation of lignin by white-rot fungi can be oxidized to form quinones. The latter are believed to be capable of reacting with amino acids to form humic substances (Flaig et al., 1975). Lignin itself has the capacity to react with NH_3. The process, called *ammonia fixation*, has been applied in industry for the production of nitrogen fertilizers by treatment of lignin and other materials rich in lignin, e.g., sawdust, and peat, with NH_3 gas. The exact mechanism of fixation is still not known, but it is believed that the NH_3 reacts with the phenolic functional groups in lignin.

4.2.2 Phenols and Polyphenols

Phenols are aromatic carbon compounds with a general formula of C_6H_5OH. They are derived from benzene, C_6H_6, by replacing one or more of the hydrogens with OH. Benzene, a flammable colorless compound, is called aromatic because of its characteristic structure marked by six carbon atoms linked by alternate single and double bonds in a symmetrical hexagonal configuration. The C_6H_5 group in phenol is called the phenyl group, from the Latin term *phene* = shining, because burning benzene produces a very bright light.

By linking several monomeric phenols together polyphenols are produced. As indicated earlier, the phenols and polyphenols can be derived from two sources, from the decomposition of lignin and from the synthesis by microorganisms. Stevenson (1994) believes that uncombined phenols are present in higher plants in the form of glucosides and tannins.

Lignin Derived Phenols and Polyphenols

Biodegradation of lignin has been implicated in producing polyphenols and phenols. Specific types of fungi have been discovered

Genesis of Humic Matter

capable of attacking lignin, compounds that are generally very resistant to microbial decomposition. In addition to the *Basidiomycetes* referred to earlier, another group, the *Ascomycetes,* has also been mentioned as important lignin-degrading organisms (Schubert, 1965). These organisms attack lignin by excreting enzymes in the phenoloxidase group, which can be distinguished into two basic types of enzymes, tyrosinase and laccase.

The mechanism of phenol formation from lignin is in essence the reverse process of lignin synthesis. Complex diagrams have been presented by a number of authors showing pages of flow sheets illustrating the degradation of lignin into its monomeric type that through a labyrinth of successive reactions is broken down into phenols (Haider et al., 1975; Schubert, 1965). A shorter and less complex diagram has been presented by Flaig et al. (1975; 1966). To avoid confusion by presenting these complex diagrams as is done in many other books, and to underscore the purpose for better comprehension by a variety of readers, a simple diagram is provided in Figure 4.5 as the present author's version of the degradation of lignin into phenols. This simplified diagram shows, what all the other authors want to imply, that lignin is broken down into its basic unit (coniferyl, sinapyl or coumaryl alcohol). The basic unit is subject to oxidation followed by demethylation and converted to a phenol compound.

Microbial Phenols

Microorganisms are reported to also contribute in producing humic precursors. A great variety of phenolic and hydroxy aromatic acids are known to be formed by microorganisms from nonaromatic hydrocarbon substances. Many fungi, actinomycetes, and bacteria have been cited to be capable of synthesizing by secondary metabolic processes simple phenols and complexed polyphenols. However, such ability is deemed to be more a characteristic of the fungi and actinomycetes than of the bacteria (Stevenson, 1994). A variety of soil fungi, including *Aspergillus, Epicoccum, Hendersonula, Penicillium, Euratium,* and *Stachybotrys* species, have been reported to produce humic acid-like substances in cultures containing glucose, glucose-

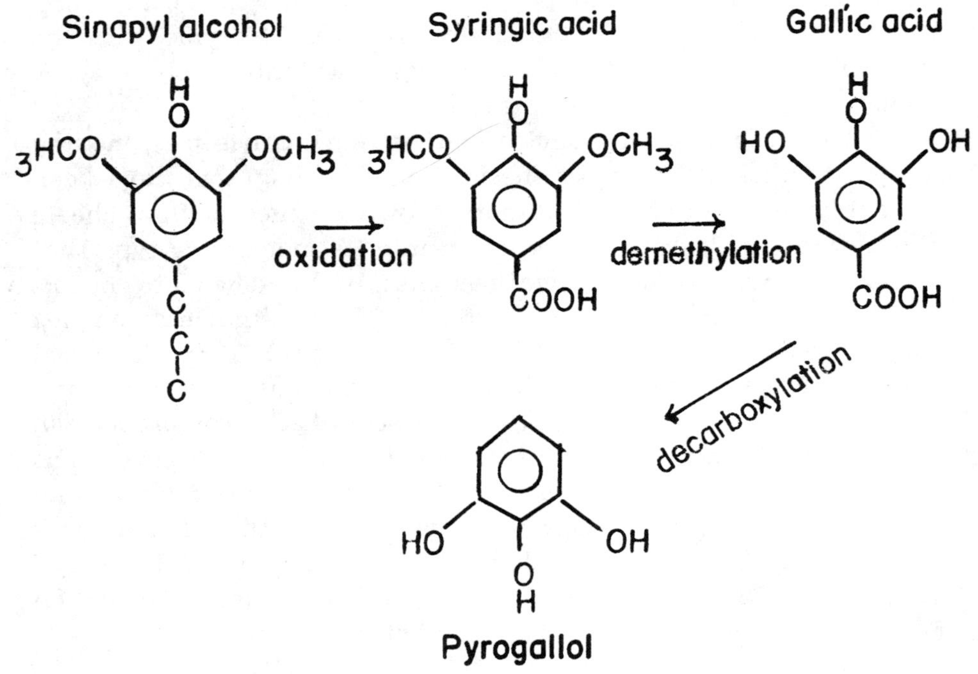

Figure 4.5 Simplified version of formation of pyrogallol by decomposition of lignin. (After Martin and Haider, 1971; 1975; Flaig et al., 1975; and Haider et al., 1975.)

$NaNO_3$, asparagine, and peptone (Filip et al., 1974;1976; Saiz-Jiminez et al., 1975). The substances formed are identified by chemical analysis to be composed of phenols, orsellinic, ρ-hydroxybenzoic, ρ-hydroxycinnamic acids, anthraquinones and melanins. Their appearance as dark-colored microbial products in the culture media is the reason for associating them with humic acids, since phenols and their derivatives are known to be building constituents of humic matter. Formation of

Genesis of Humic Matter

humic acid-like substances by mycorrhizal fungi has also been reported by Tan et al. (1978). A brownish substance is produced by the ectomycorrhiza *Pisolithus tinctorius*, grown in a Melins-Norkrans liquid culture with either sucrose or a mixture of L-malic and L-succinic acid as the C source. The brown colored substance behaves similarly to fulvic and humic acid when subjected to extraction procedures with NaOH and HCl. The substance, which is soluble in base and insoluble in acid, exhibiting infrared absorption features similar to humic acid, is believed to be composed of uronic acids. These acids are known to be waste products of microorganisms, and many authors are of the opinion that they contribute to formation of humic matter (Flaig et al., 1975).

The most probable mechanisms for the microbial synthesis of these humic precursors appear to be processes similar to those for the synthesis and/or decomposition of lignin. Two most probable mechanisms cited are the acetate-malonate and shikimic acid pathways. The data presented by Haider at al. (1975) suggest that in the acetate-malonate pathway, glucose is converted in orsellinic acid. Demethylation of the latter, followed by decarboxylation, yields resorcinol, a dihydroxyphenol. On the other hand, the shikimic acid pathway may produce pyrogallol as the end product. It is apparently a shorter pathway, since gallic acid is reported to be formed directly by aromatization of shikimic acid, which by decarboxylation produces pyrogallol, a trihydroxyphenol. Both resorcinol and pyrogallol are prominent microbial phenols, or the phenols typically produced by microorganisms. Pyrogallol is also an important product in the synthesis and in the degradation of lignin, as discussed earlier. Polymerization of these simple phenols yields polyphenols. A simplified version of the formation of resorcinol and pyrogallol is given below as illustrations (Figure 4.6). Although the two theoretical pathways have been designed to illustrate formation of different intermediate products, often both mechanisms may end up yielding a similar phenol, e.g., pyrogallol, as the final product. It is only a simple matter of hydroxylation of resorcinol to convert it into pyrogallol.

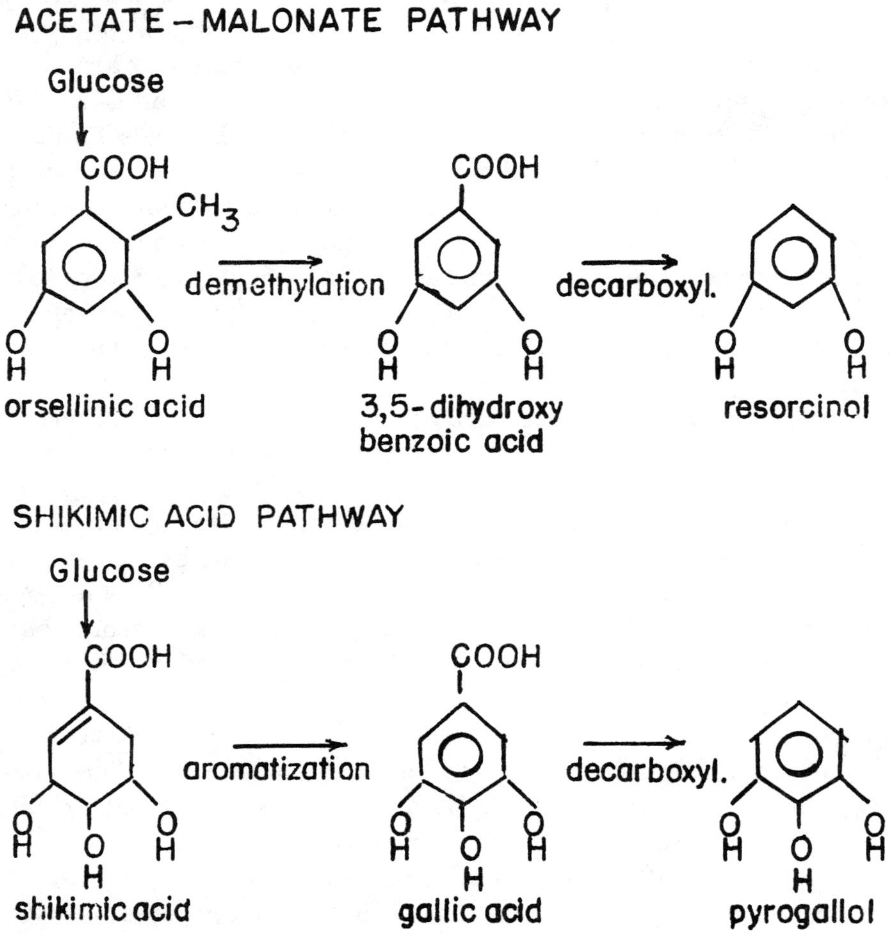

Figure 4.6 Bioformation of resorcinol and pyrogallol, according to the acetate-malonate and shikimic acid pathway, respectively.

4.2.3 Quinones

Quinones are hydrocarbon substances with a formula of $C_6H_4O_2$. These compounds are usually yellowish to red in color and biologically important as coenzymes, as hydrogen acceptors, and as key constituents of vitamins. They are derived from phenols and are diketo derivatives of dihydrobenzene. Phenols formed by decomposition of lignin or by microbial syntheses are released in soils. They can be spontaneously oxidized in alkaline solutions, a reaction called *auto-oxidation* by Ziechmann (1994), and converted into quinones. The author indicates that the formation of quinone can be explained by the electron donor-acceptor theory. Ziechmann is of the opinion that the transformation is caused by intermolecular electron transfer, by which quinone is accepting 4-π-electrons donated by the phenol molecule. However, in a natural environment, enzymes are considered required in the oxidation of phenols. In this case, the transformation into quinones is not limited to oxidation of free phenols in soils, but can also take place with phenol compounds within the microbial tissue. The quinones formed can be secreted into the soil or can be released after the microbes die. Two groups of enzymes, phenolase and laccase, are considered to play an important role in the aerobic oxidation of phenols into quinones. Schubert (1965) reports that phenolase is capable of attacking mono- and dihydric phenols, whereas laccase catalyzes the oxidation of the polyhydric phenols. To illustrate the enzymatic oxidation of a phenol yielding a quinone, a simplified diagram of reactions involved is given below (Figure 4.7). The orcinol in the figure above is formed from the decarboxylation of orsellinic acid, an acid produced in the acetate-malonate pathway as shown in Figure 4.6. Demethylation of orcinol yields catechol, which in the presence of a suitable enzyme, e.g., phenolase, will be oxidized and converted into o-quinone. It should be realized that this is not the only method for formation of quinone and many other methods are possible. For example, decarboxylation and oxidation of dihydroxybenzoic acid may also yield quinones (Flaig et al., 1966; 1988). A revised reaction, made by the present author to enhance comprehension, is given in Figure 4.7 for comparison with the oxidation reaction of orcinol and the electron donor-acceptor concept in the conversion of phenol into quinone.

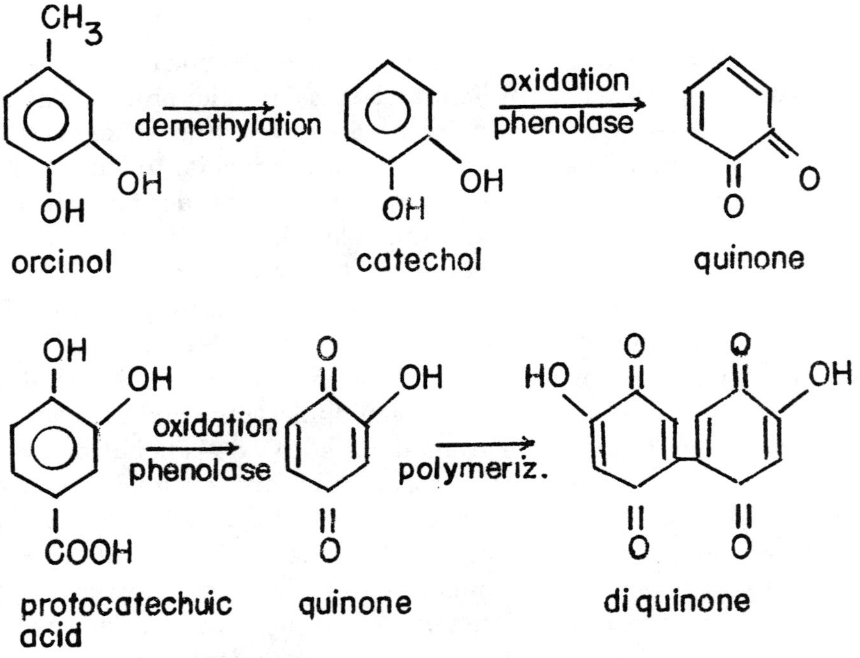

Figure 4.7 Simplified versions of catalytic oxidation of orcinol and benzoic acid, respectively, yielding quinone. (After Schubert, 1965; Flaig et al., 1975; Stevenson, 1994.)

4.2.4 Protein and Amino Acids

In the early days, protein and amino acids were not considered compounds making up humic matter. Many scientists believed humic acid to be a plain hydrocarbon substance and information has been presented off and on providing the argument for humic acid-like substances to be formed without protein. Even today, the idea still prevails that humic substances do not include peptides, nucleic acids,

Genesis of Humic Matter

sugars, and fats (Hayes and Malcolm, 2001). These biomolecules are believed to be sorbed or coprecipitated at pH 1 or 2 during the isolation procedures. However, the majority today considers humic matter to be characterized by an elemental composition showing a nitrogen content ranging from 1-5%. The latter is assumed to be contributed by amino acids and/or protein compounds (Schnitzer and Khan, 1972; Stevenson, 1994), also called peptides, as will be discussed below. To people advancing the ligno-protein theory, protein and amino acids are considered important humic precursors (Kononova, 1961; 1966; Flaig et al., 1966). Some scientists even try to make a distinction between fulvic and humic acids on the basis of the types of amino acids present in their molecular structure. Sowden et al. (1976) indicate that fulvic acids contain higher amounts of basic amino acids, whereas humic acids contain more of the acidic types of amino acids.

By definition, proteins are complex combinations of amino acids. These acids are given the name amino acids because the nitrogen in their molecules occurs as an amino (NH_2) group attached to the carbon chain. The acid part consists of a terminal C linked to an O atom and an OH group, often written as -COOH. The latter, called a carboxyl group, exhibits acidic properties, because the H of the OH radical can be dissociated. The protein is formed by the linkage of amino acid molecules through the carboxyl and amino groups:

$$\begin{array}{cccc}
\text{H} & \text{O} & \text{H} & \text{H} & \text{O} & \text{H} & \text{O} & \text{H} & \text{O} \\
| & \| & | & | & \| & | & \| & | & \| \\
H_2N-C-C-OH & + & H-N-C-C-OH & \rightarrow & H_2N-C-C-N-C-C-OH + H_2O \\
| & & | & & | & | \\
\text{H} & & \text{H} & & \text{H} & \text{H} \\
\text{Glycine} & & \text{Glycine} & & \text{Dipeptide} &
\end{array} \quad (4.1)$$

The bond linking the two groups is called the *peptide bond*, and the compound formed is called a *peptide*, or protein. Under refluxing with 6 N HCl for 18-24 hours, the protein may be hydrolyzed into its constituent amino acids. Twenty-one amino acids are usually obtained as protein constituents, but in natural environments many other types of amino acids have been identified, which according to Stevenson (1994) do not belong to proteins. Over 100 amino acids and their derivatives are reported by Stevenson to be confined as constituents or products of soil microorganisms.

Both amino acids and protein are major sources of nitrogen compounds in soils. They are perhaps less difficult to break down than lignin, but more difficult than the carbohydrates. The ease of decomposition depends on the size and their molecular structures, which appear to increase in complexity with the type of compounds. The size and complexity in molecular structure increase from aliphatic, to aromatic, and heterocyclic amino acids. In addition, many of the proteins also occur in nature in complex combination, called *conjugated*, with other compounds, complicating further the decomposition of these compounds. For example, glycoproteins in plant and animal tissue are protein conjugated with glycogen. Glucoprotein is a protein present in combination with the carbohydrate glucose, whereas lipoprotein is protein conjugated with lipids. Mucoprotein, a very important form of protein in the mucous layers of plants and animals, is supposed to be protein combined with uronic acids and other sugars. All of these factors will, of course, affect the rate of decomposition of protein and amino acids. For more details on the basics of amino acids and protein, see Tan (1998; 2000).

Decomposition of Protein and Amino acids

In contrast to lignin and phenols, protein and amino acids are major food sources for microorganisms. The nitrogen in these substances is an essential element for the growth of microorganisms as well as for the higher plants. Hence, it is expected that protein and amino acids will be subject to immediate attack by a host of microorganisms. These processes are part of the nitrogen cycle in soils and the environment. From the array of decomposition products produced, some will be adsorbed by clay minerals whereas others will be used in formation of humic substances. This part of the degraded protein and amino acid is considered temporarily resistant to further mineralization into NH_3.

Genesis of Humic Matter

The main reaction process for the decomposition of protein and amino acids is *hydrolysis*. Hydrolysis of protein, brought about by the enzymes *proteinase* and *peptidase* of soil microorganisms, results in cleavage of the peptide bonds, releasing in this way the amino acid constituents. The latter substances are broken down further into NH_3 by the enzymes called *amino acid dehydrogenase* and *oxidase*. Schematically the main pathway of decomposition can be illustrated as follows:

$$\text{Proteins} \rightarrow \text{peptides} \rightarrow \text{amino acids} \rightarrow NH_3 \quad (4.2)$$

The decomposition reactions above involve processes called *deamination* causing the destruction of the amino group or its conversion into NH_3 gas as part of the nitrogen cycle. Deamination can take place in aerobic as well as in anaerobic conditions, hence can be distinguished into *oxidative* and *non-oxidative deamination*, respectively (Gortner, 1949; Stevenson, 1986).

The reaction for oxidative deamination can be written as follows:

$$R\text{-}CH(NH_2)COOH + O_2 \rightarrow RCOOH + CO_2 + NH_3 \quad (4.3)$$
amino acid

Anaerobic deamination may result in (1) deamination and reduction and (2) decarboxylation, as can be noticed from the reactions below:

1. *Deamination and reduction*:

$$R\text{-}CH(NH_2)COOH + H_2 \rightarrow RCH_2COOH + NH_3 \quad (4.4)$$
amino acid $\qquad\qquad\qquad$ acetic acid

2. *Decarboxylation*:

$$R-CH(NH_2)COOH \rightarrow R-CH_2NH_2 + CO_2 \quad (4.5)$$
amino acid → amine

Reaction (4.4) indicates that deamination is characterized by the destruction of the amino group and its transformation into ammonia, NH_3, gas. In contrast, reaction (4.5) shows that decarboxylation involves the decomposition of the carboxyl, COOH, group into CO_2, and the subsequent transformation of the amino acid into an amine compound. The enzyme required for decarboxylation, called *amino acid decarboxylase*, is produced by *Clostridium* bacteria. When formed in animal bodies, some of the amines produced are reported to have important physiological effects. For example, *histidine decarboxylase* in animal tissue can produce histamine, an amine that can stimulate allergic effects and/or gastric secretions. Another enzyme, *tyrosine decarboxylase*, is an intermediate in the formation of *adrenaline*, an amine functioning as a *vasoconstrictor*. It is usually released in the bloodstream when a person or animal is startled or frightened (Conn and Stumpf, 1967).

All of the proteinaceous substances in their slightly or highly degraded forms are considered by many scientists to play an important role in the formation of humic matter. Most of the N content in fulvic acid is contributed by amino acids, whereas at least one-half of the N content in humic acids can be accounted for as amino acids. Lower percentages of the N in humic acids are present as NH_3, a compound apparently derived from the deamination reaction as shown in reaction (4.4). The nitrogen compounds associated with humic acids are assumed to be linked to the central core of the humic molecules.

4.2.5 Carbohydrates

Carbohydrates are perhaps the most important constituents of plants. They are considered as one of the three major groups of food substances, with the other two being protein and oil. They are

Genesis of Humic Matter

synthesized first by green plants by a process called photosynthesis, after which production of protein and oil then begins. In living plants, carbohydrates serve as sources of energy for many biological functions, and play an important role in the synthesis of nucleic acids, lignin, and other structural components in the plant tissue, in addition to protein and oil.

The carbohydrate compounds are more controversial than protein and amino acids in the issue of humic matter formation. For a long time they were regarded as contaminants rather than as precursors of humic matter. In the beginning of the twentieth century Maillard's (1916) revelation that humic matter can be synthesized from simple sugars, e.g., sucrose, compelled many scientists to start reviewing the idea of carbohydrates as possible building constituents of the humic molecule. Maillard's abiotic theory of the synthesis of humic matter from sugar is known today as *Maillard's reaction*. However, it was the discovery of aquatic humic matter that has propelled the role of carbohydrates as major contributors in the formation of humic matter. The concept of aquatic humic matter, and in particular of marine and autochthonous aquatic humic matter, is based on a carbohydrate-protein combination (Nissenbaum and Kaplan, 1972; Hatcher et al., 1985). The hypothesis was presented that this aquatic humic matter is a sugar-amino acid condensation product, though some regard it as being derived by autoxidative cross-linking of unsaturated lipids from plankton (Harvey and Boran, 1985). In terrestrial humic matter, polysaccharides have been identified earlier as important components of fulvic acids, whereas hymatomelanic acid is believed to contain polysaccharides bonded by ester linkages (Tan and Clark, 1968; Clark and Tan, 1969; Tan, 1975). To these carbohydrates are currently added amino sugars as possible precursors of humic acids. Biologically resistant complexes are formed by reaction with lignin and phenols.

Saccharides

Sugars are formed from carbohydrates, which are compounds yielding polyhydroxyaldehydes or ketones upon hydrolysis. The sugar glucose is an example of an aldose, whereas fructose is an example of a ketose. The carbohydrates, also called saccharides, are scientifically distinguished into three groups of saccharides: (1) monosaccharides, (2) oligosaccharides (Greek *oligos* = few), and (3) polysaccharides. The monosaccharides are the simple sugars, e.g., glucose and fructose, whereas the oligosaccharides are compound sugars composed of two to ten monosaccharides. Like our table sugar, a disaccharide, they are soluble in water and sweet in taste. On the other hand, polysaccharides are complex carbohydrates and are composed of many (ten or more) types of sugars or monosaccharides. They are sometimes distinguished into homo- and heteropolysaccharides. Homopolysaccharides are composed of repeating units of the same monosaccharides, whereas heteropolysaccharides are made up of different monosaccharides. Some of the units, bonded together by glucosidic bonds, are glucose, xylose, and arabinose. Starches, cellulose and hemicellulose are examples of polysaccharides, and as such are not called sugars. They are usually amorphous and tasteless, and disperse in water to form colloidal suspensions. For more details on the basics and chemistry of saccharides or carbohydrates reference is made to Tan (1998).

Mono- and Oligosaccharides. - Since carbohydrates are also the principal foodstuffs for soil microorganisms, they are rapidly attacked by the microbial population in soils. The simple sugars and the disaccharides are the preferred source of materials, and are subject to anaerobic and aerobic decomposition reactions. In the aerobic process, the sugars are broken down completely into CO_2 and H_2O, while the energy released is used by the microbes for growth and other biological processes. In the anaerobic process, the sugar is broken down into CH_4, methane, and CO_2. The decomposition processes can be illustrated by the reactions below:

$$C_6H_{12}O_6 + 6O_2 \rightarrow 6CO_2 + 6H_2O + energy \tag{4.6}$$

$$C_6H_{12}O_6 \rightarrow 3CH_4 + 3CO_2 \tag{4.7}$$

Genesis of Humic Matter 97

A partial decomposition is also possible by microbial fermentation, resulting in the production of ethyl alcohol. This process can be illustrated as follows:

$$C_6H_{12}O_6 \rightarrow 2C_2H_5OH + 3CO_2 \qquad (4.8)$$

The relatively rapid decomposition as discussed above may indicate that most of the simple sugars have been broken down before they can be used for formation of humic matter. Though it appears that in the competition for sugars, between microorganisms and humic acid synthesis, microorganisms have the advantage, a substantial amount of the sugars may in fact escape decomposition. Some may be adsorbed in intermicellar spaces of expanding clay minerals rendering them inaccessible to enzymatic attack, whereas others may enter into complex combination with toxic metals making them less susceptible to microbial attack. Additional mono- and oligosaccharides can also be produced by the decomposition of polysaccharides that are next in line in the degradation process. The resistance of polysaccharides to enzymatic attack by microorganisms depends on a number of factors. Polysaccharides are known to be able to form branch-like structures, and the greater the amount of branching, the greater will be the resistance to enzymatic degradation.

Soil Polysaccharides. – Soil polysaccharides may be different from the original plant polysaccharides discussed above. Some of them can be produced by soil microorganisms, whereas others are believed to be formed in situ (in the soil) from the partial degradation products of plant polysaccharides and free monosaccharides. The latter are derived from the decomposition of plant and microbial residues. Polymerization of these degradation products and of the free monosaccharides is reported to yield polysaccharides that are very heterogeneous and highly branched in structure. Linkage is believed to be induced by enzymes released during autolysis of microbial cells, and the 'new' polysaccharides are considered even less susceptible to biodecomposition than their plant counterparts (Stevenson, 1994; Martin et al., 1975). However, some people feel that the resistance of soil polysaccharide to microbial attack is due more to adsorption by clay minerals and chelation with toxic metals than to complex molecular structures (Cheshire et al., 1977). Regardless of the differences in opinion, this resistance is one of the reasons why polysaccharides can accumulate in soils, though their concentrations rarely amount for more than 20% in soil humus. These soil polysaccharides then serve as additional building materials for the synthesis of humic compounds. However, the opinion is present that all these carbohydrates are not considered parts of the humic molecule core. Several scientists believe that they are important only as attachments to peripheral side chains of the humic molecule.

Amino Sugars

These compounds are simple sugars with substituted amino groups in their carbon chains. The most common form of an amino sugar is *glucosamine*, found as a component of *mucopolysaccharides* and *glycoprotein* present in saliva and eggs (Conn and Stumpf, 1967). Glucosamine-like substances have also been detected in the mucous layer encasing bacteria cells. *Galactosamine* has also been mentioned as an important amino sugar in soils. It is an epimer of glucosamine, differing from the latter only in placement of an OH group in the carbon chain.

According to Stevenson (1994) amino sugars have often been mistakenly referred to as *chitin*, the material of the hard shell of insects and crustaceans. Though chitin exhibits a basic molecular structure almost the same as glucosamine, it is in fact a polymer of N-acetyl-d-glucosamine. Perhaps the name chitin is confused with the term *chitosan*, which is indeed a polymer of glucosamine (Martin et al., 1975). This then may provide some justification why chitosan can be used as a general name for amino sugars. To explain more clearly the differences and similarities between glucose, glucosamine, chitosan, and chitin, the following molecular structures are presented as illustrations in Figure 4.8. Since the structures of simple sugars can

Genesis of Humic Matter

be written in several ways, three types of structures for glucose are given in the figure: (1) the open-chain, (2) ring, and (3) cyclic structures. In aqueous solutions, it is noted that an equilibrium exists between the forms with an open-chain and a ring or cyclic structure (Gortner, 1949; Tan, 1998).

The amino sugars are believed to serve several functions in soils. They serve as an important source of N for plant and microbial life, and affect the physical and chemical conditions of soils. From the standpoint of soil physics, mention has been made in the literature on interaction reactions between amino sugars or polysaccharides and soil mineral particles encouraging soil aggregation, hence formation of stable soil structures beneficial for plant growth and the environment (Greenland et al., 1961;1962; Baver, 1963). Currently, amino sugars are also considered as important components for the synthesis of humic matter. They can enter into reactions with phenols and quinones to form a basic humic molecule. In the abiotic Maillard's reaction, glucosylamine is produced first, leading to formation of *melanin*, a dark brown to black aromatic plant pigment found widespread in the natural environment (Ziechmann, 1994). The disintegration products are called *melanoids*. Some biochemists consider melanin to be a *chromoprotein*, the colored protein of certain seaweed and the material in black wool and hair of animals (Gortner, 1949). Whatever the nature is, melanin and melanoid are assumed to be very important precursors in the synthesis of humic acids by a process sometimes also called the melanoidin pathway (Nissenbaum and Kaplan, 1972; Hatcher et al., 1985).

4.2.6 Miscellaneous Humic Precursors

Other biochemical compounds of importance in the synthesis of humic matter present in soils are lipids, nucleic acids, chlorophyl, vitamins, and hormones. To this list should be added today also pesticides and their degradation products in view of the increased influence of agricultural and industrial operations on the soil ecosystem.

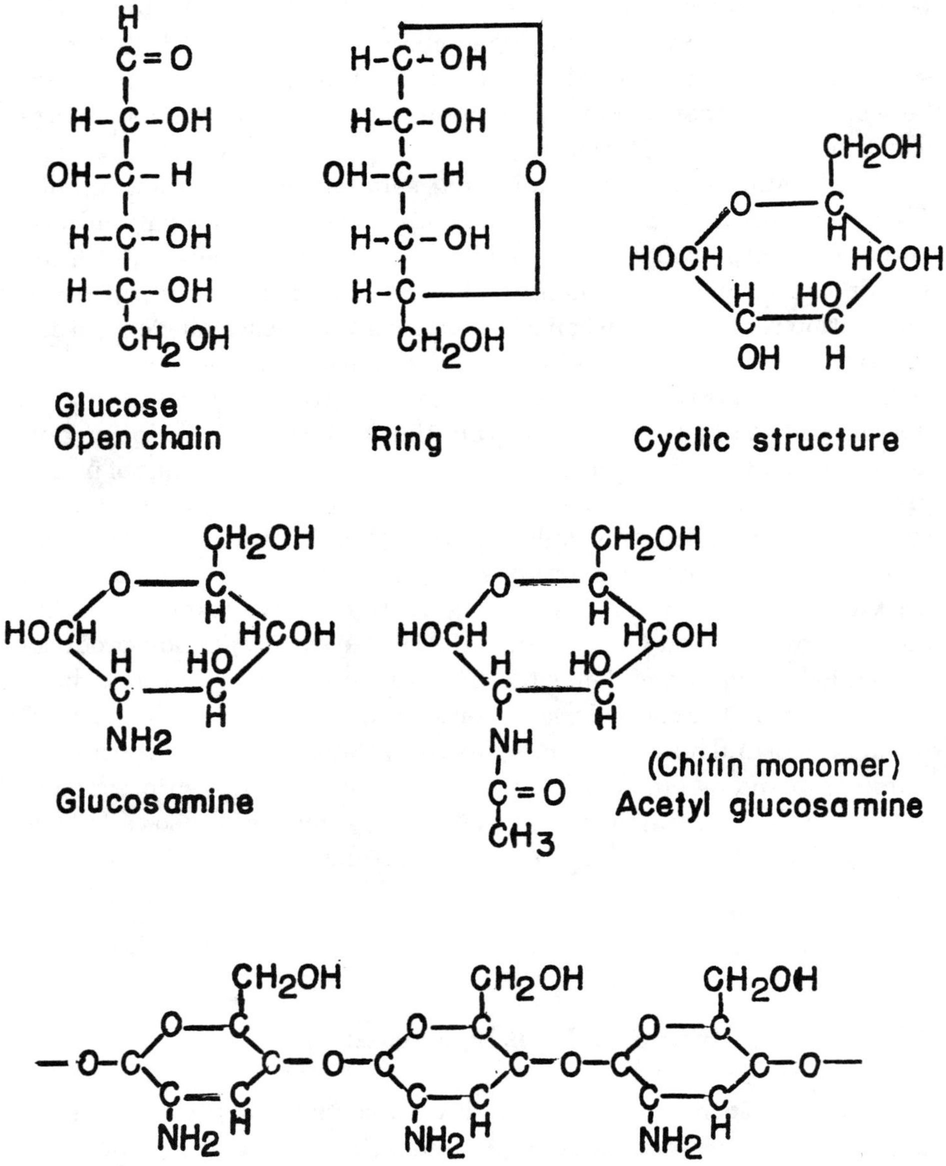

Figure 4.8 Molecular structures of glucose, glucosamine, chitin, and chitosan (some of the H and OH are not drawn due to space limitations).

Lipids

Lipids are heterogeneous compounds of fatty acids, waxes, resins, and oils. The term lipid does not imply a particular chemical structure, as with amino acid, but the name is used for substances that are soluble in fat solvents, such as ether, chloroform, or benzene. Chlorophyl and carotenoids are included in the group of lipids. For more details on the basics and types of lipids reference is made to Tan 1998; and Stevenson, 1994. Some of the soil lipids have found their origin from the higher plants, such as waxes covering leaves and fruits. Important sources of soil lipids are also terpenoids contributed by *Coniferaceae* and *Myrtaceae* plants. However, many of the soil lipids may have been derived from microbial tissue. Bacterial cells contain 5 to 10% lipids, whereas fungi may contain 10 to 25%. The microbial products are the main sources of the glycerides and phosphatides in soil lipids.

Lipids are known to affect the physical properties of soils, though not much information is available on this topic. They are hydrophobic compounds, and, therefore, will reduce the degree of wetting of soils. A high content of wax in soil humus is expected to make the soil *water-repellant*, whereas high concentrations of stearic acids, a lipoidic compound, have been implicated in causing *soil fatigue*.

Resistance to decomposition may vary considerably among the different types of lipoidic compounds. Many will decompose rapidly in well-drained soils, such as fatty acids, whereas others are relatively more resistant, like the wax, terpenoids, sterols, and paraffinic compounds. In general, it is assumed that in most soils sufficient amounts of microorganisms are present that can attack even the strongest lipid. Soil microorganisms are even available that can disintegrate oil spills, which are lipid-like compounds. The decomposition products may be in the form of lipoidic acids, which are high molecular weight organic acids containing hydroxyl and carboxyl functional groups, e.g., palmitic acid and stearic acid. These acids are important agents in the weathering of rocks and minerals, in the dissolution of plant nutrients, and in the mobilization and transportation of elements, critical in plant nutrition and soil genesis (Tan, 1986; 1998). Currently, lipids are also considered to contribute to the synthesis of the humic molecule. Paraffinic compounds have been discussed earlier as essential components of humin and kerogen (Hatcher et al., 1985).

Sterols and Steroids. – These compounds, classified as *derived lipids*, have received increased attention as additional potential components of the humic molecule. The name sterol is derived from *stereos* (meaning solid in Greek) and *ol* from alcohol, hence sterol means solid alcohol. The best-known sterol is *cholesterol*, which upon radiation will form vitamin D. These compounds are produced by animals and plants including microorganisms. *Phytosteroids* have been detected in fungi, *Chlorella*, *Chrysophycea*, *Diatomycea*, seaweed, and sphagnum. Soybean, rape seed, spinach, cabbage, and palm trees are some of the plants known to contain steroids. In particular, the vegetation of peat and peat bogs is believed to be rich in steroids (Ziechmann, 1994; Gortner, 1949). Many of these sterols and steroids are known today to exhibit medicinal properties and this is the reason why peat is often used for medical purposes. Peat baths were taken in the old days for therapy of gynecological and rheumatic diseases, and even today mud baths are offered in many European health spas and clinics. These therapeutic properties of peat, known for a long time, have currently been traced to humic acid as the dominant component of peat. Since the phytosteroids are present in soils as decomposition products of plant residues, the chances are that they may also participate in building up the humic molecule.

Nucleic Acids

Nucleic acids, first isolated in 1869 by F. Miescher from the nucleus of plants, are polymers of high molecular weights. Their repeating units are *mononucleotides*, rather than amino acids. Two types of nucleic acids are generally recognized: (1) *deoxyribonucleic acid (DNA)*, and (2) *ribonucleic acid (RNA)*. Both DNA and RNA consist of long chains of alternating sugar and phosphate residues twisted in the form of a helix. The strands are bonded together by the

Genesis of Humic Matter

purine bases *adenine* and *guanine*, and by the pyrimidine bases *thymine* and *cytosine*. For the basics of nucleic acids in soils, see Tan (1998).

Upon decomposition of plants, these nucleic acids will be released into the soil. Though, they are one of the organic compounds from living cells that are apparently broken down rapidly in soils, they are also potentially capable of being incorporated in the humic molecules. As shown in Figure 4.9, the molecular structures of especially the nitrogenous bases lend themselves very well to rapid in-

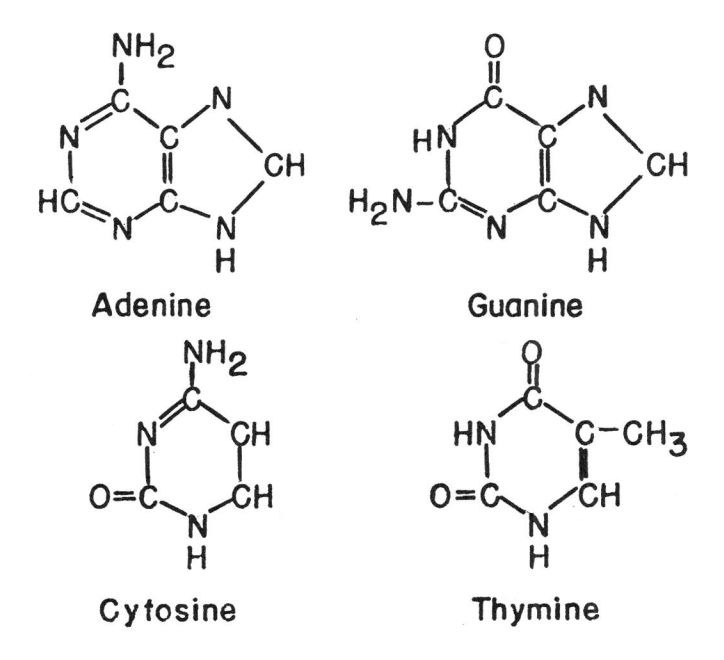

Figure 4.9 Molecular structures of nitrogenous bases of nucleic acids showing functional groups, potentially active for interactions with other humic precursors.

teraction with other humic precursors.

Because of their N and P content, nucleic acids are expected to be important sources for soil N and P. Humic acid contains a considerable amount of N that cannot be accounted for in analysis. This unaccounted for N content, called *HUN* for *hydrolyzable unknown N* or just *simply unknown* N, is believed to be associated with the N in nucleic acids or their derivatives (Schnitzer and Hindle, 1980).

Growth Promoting Substances

Humic matter has often been implicated in promoting hormone-like effects in plant growth. Stimulated seed germination, rapid root elongation, elongation of young seedlings, and accelerated shoot growth have frequently been assumed to be caused by a hormonal growth effect of humic matter (Poapst et al., 1970; 1971; Guminski et al., 1977). Some truth may be present in these allegations, since hormones can be released from plants and incorporated in the humic molecular structure. They are released in the soil upon plant and microbial decomposition. However, growth promoting substances are not confined only to hormones, since a number of other compounds are currently noted to possess similar effects as hormones. Degradation of lignin due to microbial decay is reported to produce substances exhibiting hormonal activity (Stevenson, 1994). Vitamins are also recognized to be able to promote plant growth. These growth promoting substances in general are synthesized by plants and soil microorganisms. They are formed by the organisms for a specific purpose that up till now remains mostly a mystery. Though their concentrations are always reported to be very low, they are expected to be present in soil humus and are potential participants in building up a humic molecule. The extremely low concentrations are perhaps one of the reasons why in many cases the effect is rather obscured.

Plant Hormones. - The term hormone is originally used for compounds affecting specific growth functions in animal bodies, and refers initially to chemical compounds secreted by endocrine glands

Genesis of Humic Matter

(Gortner, 1949). In plant science, it was Charles Darwin who started the idea of a substance present in plant tissue causing seedlings to bend when exposed to light. Darwin's theory, discussed in his book entitled the *Power of Movement in Plants* (1881), was later supported by results of light experiments in growing plant cells, conducted in 1910 and the following years by Boysen-Jensen (1936). Since then a large number of investigations have been carried out on the subject of growth promoting substances in plants, which finally resulted in the term hormone being adopted for compounds showing the capacity to stimulate the growth of plants. The name *auxin* from the Greek term to *increase* was later suggested by F. Kögl for this group of plant regulatory compounds (Gortner, 1949).

Many types of natural and artificially produced hormones are known today, e.g., auxin-a, auxin-b, *heteroauxin*, and *β-indoleacetic acid*. Auxin-a and b can be isolated from the oil extracts of corn, mustard, sunflower, and flax plants. Human urine also contains substantial amounts of these types of auxins. Indoleacetic acids are artificially produced auxins.

Substances regulating plant growth can also be produced by microorganisms. A fungus, called *Gibberella fujikuroi*, is noted to produce a compound, named *gibberellin*, that in low concentrations promotes root growth. The artificially prepared product, called gibberellic acid, a crystalline acid with the formula $C_{19}H_{22}O_6$, is associated with and similar in effect to gibberellin. Another example is the hormone mentioned above under the name heteroauxin. It is reported to be formed by a variety of fungi, including yeast, and has been isolated from *Aspergillus niger* and *Rhizopus sp*. Heteroauxin is considered highly effective as a growth promoting substance especially in lower plant life. This hormone has found today practical applications in industry and horticultural operations. Synthetic heteroauxins are for example β-indolebutyric acid and α-naphthaleneacetic acid. They are presumably active in root formation, and have been noted to induce formation of roots on plant cuttings, making possible commercial propagation of plants that normally cannot be propagated by this method.

As indicated above, this group of compounds may be incorporated in the synthesis of the humic molecule and is then the reason for the alleged hormonal effect of humic matter. The molecular structure of auxin (Figure 4.10) shows the presence of carboxyl and hydroxyl groups for easy hook-up to phenols, quinones, lignin, and other humic acid precursors by chelation or other interaction processes.

Vitamins. – These compounds have hitherto been considered only of importance in animal and human nutrition. The term *vitamine*, from

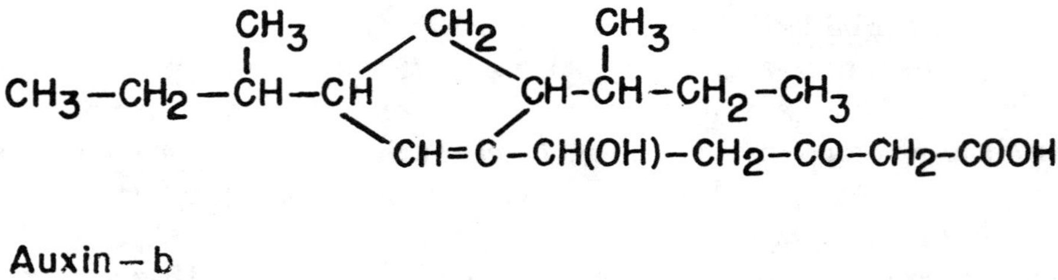

$$CH_3-CH_2-\underset{CH_3}{\overset{|}{CH}}-CH\underset{CH=C-CH(OH)-CH_2-CO-CH_2-COOH}{\overset{CH_2}{\diagup\diagdown}}CH-\underset{CH_3}{\overset{|}{CH}}-CH_2-CH_3$$

Auxin – b

Figure 4.10 Molecular structure of auxin showing carboxyl and hydroxyl functional groups for possible interactions with humic precursors.

which the present-day vitamin is derived, has been used to designate chemical substances essential in the diet of animals and humans that exhibit a hormone-like or enzyme action affecting the control and coordination of specific reactions in the animal or human body (Gortner, 1949).

The vitamins are usually distinguished into the (1) fat-soluble vitamins, e.g., vitamin A, D, E, and K, and (2) water-soluble vitamins,

Genesis of Humic Matter

e.g., vitamin C and vitamin B complex. At the present, it is especially the vitamin B complex that has attracted much attention in plant physiology and humic acid chemistry. It is called the B complex, since it is composed of several types of vitamin Bs, each responsible for specific functions in the plant body. Thiamin, riboflavin, nicotinic acid or niacin, vitamin B_6, biotin and pantothenic acid are examples of vitamin B. This B complex also includes aminobenzoic acid and inositol.

Since these vitamins are produced by plants, they are suspected to also play an important role in plant life though little is known on this aspect. The best-known vitamins produced by plants are perhaps vitamins A and C. Vitamin A is formed by carrots in the form of *β-carotene*, whereas vitamin C is produced especially by red pepper, citrus and pineapple plants. It is also called *ascorbic acid* due to its anti-scorbutic effect. For commercial purposes, vitamin C is synthesized from glucose, which is why its molecular structure shows close similarities to a pentose or sugar (Gortner, 1949). It lacks a carboxyl group, but perhaps the OH groups (Figure 4.11), behaving as Arrhenius or Brønsted acids, are the reasons for considering it as an acid. Scientifically it is an acetone, since it is classified by Gortner (1949) as a 2,3-dienol-L-guluforanol-acetone. The role of vitamin C as a redox agent or antioxidant in plant tissue is still under investigation.

In contrast, more is known on the effect of the vitamin B complex on plant life. For example, biotin and the related vitamin H, essential for normal skin growth in humans, are believed to affect the growth and respiration of *Rhizobium trifolia* and other microorganisms. Aminobenzoic acid, another vitamin B, is noted to stimulate bacterial growth. Thiamin is known to be produced by higher plants and microorganisms for use in their metabolism. In higher plants, thiamin is concentrated in seeds, and especially in the embryos. Nicotine acid, also called niacin, is found in cereals, fruits and vegetables. Of course, meat, milk, and egg are the primary sources of niacin. By oxidation and methylation it is believed to be converted into nicotine, an alkaloid, present in tobacco plants (Gortner, 1949).

As is the case with hormones, these vitamins may also be included in the synthesis of humic matter and are parts of the reasons for the growth-stimulating effect shown by humic substances. The molecular structures shown for some of the vitamins (Figure 4.11) suggest again the presence of chemically active functional groups for easy hook-up to phenols and other humic precursors.

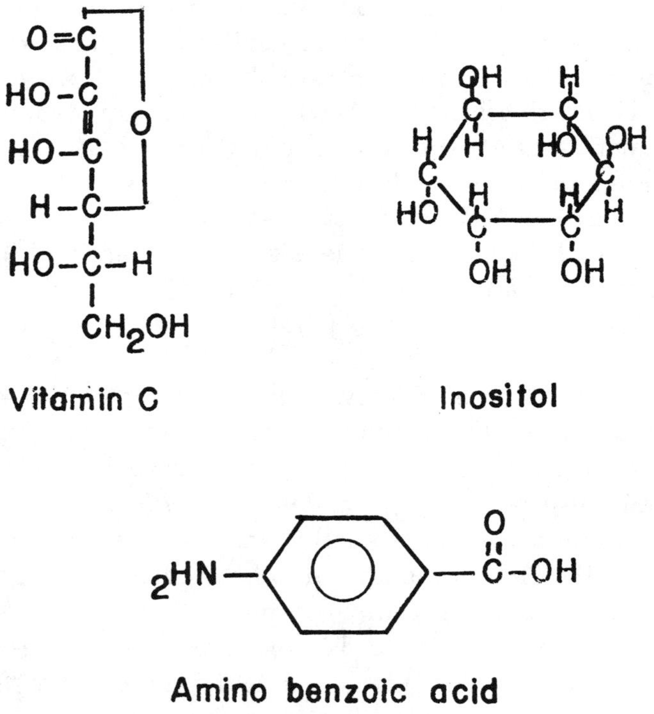

Figure 4.11 Molecular structures of vitamin C, aminobenzoic acid, and inositol showing functional groups, potentially active for interactions with other humic precursors.

Genesis of Humic Matter

Xenobiotics

The name xenobiotic refers to foreign organic substances, here meaning foreign to the soil. They are introduced into soils by the modern and extensive agricultural and industrial operations of today. The major xenobiotics expected to participate in the synthesis of humic matter include pesticides and their degradation products, e.g., the triazines, substituted ureas, and phenylcarbamates. These substances have the capability of forming stable complexes with soil organic constituents, which greatly affect their chemical behavior in soils in several ways. Such interactions may increase their persistence in soils, bringing with it a variety of effects on the environment as discussed earlier by Tan (2000). The behavior of the pesticide incorporated in the humic molecule may go one way or the other. It may remain chemically active and toxic or it may often lose its identity and behavior as a pesticide. The latter process is called biodetoxification by Tan (2000). Biodegradation of pesticides is reported to yield chemically active substances that can link with carbonyl, carboxyl, phenolic-OH, and amino groups of other soil organic substances to form humic acid-like compounds. The redox property of humic acids, as discussed before, can in turn play a definite role in the transformation of pesticides. For example, s-triazine is reported to be converted into its cationic form by humic acid, and the protonation process causing the development of the positive charge occurs on the ring-nitrogen atom (Weber, 1970), as illustrated in Figure 4.12. This process can also be explained by the *electron donor-acceptor concept* (Ziechmann, 1994). The triazines are considered the electron donors, whereas humic acids are the electron acceptors. A positively charged triazine is produced when an electron can be transferred from the electron-rich triazine to an electron-deficient quinone in the humic molecule. The transfer also results in humic acid becoming negatively charged.

Of interest perhaps is also the chemical compound 2,4-D, which in low concentrations can function as a growth promoting substance, but at relatively higher concentrations will act as a herbicide. The formal use of 2,4-D today is as a weed killer. However, in the past it has found application in controlling blooming and fruiting of crops. It is also used in rubber (*Hevea brasiliensis*) cultivation to increase the

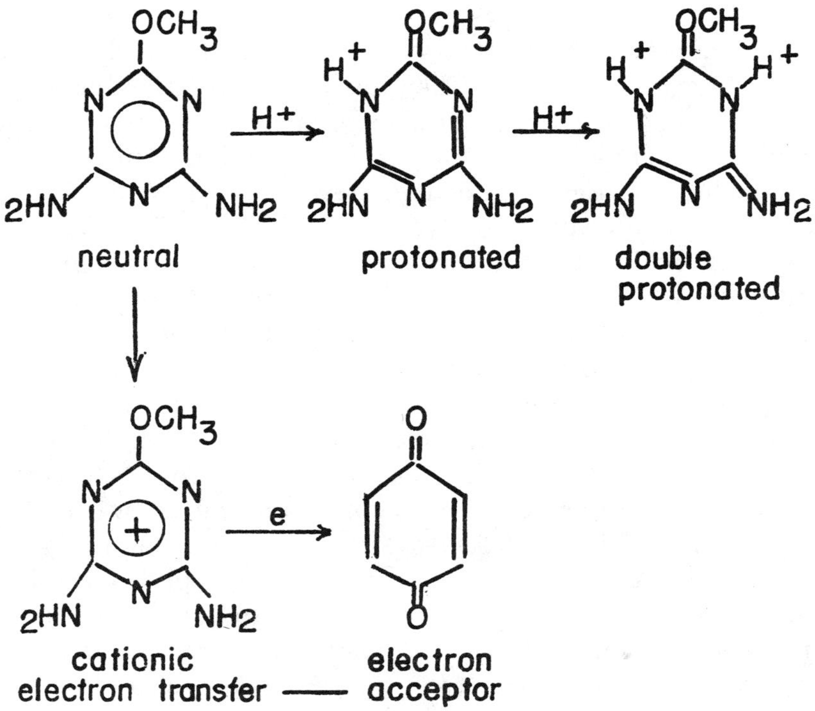

Figure 4.12 A schematic illustration of the transformation of a neutral triazine molecule into a cation by induced protonation and by the electron donor-acceptor concept.

latex production of the trees. The structure of 2,4-D (Figure 4.13) shows an active carboxyl group, whereas the Cl on the aromatic ring can be easily replaced by organic substances including humic acids. By incorporation into a humic acid molecule, the chemical activity of 2,4-D

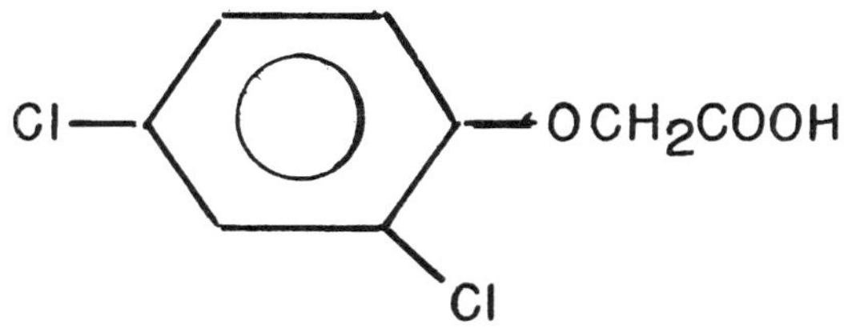

Genesis of Humic Matter

Figure 4.13 The molecular structure of 2,4-D showing the presence of Cl and a carboxyl group providing locations for potential reactions in the synthesis of humic matter.

can be preserved. It can remain active as a weed killer and this is reflected by a process called *residue effect*, or it may contribute to the hormone-like behavior of humic acid. Since in all likelihood only extremely small concentrations are chelated in the humic molecule, the hormonal effect is more likely to prevail over the herbicidal effect. However, it is also possible that these activities can be erased, since as an integral part of the humic structure the identity of 2,4-D has faded completely. Persistence of this compound in soils is usually not a big issue, since it is known to exhibit only a short half-life or mean residence time.

4.3 THEORIES OF HUMIFICATION

It is perhaps apparent from the discussions in the preceding sections that a variety of organic compounds take part in the humification processes. In fact, not a single organic compound can theoretically be ignored, and a variety of substances is expected to react in a variety of combination, which is perhaps creating all the consternations about humic acids and fulvic acids being *artifacts* or *operational compounds* and the like. The biopolymers produced by plants, when released into the soil upon decomposition of the plant tissue, can be used as such or in slightly degraded forms to form humic substances. Lignin or slightly degraded lignin may react with decomposed protein or peptide polymers and other large organic polymers, e.g., polysaccharides and phenol polymers. This type of humification is reflected by the ligno-protein theory. A second group of theories argues that the large plant biopolymers are decomposed first into their monomers or smaller molecular compounds prior to humification. According to this idea, single phenols, quinones, monomeric lignin, monomeric amino acids and ions, such as NH_3, are considered the basic molecules, instead of lignin, hence names such as polyphenol or phenol theory have been advanced for this type of humic matter synthesis. The reaction between the small molecules can take place perhaps abiotically, meaning purely chemically, but it should be realized that the phenols, quinones, and monomeric amino acids have been derived biologically. Not only can they be formed from the degradation of large biopolymers such as lignin, but they can also originate from metabolites of microorganisms. In addition to the ligno-protein and phenol theories above, the *sugar-amino condensation theory* is perhaps a third theory of major importance. It is sometimes called by different names, such as the melanoidin pathway or Maillard's reaction. Sugar and amino acid form the starting points, instead of lignin or phenol and quinone, as is the case in the ligno-protein and phenol-protein theories, respectively. These are the three major concepts on the synthesis of humic matter, though perhaps other theories are available but not known by the present author. The three theories on humification will be discussed below in more detail.

4.3.1 The Ligno-Protein Theory

This is the theory that has dominated for years the view of many

Genesis of Humic Matter 113

scientists on the synthesis of humic matter. It has been promoted by renowned scientists in the subject worldwide, such as Waksman (1932) in the USA, Kononova (1961) in Russia, and Flaig (1975) in Germany. The concept is called the lignin theory by Stevenson (1994), though the author failed to state his reason for this. Normally, lignin and amino acids (protein) are considered to form the core of humic acid, and not lignin alone. Perhaps, it is because the synthesis of humic matter does not only involve reactions of lignin and protein, but also reactions between lignin derivatives and simple nitrogenous compounds, such as NH_3, which is not a protein. Stevenson is also of the opinion that this ligno-protein or lignin theory is currently obsolete and has been replaced by the polyphenol theory. However, the idea that lignin and protein derivatives are playing an important role in the synthesis of humic substances has gained considerable support from reports recently published by Chefetz et al. (2002). Using advanced techniques of ^{13}C-NMR and thermochemolysis-gas chromatography/mass spectrometry, large amounts of lignin, protein, and cuticular material are detected in the humic acid structures by the authors.

The ligno-protein theory envisages lignin units entering into reactions with amino acids forming the core of the humic molecule, hence the name *lignin* and *protein* concept. To this core can be attached other organic substances, such as polysaccharides. A simple illustration of these reactions is provided in Figure 4.14. For simplicity, the lignin fraction is represented by coniferyl alcohol, whereas the protein fraction is shown by a monomeric amino acid. The reaction product formed should be regarded as a prototype, or humic-like substance. Addition of COOH and OH groups to the aromatic core is required for the development of the smallest possible monomer of a humic molecule. According to Flaig's theory the propane section is oxidized to yield by carboxylation a COOH group, whereas the methyl group can be converted into an OH group by hydroxylation producing in this way the phenolic-OH group.

In the biopolymer degradation concept, these constituents – coniferyl alcohol and amino acid – should be visualized as large polymeric compounds producing first large molecules of humic matter. The largest possible humic substance is humin, which upon depolymerization or degradation can be converted into humic acid, followed by

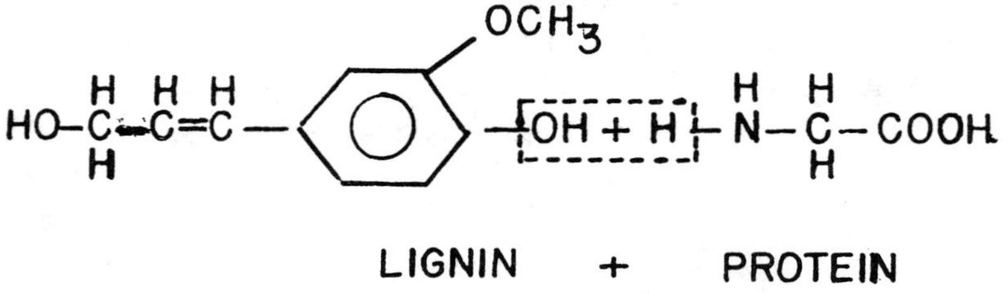

$$\text{HO-}\underset{\underset{H}{|}}{\overset{\overset{H}{|}}{C}}\!\!-\!\!\overset{H}{C}\!\!=\!\!\overset{H}{C}-\!\!\!\left\langle\!\bigcirc\!\right\rangle\!\overset{OCH_3}{}\!-[\text{OH} + \text{H}]\!-\!\text{N}\!-\!\underset{\underset{H}{|}}{\overset{\overset{H}{|}}{C}}\!-\!\text{COOH}$$

LIGNIN + PROTEIN

Figure 4.14 Schematic formation of a humic molecule according to the ligno-protein concept.

fulvic acid. In turn, fulvic acid will be broken down further into its constituents, e.g., phenols and quinones, referred to as small labile molecules by Hedges (1988).

The concept that fulvic acid can be formed from the decomposition of humic acid is perhaps reasonably acceptable, and so is its decomposition into the *'small labile molecules.'* However, as raised earlier, the issue is now whether a degradation of humin into small labile nonhumic substances can be called a humification process? Is it correct for a humification process to yield nonhumic substances as end products? How does this relate to the soil carbon sequestration theory?

On the other hand, the monomers, as noted in Figure 4.14, can also react together as such. They are then expected to form the smallest possible molecule of a humic substance. The smallest humic substance known today is fulvic acid, which upon polymerization can be converted into humic acid. Further polymerization or condensation will form humin and kerogen, which constitute the end products of the reactions. This is the process referred to as an abiotic polymerization process by Hedges (1988). Perhaps, no one will argue that this is a

Genesis of Humic Matter 115

humification process and the rate of humification and condensation increases in the direction from fulvic acid to consecutively humic acid, humin and kerogen.

However, recently a slightly different version has been presented on the diagenetic transformation of the humic substances, which clouds the issue on biopolymer degradation and polymerization processes in the synthesis of humic substances. Stevenson (1994) is of the opinion that lignin will yield humic acid, which can be directly transformed into coal. Stevenson's diagram, given below (Figure 4.15) for the sole purpose of explaining the confusion it creates, suggests that

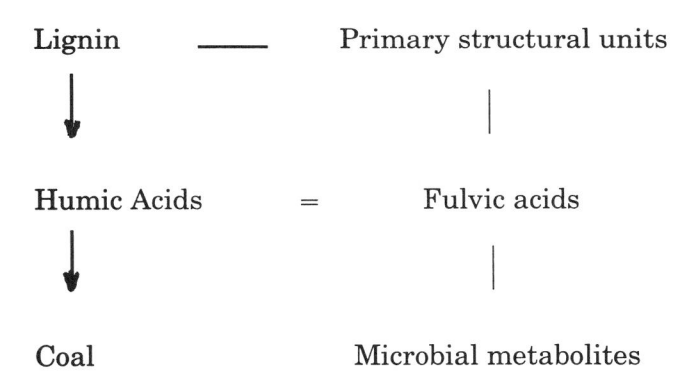

```
Lignin          ─────     Primary structural units
  ↓                              │
Humic Acids      =        Fulvic acids
  ↓                              │
Coal                      Microbial metabolites
```

Figure 4.15. Diagenetic changes of humic matter according to Stevenson (1994).

formation of humic acid directly from lignin is a biopolymer degradation, but the transformation of humic acid into coal is a condensation process. Stevenson is of the opinion that a biopolymer degradation process starting with lignin to humic acids will end with the formation of fulvic acid. No indications are provided that the latter can also be broken down further into even smaller units. In this process the fulvic acids are assumed to be created only as byproducts, since the main route is from lignin ⇢ humic acids ⇢ coal, as indicated by the bold-shaded arrows. Polymerization of the small units derived either from the degradation products of lignin or from metabolites produced by microorganisms will form fulvic acids. However, the way the diagram is constructed, it can be construed that fulvic acids are functioning only as intermediates in the diagenetic process.

Another possibility or hypothesis inserted here for further contemplation is the formation of humic compounds through transamination reactions of lignin monomers. Lignin monomers were indicated earlier to be formed from carbohydrates by the shikimic acid pathway. Decarboxylation of the end product, tyrosine (see Figure 4.3), may form a substance with the structure of an ammoniated coumaryl alcohol, or an ammoniated lignin monomer. Such a compound falls into the category of a *ligno-protein compound*, hence can be considered as a prototype of a humic substance (Figure 4.16). Again a COOH and an OH group should be added to the phenyl-core before the substance qualifies to be called a humic molecule. The ammonia, NH_3, has been derived from the decomposition of amino acids that can occur in both aerobic and anaerobic conditions, as illustrated previously by the deamination reactions (4.2), (4.3), and (4.4).

4.3.2 The Phenol-Protein Theory

This theory is called the *polyphenol theory* by Stevenson (1994) or the *phenol autoxidation theory* by Ziechmann (1994). However, in view of current concepts considering the humic acid core to be composed generally of an aromatic and nitrogenous substance, usually in the form of protein or amino acid, the present author proposes using the name *phenol-protein theory* in analogy to the ligno-protein theory. All these names can also be considered misnomers, since not only phenols, but also quinones are involved in the reactions forming humic matter. Quinones are phenol derivatives, and often play a dominant role in the formation of humic matter. Parts of the phenol-protein theory also tend to overlap with the ligno-protein theory. Many of the precursors and their reactions in humic matter formation are closely

Genesis of Humic Matter 117

Figure 4.16 An abbreviated version of the synthesis of a ligno-protein complex through formation of lignin monomers according to the shikimic acid pathway. Adapted from Schubert (1965) and Haider et al. (1975).

related and make one wonder about the necessity for advancing two different theories in the synthesis of humic matter. Perhaps only the sugar-amine condensation theory needs to be distinguished from the two above, since it provides an explanation for the formation of humic matter in environments in which lignin and its derivatives are absent. However, since phenols and quinones are not derived only from the enzymatic degradation of lignin, but can also be formed by microorganisms from sugars through the shikimic acid and acetate-malonate pathways, the feeling is that there is a need to also recognize a phenol theory as a distinct humification process.

In the phenol-protein theory, phenols, quinones, and amino acids are the key components for formation of the humic molecule core. A simple reaction between gallic acid and a small amino acid molecule is shown in Figure 4.17. For comparison, a reaction is also provided between a hydroxyquinone and an amino acid. The formation of humic matter with the constituents as shown in Figure 4.17 can be used as a model for the abiotic condensation theory, since the phenol and other reacting molecules are normally small in sizes. As stated earlier, the abiotic prescript applies only to the chemical reactions. The phenols and other participating components have been formed by biotic enzymatic reactions. This condensation theory assumes then that polymerization of the reaction products above is expected to yield first fulvic acid, which upon further condensation produces humic acid, and ultimately humin and kerogen. However, large polymers of phenols, quinones, and amino acids (protein) can also be formed prior to formation of humic matter. When these large biopolymers enter into reactions, large humic molecules are expected to be formed first, such as humin, which upon degradation can be converted into humic acid, which in turn yields fulvic acid. This is a model example of a biopolymer degradation concept showing a principal difference from the original proposed by Hedges (1988). In the reactions above as illustrated by Figure 4.17, the phenol biopolymers are not degradation products of lignin, but have been synthesized by polymerization of the single phenol molecules.

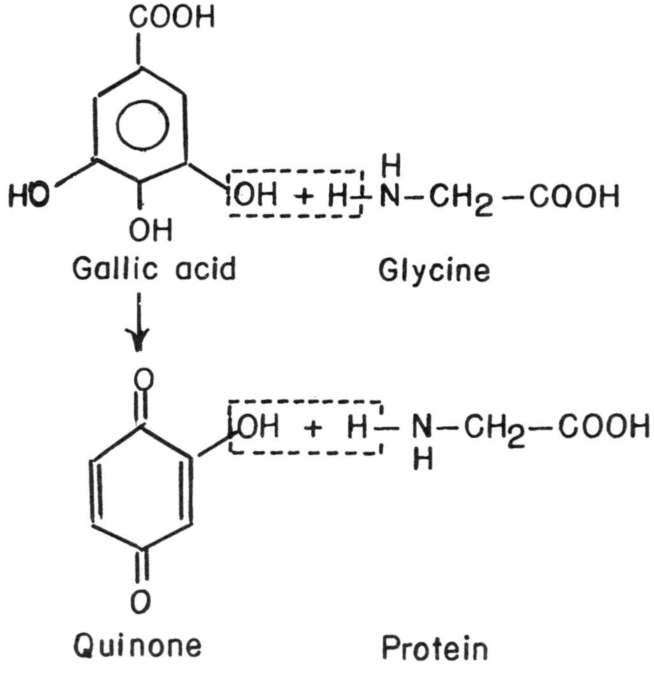

Figure 4.17 Possible reactions for humification according to the phenol-protein theory.

4.3.3 The Sugar-Amine Condensation Theory

This theory is also reported under different names and as discussed earlier has been called the melanoidin pathway by some (Hatcher et al., 1985; Nissenbaum and Kaplan, 1972), and the Maillard reaction by others (Ziechmann, 1994; Stevenson, 1994). It is a hypothesis using sugar and amine as starting and key components for the synthesis of humic matter, and no lignin derivatives are required.

The reaction, reported for the first time by Maillard in 1911, can take place nonenzymatically or abiotically between glucose and amine, producing melanins as end products. Ziechmann (1994) reports that glycine can substitute for the amine, which in the first step of the reaction with glucose yields glucosylamine. The latter, also called glycosylamine by other authors (Stevenson, 1994), will eventually be converted into melanoidins, and subsequent polymerization or condensation of the melanoidins produces humic matter. Melanins or melanoidins are pigments that are used in the food industry for coloring and flavor, because of the aroma they carry. They also occur in nature and are believed to exhibit chemical properties characteristic of humic substances, especially of humin and kerogen in marine sediments (Hatcher et al., 1985; Nissenbaum and Kaplan, 1972). The name melanin has been applied in the U.S. Soil Taxonomy for the development of *melanic* (Greek melas-anos = black) *epipedons*, meaning surface soils, rich in humic acids, hence exhibiting intense black colors as found in andosols (Soil Survey Staff, 1990).

The Maillard reaction as reported by Stevenson (1994) involves a very complex series of chemical processes. Sugar, the starting point, is converted first into a *Schiff base* by reaction with an amine. The latter is then converted into an η-substituted glycosylamine, which subsequently undergoes complicated *Amadori rearrangement* processes. The rearrangement products are then subjected again to a series of additional reactions, which make the concept very confusing and mind boggling. For those interested, reference is made to Stevenson (1994) and Ziechmann (1994) for details and specifics of this chain of reactions. However, for the purpose of increasing comprehension, a simple reaction is provided in Figure 4.18 to illustrate the interaction between glucose and an amine. For comparison a reaction between glucose and glycine is also provided. The basic molecular structures shown in the figure, composed of carbohydrates and nitrogenous compounds, correspond closely to basic structures advanced for autochthonous aquatic humic matter. As discussed earlier, plant materials in marine environments do not require lignin and are composed mostly of carbohydrates. Therefore, this theory indicates that marine humic matter, not affected by terrestrial material, is composed mainly of carbohydrate-protein complexes.

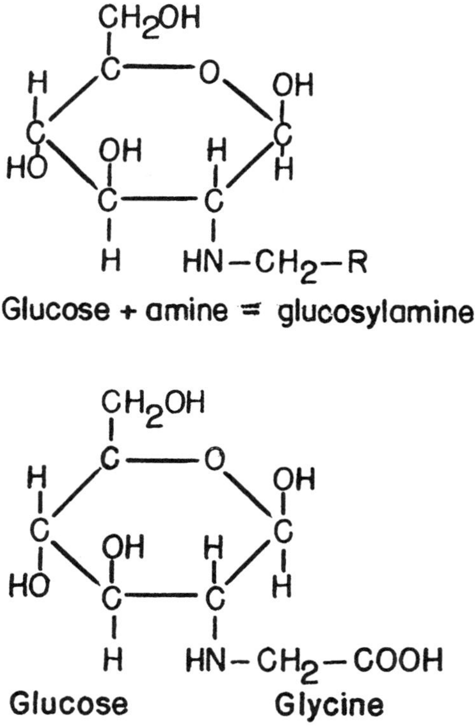

Glucose + amine ⇌ glucosylamine

Glucose   Glycine

Figure 4.18 A simplified version of the reactions between glucose and an amine (top), and glucose and glycine (bottom).

4.4 STATISTICAL MODELING OF HUMIFICATION

In today's computer age, statistical modeling of almost any process in nature is becoming a great fashion. Statistical formulas abound in soil chemistry and other sciences, and many are so complicated that the message they want to convey is often obscured by the maze of symbols and pages of computations. Though not much is known yet on these efforts in humic acid chemistry, the limited information available shows a similar tendency in statistical modeling of the humification process. Some are reasonably easy to follow, whereas it is difficult to fathom the rationale and applicability of others. To make more sense of the complicated statistics, the following attempt is made to present and explain in simpler terms some of the theories available on statistical modeling of the humification process.

4.4.1 Humification Indexes

Attempts have been made to express degree of humification in terms of optical density values, extinction or absorbance. By plotting the logarithm of the absorbance against the wavelengths in the visible light range, an absorption spectrum of humic or fulvic acid is usually produced in the form of a straight line. The slope of such a line or curve is taken as a characteristic for differentiating the humic substances. The lighter the color of the humic substance, the steeper the slope, whereas the darker the color the more horizontal tends to be the spectral line. Fulvic acids are usually characterized by spectra with the steepest slopes, compared to humic acids (Tan, 2000; Tan and Van Schuylenborgh, 1961; Kumada, 1955; 1987). Hence, the degree of inclination of the spectral curves may translate into rate or degree of the humification process. This inclination or slope of the spectral curve can be expressed statistically in several ways. It can be formulated in terms of a ratio or quotient of the absorbance at two arbitrary selected wavelengths, e.g., absorbance or extinction at 400 and 600 nm, called the color ratio E_4/E_6 or $Q_{4/6}$ (Springer, 1934; Tan and Van Schuylenborgh, 1961; Flaig et al., 1975; Kononova, 1966):

$$E_4/E_6 \text{ or } Q_{4/6} = \frac{\text{Log absorbance at 400 nm}}{\text{Log absorbance at 600 nm}} \qquad (4.9)$$

Genesis of Humic Matter

A high color ratio, 7-8 or higher, corresponds to steep curves, usually observed for fulvic acids. On the other hand, a low color ratio, 3-5, indicates curves that are less steep, normally exhibited by humic acids. Consequently, the smaller the value of the color ratio the greater will be the degree of humification (Tan and Giddens, 1972). This conclusion seems to be in close agreement with the contention of Chen et al. (1977), indicating the presence of an inverse relationship between E_4/E_6 ratios and degree of condensation of humic matter. An inverse relationship has also been cited by Stevenson (1994) between the color ratio and the mean residence time of humic matter. Humic substances exhibiting the highest E_4/E_6 ratios are the lowest in mean residence time. Consequently, fulvic acids tend to possess a shorter life time in soils than humic acids.

The slope of the spectral curve can also be expressed as the difference between the log values of the respective extinctions, and its value can also be used as a humification index (Tan and Van Schuylenborgh, 1961; Kumada, 1987):

$$\Delta \log E = \log E_{400} - \log E_{600} \qquad (4.10)$$

where E = extinction or absorbance. The values for $\Delta \log E$ may vary from 1.1 or higher for less humified material, e.g., stable manure, to 0.6 or smaller for humic acids in andosols.

4.4.2 Stability Coefficient of Humus

The absorbance or extinction of humic solutions has also been applied by Hargitai (1955) for the formulation of a stability coefficient of humus as follows:

$$K = \frac{E_{NaF}}{E_{NaOH} H_c} \qquad (4.11)$$

in which K = the stability coefficient of humus, E_{NaF} = extinction coefficient of humic acid extract in NaF, E_{NaOH} = extinction coefficient of humic acid solution in 0.5 % NaOH, and H_c = humus content in %. Hargitai (1997) assumes that the K value relates the rate of solubility of humic acid in NaOH and the extent of mobilization in NaF of long chain humified components. However, in the opinion of the present author, this is not what equation (4.11) wants to convey. The use of the ratio $E_{NaF}/E_{NaOH}H_c$ refers to the amount of humic acid soluble in NaF in relation to that dissolved in NaOH times the humus concentration, in other words the amount of humic acid soluble in NaF *per* the amounts of NaOH soluble humic acid and humus concentration. Consequently, the ratio indicates that the higher the value of E_{NaF}, the larger will be the amounts of humic acid dissolved by NaF, hence the smaller will be the values for the denominators E_{NaOH} and H_c. This translates into high K values, indicating that the humus is unstable. The latter is perhaps being used or 'destroyed' by humification. The relationship between high K values and humification seems to be supported by the author's experiments on C/N ratios, where an increase in K values is noted to be closely related to a decrease in C/N ratios. Nonhumified organic residue usually exhibits a high C/N ratio (= 80 for straw), and humic acids are characterized by low C/N ratios (10-15). Therefore, the increase in K value may well mean an increased rate of humification.

4.4.3 Humification Model

A statistical model on humification is presented by Ziechmann (1994) and Kappler and Ziechmann (1969). Though at first glance the model looks reasonably simple, it turns out to be a complicated model. It is apparently developed on the basis that a system of humic substances can be separated into several fractions: F_1, F_2, F_3 to F_n. The molecular weight values increase from F_1 to F_n, implying that humification increases in the same direction. The authors above claim that the humification process, H, of such a system can be defined as the negative sum of the products of the possible transition fraction and its logarithm. In their opinion, this definition assumes a *Markhoff*

Genesis of Humic Matter

model which can be written as follows:

$$H = -\sum p_i \cdot \log p_i \tag{4.12}$$

in which p_i = the relative mass of fraction i, $p_i \geq 0$ and $\sum p_i = 1$. Such a formulation for humification appears to raise a number of questions. In many other books, authors often consider that the reader should know the basic of statistics, hence offer only sparse explanations of symbols and declarations of rules or assignments of limits, and the like. They perhaps forget to realize that there are a variety of readers examining the book, who may not necessarily have the same expertise. In this case, the use of a negative sign in equation (4.12) has to be justified, and assigning the sum of p_i to equal 1 explained properly. It is difficult to express degree or rate of humification by negative values. Secondly, the definition does not correspond to or is in conflict with the assignment '$\sum p_i = 1$' and its application. As will be explained below, calculation of an application provides strong evidence that the humification model should perhaps be written more properly in the following form:

$$H = -(\sum p_i) \log p_i \tag{4.13}$$

When the masses of humic substances are distributed uniformly over all the fractions, the authors state that a maximum has been reached in the humification process, and $p_i = 1/n$. The H function is accordingly solved by the authors as follows:

$$H = -\sum \frac{1}{n} \log \frac{1}{n} = -\log \frac{1}{n} = \log n \tag{4.14}$$

The answer, log n, can only be obtained by using the formula as written in equation (4.13), and in addition when $\sum 1/n = 1$, since by definition $\sum p_i = 1$. Another issue is the selection of the value 1/n at a maximum rate of humification. In view of the sum of p_i being defined by the authors to equal 1, whereas n is the number of fractions, the value 1/n would more reasonably refer to the *average* or *mean value* of H, instead of to the maximum value.

CHAPTER 5

CHEMICAL COMPOSITION OF HUMIC MATTER

5.1 ELEMENTAL COMPOSITION

5.1.1 Weight Percentage

As indicated in Chapter 1, the elemental composition of humic matter is a very big issue among scientists, with one group being very critical about the presence of an elemental composition, and the other group proclaiming the existence of an elemental composition characterizing humic matter. A number of scientists in the first group above fail to see the significance of an elemental composition, indicating that extraction procedures may have effected changes in the elemental composition of humic substances. However, Steelink (1985) declares that this issue is not clear yet and needs to be resolved by more detailed research. Except for a generous number of criticisms, no further information of interest is available from this group, and as stated by Ziechmann (1994), elemental composition is a dead issue among these scientists. In contrast, a lot of information has been supplied by the second group. Many of them have analyzed the C, H, O, N, and S contents of humic and fulvic acids extracted from a variety

Table 5.1 Summary of Elemental Composition of Humic Acids (HA) and Fulvic Acids (FA) Extracted from Tropical and Temperate Region Soils and Miscellaneous Environments. (The figures are averages of weight percentages from multiple analyses of various sources.)[a]

	C	H	O	N	S	C/N
	----------------- % -----------------					
Tropical Region Soils						
HA-Alfisols	52.3	5.2	37.2	3.6	n.d	14.5
HA-Andosols	58.5	3.9	34.8	3.8	0.3	15.4
HA-Oxisols	54.5	4.4	38.0	3.1	n.d	17.5
FA-Andosols	48.9	4.3	44.5	2.3	n.d	21.3
Temperate Region Soils						
HA-Alfisols	56.8	5.0	33.6	4.6	n.d	12.3
HA-Aridisols (Solonetz)	54.5	4.1	36.4	5.0	n.d	10.9
HA-Histosols (Peat-bogs)	58.7	5.0	32.9	3.4	n.d	17.3
HA-Inceptisols	51.4	5.4	39.1	4.1	n.d	12.5
HA-Mollisols	53.7	4.3	36.3	3.7	n.d	14.5
HA-Spodosols	53.1	4.9	39.2	3.5	0.4	15.2
HA-Ultisols	50.5	5.2	40.0	3.9	0.5	12.9
FA-Inceptisols	47.9	5.2	44.3	2.6	n.d	18.4
FA-Mollisols	41.6	4.0	51.9	1.1	n.d	37.8
FA-Spodosols	50.6	4.0	44.1	1.8	0.3	28.1
FA-Ultisols	45.1	4.5	48.0	2.2	0.3	20.5
Geologic Deposits						
Lignite	52.6	2.8	31.8	2.0	0.7	26.3
HA-lignite	57.4	3.9	36.8	1.2	0.8	47.8
FA-lignite	46.4	4.4	45.8	1.5	0.6	30.9
Rivers and Swamps						
HA-aquatic	49.6	4.8	43.5	2.1	n.d	23.6
FA-aquatic	50.8	4.3	43.1	1.6	n.d	37.8
Merck Chemical. Co.						
HA-artificial	59.8	4.8	33.6	1.8	n.d.	33.2

Chemical Composition of Humic Matter

Table 5.1 *Continued*

	C	H	O	N	S	C/N
			%			
'Reference' humic acid	56.2	4.7	35.5	3.2	0.8	17.6
Peat	55.5	5.4	33.1	3.0	n.d	18.5
Plant Residue	49.6	6.3	41.6	2.5	n.d	19.8

^aSources: Lobartini et al. (1997; 1992; 1991); Tan et al. (1991); Kumada (1987); Thurman and Malcolm (1981); Orlov (1985); Steelink (1985); Schnitzer et al. (1991); Schnitzer and Mathur (1978); Schnitzer and Khan (1972); Cranwell and Haworth (1975); Tan and Van Schuylenborgh (1959). n.d = not determined.

of soils in tropical and temperate regions. These are considered the major elements in humic matter, and a summary of the data is given as examples in Table 5.1. The lignite samples listed in the table above were from the deposits in North Dakota, and the data for the lignite-humic and fulvic acids are the average figures of data reported by Mathur and Parnham (1985), Steelink (1985), Tan et al. (1991), Lobartini et al. (1992), and from unpublished data of the present author. The elemental composition of aquatic humic and fulvic acids are also the average figures of data supplied by Lobartini et al. (1991), Thurman and Malcolm (1981), and from unpublished data of the present author. The black water samples for extraction of these humic substances are from the Okefenokee swamps, Satilla, Ohopee, and Suwannee rivers in the southeastern United States.

Historical data reported as early as 150 years ago by Mulder and other pioneers in humic acid science, for %C (56-62%), %H (2.0- 5.5%), and %N (2-8%), are remarkably within range of those listed in Table 5.1. However, Orlov (1985) cautions their use for comparison, because they are in the higher ranges, due perhaps to the habit in the old days of drying humic acids at 140°C or higher, and the use of different values in atomic weights.

Both the old and especially the modern data show the elemental composition to be within relatively fixed limits, meaning that it does not exhibit an erratic or a very wide range of variation, as would have been expected with fake materials or artifacts. The differences noticed may be due to differences in origin or to the types of humic substances. Orlov (1985) believes that variation in the elemental composition is affected by (1) variability in soils, (2) variability of humic substances in time and space, (3) different isolation techniques, and (4) errors in sampling and analyses. Nevertheless, the general composition of humic matter, as listed in Table 5.1, is still displaying a close relation with that of the plant material from which it has been derived. The observed divergence from the plant residue is apparently the result of the humification process and other soil factors. The composition is also in the range of that of humic acid, listed as a *'reference'* in the table. This particular sample is considered the ideal example of humic acid in soils by Steelink (1985) and Schnitzer and Mathur (1978).

With few exceptions, the carbon content of humic acids is similar to slightly higher than that in plant residue. On the other hand, fulvic acids exhibit carbon contents almost similar to slightly lower than that in plant residue. The differences in carbon contents between the two types of humic substances are in agreement with data supplied by Steelink (1985), showing carbon contents of 53.8 to 58.7% in humic acids and that of 40.7 - 50.65 % for fulvic acids. This may indicate that fixation of carbon or carbon sequestration takes place during the synthesis of humic matter and goes up slightly with an increased rate of humification. The decomposition of organic residue is characterized generally by a loss of C in the form of CO_2, but when humification steps in, some of the C will be incorporated into humus and humic matter. It is estimated that approximately one-third of the C from plant residues is retained in the soil and stabilization of nonaromatic C is expected to take place through microbial transformation into aromatic substances used in the synthesis of humic matter. By comparison with the elemental composition of the plant residue listed at the bottom of the table, the carbon contents of humic and fulvic acids indicate that losses of C from decomposition of plant residue should have been

Chemical Composition of Humic Matter

minimal. Most of the liberated carbon is apparently retained in the soil in the form of humic matter, since both humic and fulvic acids exhibit C contents in the range of that in plant residues.

The hydrogen content of humic and fulvic acid does not differ much from each other. The data for both of the humic substances are also in the range of the average hydrogen values of 3.2 - 6.2% and 3.8 - 7.0% in humic and fulvic acids, respectively, as reported by Steelink (1985). The oxygen contents listed in the Table 5.1 above for fulvic acids tend to be slightly higher than the average values of 39.7-49.8% as reported by Steelink, but the big difference is in the nitrogen contents. The data in Table 5.1 display nitrogen contents of 2.6 to 5.05 % for soil humic acids, which are considerably higher than Steelink's average values of 0.8 -4.3% N, whereas those of fulvic acids, ranging from 1.1 -2.6%, tend to be lower than the average values of 0.9 -3.3 % as reported by Steelink. It is conspicuous that soil humic acid is substantially higher in N than soil fulvic acid, which is not apparent from Steelink's average ranges. The exceptions to the above are the geologic and aquatic humic substances, which are characterized by low N contents. The differences in N content are also not too obvious between aquatic humic and fulvic acid.

5.1.2 The C/N Ratio

The carbon to nitrogen ratio is often considered as an index of a decomposition process of crop residue. Its value varies from 13 to 20 in legume crops, to 40 in cornstalks, or 80 in straw of cereal crops, and has been reported as high as 500 to 800 in sawdust (Brady, 1990; Miller and Gardiner, 1998). The decomposition of plant residues results in some of the carbon and nitrogen being lost, processes considered part of the carbon and nitrogen cycles. However, some of the C and a considerable amount of the N liberated are incorporated into microbial cells or fixed in substances used for formation of humic matter. These processes lead to decreasing the values of the C/N ratios, eventually reaching relatively constant values in soils. In most soils, the C/N ratio falls within narrow limits to about 10 to 15, when decomposition is virtually 'completed', meaning organic matter decomposition is in equilibrium with the synthesis and accumulation of new organic materials.

This decrease in C/N ratio, reaching a constant value, is now extended to also indicate degree or rate of a humification process. As indicated above, part of the liberated carbon and a substantial amount of the organic nitrogen are 'sequestered' by the humic molecule. Hence, it is generally believed that the C/N ratio will also decrease with increased rate or degree of humification, and C/N ratios between 10 to 15 are often considered to be characteristic for well-developed humic acids. The data listed in Table 5.1 show some support for the opinion above by displaying C/N ratios of soil humic acids between 12.3 and 17.3. Higher C/N ratios are exhibited by aquatic and geologic humic acids, which is perhaps caused by disturbance in the humification process due to the different environments in aquatic or geologic systems from that in the soil ecosystem, e.g., lower N contents and reduced condition in aquatic systems.

Of interest to note is the high C/N ratios of soil fulvic acids, ranging from 18.4 to 37.8, in comparison to those of humic acids. As discussed above, the differences in carbon contents between fulvic and humic acids were rather small and would not have caused the values of C/N to differ that much. In contrast, the nitrogen content is approximately 2 to 3 times higher in humic acids than in fulvic acids, which may indicate that fixation of N increases with increased humification from fulvic to humic acid. Increasing amounts of nitrogenous compounds are apparently being sequestered in the process of the synthesis of a humic acid molecule. This raises the possibility that the polymerization or condensation theory of formation of humic acids is not quite correct. Polymerization only of fulvic acids is not adequate, since it would not be able to increase the N content so that it causes the C/N ratio of the polymerized product (humic acids) to decrease that much. The polymerization of two moles, 10 moles, 100 moles (or even higher) of fulvic acids will change neither the composition nor the C/N ratio. Therefore, in addition to polymerization, other reactions, such as interaction, adsorption, and chelation of nitrogenous substances are apparently also involved. For raising the nitrogen content in humic acid, it is necessary to also invoke these reactions for the inclusion of the needed nitrogen sources in the humic

Chemical Composition of Humic Matter

acid molecule.

The differences in N content between fulvic acids and humic acids can perhaps be explained partly by invoking the biopolymer degradation theory. Since in this theory humic acid is formed first, it is more likely that the degradation process yields humic acids with less N than the plant residue, or maximal with N contents similar to that of the parent material. This will not explain the N contents of humic acids, which are higher than that in the plant residues. However, by further degradation of humic acids into fulvic acids, the possibility arises that in this process large amounts of the nitrogenous constituents have been broken down and removed from the humic acid molecule. The losses of N by comparison to C must be substantial, causing the N contents to be very low and the C/N ratio to become high in fulvic acids.

5.1.3 Atomic Percentage

A number of scientists believe that elemental composition based on weight percentages cannot be used to explain the molecular structure of humic substances. For the purpose of studying and devising structural formulas for humic substances, they suggest the use of atomic percentages. This method of expressing the elemental composition of humic matter is especially popular in eastern European countries (Cieślewicz et al., 1997; Dębska, 1997). Orlov (1985) is of the opinion that atomic percentages give a better picture than weight percentages on the composition of these substances and on the role they play in molecular structure. He has used them for the determination of molecular weights and formulas, and in distinguishing two groups of humic acids, one with an atomic percentage of 40-42% C, and the other with an atomic percentage of 37-38% C. Steelink (1985) also believes that atomic percentages and atomic ratios are useful as guides in the identification of different types of humic acids, and for drafting structural formulas of the humic substances.

Table 5.2 Summary of Elemental Composition of Humic Acids (HA) and Fulvic Acids (FA) Extracted from Tropical and Temperate Region Soils and Miscellaneous Environments in Atomic Percentages

	C	H	O	N	ω
	\-\-\-\-\-\-\-\-\-\- atomic % \-\-\-\-\-\-\-\-\-				
Tropical Region Soils					
HA-Alfisols	35.9	42.8	19.2	2.1	0.053
HA-Andosols	43.4	34.8	19.4	2.4	0.258
HA-Oxisols	39.4	38.1	20.6	1.9	0.223
FA-Andosols	36.0	38.0	24.6	1.4	0.428
Temperate Region Soils					
HA-Alfisols	38.9	41.1	17.3	2.7	0.041
HA-Aridisols (Solonetz)	40.3	36.4	20.2	3.1	0.330
HA-Histosols (Peat-bogs)	40.1	41.0	16.9	2.0	−0.080
HA-Inceptisols	34.5	43.5	19.7	2.4	0.090
HA-Mollisols	39.6	38.0	20.1	2.3	0.230
HA-Spodosols	36.8	40.7	20.4	2.1	0.174
HA-Ultisols	34.5	42.7	20.5	2.3	0.151
FA-Inceptisols	32.9	42.8	22.8	1.5	0.222
FA-Mollisols	32.1	37.1	30.1	0.7	0.785
FA-Spodosols	38.0	36.0	24.8	1.2	0.453
FA-Ultisols	30.9	37.4	30.8	0.9	0.871
Geologic Deposits					
Lignite	47.1	30.1	21.3	1.5	0.361
A-lignite	43.2	35.2	20.8	0.8	0.204
FA-lignite	34.4	39.1	25.5	1.0	0.433
Rivers and Swamps					
HA-aquatic	35.0	40.7	23.0	1.3	0.263
FA-aquatic	37.3	37.9	23.8	1.0	0.340
Merck Chemical Co.					
HA-artificial	41.5	40.0	17.5	1.0	−0.048
'Reference' humic acid	39.6	39.7	18.8	1.9	0.091

Chemical Composition of Humic Matter

Table 5.2 *Continued*

	C	H	O	N	ω
	---------- atomic % ---------				
Peat	37.6	43.9	16.8	1.7	−0.138
Plant residue	29.4	44.7	24.6	1.3	0.286

Hence, the data in Table 5.1 are converted by the present author into atomic percentages and the results are listed in Table 5.2.

By comparison with weight percentages, the variation in elemental composition expressed in terms of atomic percentages is also relatively small. The humic and fulvic acids from soils and aquatic environments exhibit C, H, O, and N atomic percentages within relatively narrow fixed limits, also providing strong credentials for the presence of real natural compounds, instead of fake or operational substances.

The data suggest that for every C atom there is at least one H atom. It is also more evident now that the composition of humic substances contains approximately one atom of O to two atoms of C. Again, one can notice that in general the N content is larger for humic acid than for fulvic acid. The nitrogen atomic percentages are approximately twice that large in humic acids than in fulvic acids of terrestrial soils.

5.1.4 Internal Oxidation of Humic Substances, ω

The parameter in Table 5.2, designated by the symbol ω, refers to the internal oxidation value of humic substances. This is considered a very important value, especially in Europe, for studying diagenetic changes of humic substances. A number of scientists believe that diagenetic changes of humic acids are closely related to degradation and oxidation reactions. This hypothesis, as discussed by Orlov (1985), starts with the degradation of plant residue to form a humic acid-like substance. The degradation process involves the loss of CH_3 groups. The transformation of the humic substance above to a humic acid characteristic – for example – in mollisols, is postulated to take place by a continuation of the degradation process, involving now also partial oxidation and further losses of CH_3 groups. An increase or decrease in the number and length of aliphatic chains in the humic molecule is considered to be inherent to this basic process of addition or loss of especially terminal CH_3- or CH_2-groups. The resulting difference between the number of oxygen and hydrogen atoms is used as a measure of the degree of the oxidation processes. This difference can be calculated by several different, though related, methods. The simplest is the method using a revised formula below, adapted from Orlov (1985):

$$\Delta = 2O - H \tag{5.1}$$

in which Δ = difference between numbers of oxygen and hydrogen atoms, O = number of oxygen atoms, and H = number of hydrogen atoms. For a water molecule (H_2O), a fundamental basis of this hypothesis, the difference equals zero:

$$\Delta = 2O - H = (2 \times 1) - (2 \times 1) = 0 \tag{5.2}$$

To compare the degree of oxidation in different organic substances, Orlov (1985) proposes to calculate the difference on the basis of 1 gram atom of carbon or per unit amount of carbon atoms:

$$\omega = \frac{2O - H}{C} \tag{5.3}$$

Chemical Composition of Humic Matter

in which ω = the difference between numbers of O and H atoms per 1 gram atom of C, and C = number of carbon atoms.

The value of ω = negative with substances in the reduced state, meaning in substances with excess H atoms. The value of ω = positive with substances in the oxidized state, or substances containing excess oxygens. The value is considered to fluctuate between a minimum of -4 and a maximum of $+4$, corresponding to the formation of CH_4 and CO_2, respectively. Methane, CH_4, is considered a C compound at the highest reduced state, whereas carbon dioxide, CO_2, is assumed to be a C compound at the highest oxidation state. Examples of calculations are given below as illustrations:

$$CH_4: \quad \omega = (0-4)/1 = -4 \qquad (5.4)$$

$$CO_2: \quad \omega = (4-0)/1 = +4 \qquad (5.5)$$

Applying formula (5.3), Orlov argues that the degree of oxidation in the formation of a carbohydrate, with a general formula of $C_x(H_2O)_y$, must equal zero:

$$\omega = \frac{2O - H}{C} = \frac{2y - 2y}{x} = 0 \qquad (5.6)$$

Orlov's formula discussed above is considered inadequate to explain the formation of humic substances. It is now an established fact that, in addition to C, H, and O, humic substances also contain nitrogen as an additional major constituent. Hence, a new hypothetical model has recently been proposed to include the significance of N in the formation and diagenetic changes of humic substances. The oxidation process, used above as the basis for reflecting formation and diagenetic changes of humic substances, is now called *internal oxidation*. The relationship, by which the degree of internal oxidation can be calculated, is formulated as follows (Ciéslewicz et al., 1997; Dębska, 1997):

$$\omega = \frac{(2\,O + 3\,N) - H}{C} \tag{5.7}$$

The rational of adding 3N is not clear and can only be guessed. Perhaps it is based on the formation of NH_3 as the most reduced N compound, since the value of $\omega = 0$ for formation of NH_3 gas.

Using equation (5.7), the values for ω have been calculated by the present author for the humic substances, using the atomic percentages in discussion as listed in Table 5.2. As can be noticed from the data, most of the values for ω are positive in sign, indicating a high degree of internal oxidation. Only HA-histosol, extracted from peat bog samples, HA-artificial, and peat have ω-values that are negative, indicating a low degree of internal oxidation. In Orlov's terms, this means that humic acids extracted from soils have excess oxygens, in other words are compounds in an oxidized state. On the other hand, humic acid from peat, peat itself, and humic acid artificially prepared by the Merck Chemical Company have excess hydrogen atoms, or compounds that are in a reduced state. The above lends support to the more recent findings by Ciéslewicz et al. (1997), who conclude from their research that positive values for ω are exhibited by humic acids extracted from well drained soils, underscoring the aerobic environment prevailing in soils affecting the oxidation of humic acids. In contrast, the negative values for ω are noticed by Ciéslewicz et al. (1997) for humic acids extracted from lagoon sediments, indicating anaerobic conditions in the transformation of humic acids.

5.1.5 Atomic Ratios

For the purpose of creating structural formulas or formula weights, the ratios of the atomic percentages H/C, O/H, O/C, and N/C

Chemical Composition of Humic Matter 139

are found to be helpful. In addition, Steelink (1985) thinks that they can also be used for the identification of types of humic substances. Hence, the atomic ratios are calculated by the present author from the figures in Table 5.2, and the results are presented in Table 5.3.

The atomic ratios of H/C of the humic substances, ranging from 0.90 to 1.35, are in agreement with those reported by Orlov (1985) and Steelink (1985). As indicated earlier by Orlov (1985), the figures suggest that for each carbon atom, there is indeed one hydrogen atom in the humic molecule. The atomic H/C ratios of aquatic humic matter are also within range of the limits stated above. However, the present data do not support the observation of Ishiwatari (1975) for higher H/C ratios for humic substances in lake sediments than those in soils. The present maximum of 1.35 for humic substances conforms fairly well with the contention of Chen et al. (1977) that H/C ratios >1.3 indicate the presence of nonhumic compounds. As can be noticed in Table 5.3, the H/C ratio of plant residue equals 1.5.

The atomic ratios of O/C show values that are in agreement with those reported by Orlov (1985) and Steelink (1985). As is the case with the other authors, the present data also clearly separate soil humic acids from soil fulvic acids. Humic acids exhibit O/C values, ranging from 0.421 to 0.657, whereas fulvic acids are characterized by O/C values between 0.653 to 0.997. This is in agreement with Steelink (1985), who also reports lower O/C atomic ratios in humic acids (averaging 0.5) than in fulvic acids (averaging 0.7). Steelink is of the opinion that the O/C ratio is the best parameter for differentiating between types of humic compounds.

The O/C ratios of aquatic and lignite humic matter also fall within the range of those of soil humic matter. These values are noted to be lower in the humic acids than in the fulvic acid compounds extracted from lignite. However, the O/C values are not much different between aquatic humic and fulvic acid, contradicting Steelink's (1985) report for higher O/C ratios in aquatic fulvic than in humic acid.

Although Steelink (1985) is of the opinion that the O/C ratio was the best, the present data in Table 5.3 indicate that the N/C ratios are equally adequate for differentiating between humic and fulvic acids. As can be noticed, soil humic acids exhibit N/C ratios ranging from 0.048 to 0.077 as opposed to those of soil fulvic acids with N/C ratios in the

Table 5.3 Atomic Ratios of Major Elements in Humic and Fulvic Acids and Miscellaneous Materials

	H/C	O/H	O/C	N/C
Tropical Region Soils				
HA-Alfisols	1.192	0.449	0.539	0.058
HA-Andosols	0.802	0.559	0.447	0.055
HA-Oxisols	0.967	0.540	0.523	0.048
FA-Andosols	1.055	0.647	0.683	0.039
Temperate Regions Soils				
HA-Alfisols	1.121	0.392	0.439	0.071
HA-Aridisols (Solonetz)	0.903	0.055	0.501	0.077
HA-Histosols (Peat-bogs)	1.022	0.412	0.421	0.050
HA-Inceptisols	1.352	0.417	0.564	0.069
HA-Mollisols	0.960	0.529	0.508	0.058
HA-Spodosols	1.105	0.501	0.554	0.057
HA-Ultisols	1.237	0.480	0.594	0.067
FA-Inceptisols	1.301	0.533	0.693	0.046
FA-Mollisols	1.155	0.811	0.938	0.022
FA-Spodosols	0.947	0.689	0.653	0.032
FA-Ultisols	1.210	0.824	0.997	0.029
Geologic Deposits				
Lignite	0.639	0.708	0.452	0.032
HA-lignite	0.815	0.591	0.481	0.019
FA-lignite	1.136	0.652	0.741	0.029
Rivers and Swamps				
HA-aquatic	1.162	0.565	0.657	0.037
FA-aquatic	1.016	0.628	0.638	0.027
Merck Chemical Co.				
HA-artificial	0.964	0.438	0.422	0.024
'*Reference*' humic acid	1.00	0.474	0.475	0.048

Chemical Composition of Humic Matter

Table 5.3 *Continued*

	H/C	O/H	O/C	N/C
Peat	1.167	0.383	0.447	0.045
Plant Residue	1.520	0.550	0.837	0.044

range of only 0.022 to 0.046. Extremely low N/C ratios are noticed for aquatic and lignite derived humic and fulvic acids.

The observations above suggest that in general the humic molecule contains more carbon atoms than oxygen atoms, which is in close agreement with the assumption made by Orlov (1985). The humic acid molecule has at least two carbon atoms for one oxygen atom. The O/H ratios indicate the presence of one oxygen atom to two hydrogen atoms, a composition similar to that of a water molecule, and according to Orlov (1985) also similar to carbohydrates. The higher O/C ratios reaching values of 0.99 in fulvic acids suggest a composition made up of one carbon to one oxygen atom. Such a ratio is common for carbohydrate molecules. For example, the carbohydrate $C_6H_{12}O_6$ has six carbon atoms to six oxygen atoms per molecule, giving an O/C ratio = 1.0. This observation increases the credibility of opinions that fulvic acids are polysaccharidic in nature. Steelink (1985) and Ishiwatari (1975) believe that the carbohydrate content increases during formation of fulvic acid. Such an assumption is perhaps correct when considered from the standpoint of the polymerization concept, where fulvic acid is the starting point in the humification process. During formation and polymerization of humic precursors into fulvic acids, more polysaccharides or carbohydrates are incorporated than phenolic or quinonic substances. However, from the viewpoint of the biopolymer degradation concept, it is unlikely that during the degradation process into fulvic acid, more carbohydrates are being sequestered. By adding more carbohydrates into its molecule, the humic substance to be formed will not be reduced in size and is expected to be similar in molecular size as humic acid. Hence, the proposal is now presented that during the transformation into fulvic acid more of the aromatic substances are broken down and released by comparison with the polysaccharide constituents. The latter increases in amount because of the larger losses of phenolic and quinonic constituents. Consequently, the fulvic acid formed becomes increasingly more polysaccharidic in nature.

Finally, the values of N/C ratios of soil humic acids suggest on the average a composition of 0.06 - 0.05 N atoms to one atom of carbon. This means that approximately one nitrogen atom is present to 16 - 20 carbon atoms in a humic acid molecule. In fulvic acid, the molecule is expected to contain at least one nitrogen atom to 50 atoms of carbon.

5.1.6 Group Composition

In addition to elemental composition, group composition can also be used in the characterization of humic substances. Instead of being composed of elements or atoms, this composition is made up of chemical units of atoms or compounds, hence the name *group composition* is used here by the present author. It plays an important role in determining the chemistry and structural properties of humic substances. Composing a structural formula and the study on chemical behavior of humic substances would be incomplete without the knowledge of group composition. The chemical reactions characteristic of humic substances are attributed for the most part to the existence of a particular group composition. It is not enough to know how many atoms are present in a humic molecule, but it is in addition necessary to learn about the units created by the atoms. Though the issue of group composition is discussed in the literature many times, often in considerable length, not much concrete data are available. At the present knowledge, it is known that it can be distinguished into a (1) functional group composition, and (2) group-compound composition. For readers interested in the methods of determination of both the functional groups and group-composition of humic matter reference is made to Tan (2000), Stevenson (1994), and Purdue (1985).

Chemical Composition of Humic Matter

Functional Group Composition

The functional groups are composed of a set of active chemical groups that gives the humic substances their unique chemical behavior. Interaction reactions characteristic for humic substances, e.g., complex formation, chelation, and ion or metal bridging, are attributed to their presence in the humic molecule. These processes will be discussed in more detail in Chapter 7.

These groups are sometimes referred to by different names. Stevenson (1994) suggests naming them *oxygen containing functional groups*, though some of the groups do not contain oxygen at all. In contrast, Purdue (1985) prefers to call them *acidic functional groups*, which also limit the groups to units that behave acidic only. The major types of functional groups are in the form of COOH, phenolic-OH, alcoholic-OH, and carbonyl groups. These units indeed contain oxygen atoms. In addition to the above, amino groups are also important functional groups, but these are neither of the oxygen containing nor of the acidic types.

Carboxyl Groups. - The carboxyl or COOH groups give to the humic molecule its acidic characteristic. Their presence is the reason why humic substances exhibit charge properties, and have the capacity to adsorb and exchange cations. As indicated in Chapter 7 and as can be noticed in Figure 7.1, these carboxyl groups will dissociate their H atoms at pH 3.0. Such behavior -donating protons - fits the concept of a Brønsted-Lowry acid. However, Purdue (1985) considers the acidity of humic substances to be attributable to their proton binding capacity. Elaborate statistical models have been presented by the author for describing proton binding that can be used as a basic foundation for the acidity of humic substances, before finally admitting that none of them were satisfactory. The latter is due to the extreme complexity of the distribution of acidic functional groups in the humic molecule.

Not much information is available in the literature on carboxyl content of humic substances. Most of the published material is often on the analytical procedures of determination (Stevenson, 1994; Purdue, 1985), and only some authors discuss the amounts present in humic matter. The most common method used for determination is the Ca-acetate titration procedure (Tan, 1995), but other chemical methods are also available. Carbon-13 NMR spectroscopy (Schnitzer and Preston, 1986) and x-ray diffraction analysis (Schnitzer et al., 1991) have also proven to be useful in the determination of COOH content in humic substances. However, the figures reported show large variations due to differences in methods of determination. Though many think that these data are in disagreement and should be interpreted with caution, the present author notices that data obtained by the same analytical procedure agree fairly satisfactorily, with variations occurring only within very narrow limits. A summary of the average values of such data is provided in Table 5.4.

As can be noticed, the carboxyl content in soil humic acids are in the range of 2.4 to 5.4 meq/g, which is substantially lower than those in fulvic acids with COOH values of 8.5 meq/g. These values, obtained by the Ca-acetate titration method, are in the range for COOH content summarized by Stevenson (1994) from data reported by Schnitzer (1977) at the International Atomic Energy Agency meetings in Vienna. However, the COOH values for humic acids are rather low when compared to the average values of 6.4 to 9.0% obtained by ^{13}C-NMR analysis (Schnitzer and Preston, 1986). Higher values for COOH content have even been reported by Hatcher et al. (1981), who obtained by NMR analysis concentrations of 10 to 11 meq/g in humic acids from inceptisols. In the opinion of Schnitzer and Preston (1986), this discrepancy is attributed to the inclusions of amides and esters in the analysis by NMR spectroscopy. On the other hand, lack of reactivity and steric hindrance are the reasons cited by the authors for the low percentages in the titration procedures. Analytical constraints on acidic functional group determination are apparently not limited to the above. Purdue (1985) indicates that the amount of oxidic groups behaving as acids is limited by the oxygen content in a humic substance. On the basis of oxygen content alone, Purdue argues that humic acid with an oxygen content of 48% will contain 15 mmol/g of COOH or 30 mmol/g in the form of phenolic-OH or other oxy-type acidic groups. Unsaturation of COOH and phenolic groups is another possibility for variation in results of determination of COOH contents.

Stevenson (1994) agrees that fulvic acids are higher in COOH

Chemical Composition of Humic Matter

Table 5.4 Summary of COOH, Phenolic-OH Contents, and Total Acidity in Soil, Geologic, and Aquatic Humic Matter

	Carboxyl	Phenolic hydroxyl	Total acidity
		---------------- meq/g ----------------	
Soil Humic Matter			
HA-Alfisols	3.9	2.8	6.8
HA-Inceptisols	2.4	3.6	6.0
HA-Spodosols	5.4	3.5	8.9
HA-Ultisols	3.7	4.4	8.1
FA-Spodosols	8.5	3.8	12.3
FA-Ultisols	8.5	1.5	10.0
Geologic Deposit			
HA-lignite	5.2	0.9	6.1
FA-lignite	6.6	4.9	11.5
Rivers and Swamps			
HA-aquatic	3.4	2.5	5.9
FA-aquatic	3.9	2.3	6.2

Sources: Schnitzer and Khan (1972); Schnitzer (1977); Tan et al. (1990; 1991); Lobartini et al. (1992; 1991).

content than humic acids. According to this author the amount of COOH groups is inversely related to molecular weights, and fulvic acids are known to be lower in molecular weights than humic acids. He is of the opinion that the oxygen-containing functional groups are the highest in fulvic acids than any of the other naturally occurring organic polymers. The COOH concentration is considered to decrease upon *carbonization*, a theory advanced by Kumada (1987) for formation and transformation of humic matter. Stevenson (1994) argues that all humic and fulvic acids surviving biological attack are eventually diagenetically transformed into kerogen or coal. If humic and fulvic acids are affected by the *coalification* process, the first to disappear are the COOH groups, followed by methoxyl, OCH_3, and C=O groups.

Hydroxyl Groups. - Humic substances contain a variety of hydroxyl groups, but for characterization of humic acids generally three major types of OH groups are distinguished:

1. **Total hydroxyls**. They are the OH groups associated with all functional groups, such as phenols, alcohols, and hydroquinones. However, often the term "total hydroxyls" refers only to the sum of phenolic- and alcoholic-OH groups. Total OH is usually measured by acetylation.

2. **Phenolic-hydroxyls.** These are OH groups attached to phenolic (aromatic) structures. Currently, there is no wet chemical method for the determination of these groups. The amount is usually determined by difference as follows:

$$\text{meq phenolic-OH} = \text{meq total acidity} - \text{meq COOH} \qquad (5.8)$$

3. **Alcoholic-hydroxyls.** They are OH groups associated with alcoholic groups or nonaromatic carbons. The amount can also be determined by difference only:

$$\text{meq alcoholic-OH} = \text{meq total OH} - \text{meq phenolic-OH} \qquad (5.9)$$

The reactivity of alcoholic-OH groups is usually considered lower than that of phenolic-OH groups, hence the latter are assumed to be the most important in humic acid reactions. These phenolic-OH groups are another reason for giving the humic molecule its charge characteristic and unique chemical behavior. In basic chemistry, hydroxyls react as bases. However, phenolic-hydroxyls react as weak acids, since they dissociate their protons at pH 9.0 (see Figure 7.1). Because of this

Chemical Composition of Humic Matter

dissociation, Stevenson (1994) prefer to call them *acidic OH* rather than phenolic-OH groups, whereas Purdue (1985) prefers to use the name weakly acidic groups.

When electrically charged, the phenolic-OH groups together with the COOH groups are providing the capacity for interaction reactions. Complex reactions, chelation, water and metal bridging are made possible as shown in Figure 7.2.

As can be noticed from Table 5.4, the phenolic-OH content in soil humic acid is in the range of 1.5 to 4.4 meq/g, which agrees with values reported by Schnitzer (1977). Not many other authors have studied concentrations of phenolic-OH groups in humic matter, since no direct wet chemical methods are available. As indicated above, its measurement can be conducted only indirectly, The amount of aromatic carbon in the humic molecule is said to also limit the phenolic-OH concentration in humic substances (Purdue, 1985), which no doubt adds to the difficulties in the determination.

Total Acidity. - This is not a functional group, but it is a very important characteristic of humic substances closely related to the functional groups. It is commonly used as a measure for the cation exchange capacity of humic substances. Though it can be determined directly by titration, the common method is to calculate it by summation as follows:

$$\text{meq total acidity} = \text{meq COOH} + \text{meq phenolic-OH} \qquad (5.10)$$

However, the use of different analytical procedures in the determination of total acidity or carboxyl groups has been noted to often yield different results, as explained earlier. This is also shown by Felbeck (1965), who demonstrates that total acidity measurements by the $Ba(OH)_2$, KOH, NaOH, or Ba-acetate procedures generally yield different results. Nevertheless, by limiting collection and comparison of data within similar methods, the present author notes that the variation in values occurs only within narrow limits. The average figures in Table 5.4 for total acidity determined by the summation method vary only from 6.0 to 8.9 meq/g for soil humic acids, and from 10 to 12.3 meq for fulvic acids. These values are in agreement with those reported by Stevenson (1994), Hatcher et al. (1981), and Schnitzer (1977). The big difference in total acidity between humic and fulvic acid is also supported by the authors above. Stevenson indicates that fulvic acids are unmistakably substantially higher in total acidity than humic acids, which is due to their higher COOH content. Extremely high values of 14 meq/g are in Stevenson's opinion caused by analytical errors, such as failure to remove inorganic acids from the samples before titration.

In summary, the conclusion can be made that soil humic acid is characterized by a carboxyl content ranging from 2.4 to 3.9 me/g, and a phenolic-OH content of 2.8 to 4.4 me/g. These differences between the carboxyl and phenolic-OH content are so small that in essence one can conclude that humic acid has equal amounts of COOH and phenolic-OH groups. On the other hand, fulvic acid appears to contain almost twice the amount of carboxyl groups as humic acids. The amount of COOH in fulvic acid averaging 8.5 me/g is also substantially higher than the phenolic-OH content, which is 2.6 me/g on the average. Such differences in functional group composition result in remarkably large differences in total acidity values between humic and fulvic acids, as noticed from the data in Table 5.4. Humic acids from lignite and aquatic humic and fulvic acids exhibit COOH and phenolic-OH contents in the range of soil humic matter discussed above. However, no marked differences have been observed between COOH and phenolic-OH contents in aquatic humic matter.

Group-Compound Composition

This composition is made up of well defined compounds, such as carbohydrates, aromatic, and carboxyl group compounds. Such information proves to be very valuable in the identification of types of humic substances. It is usually determined by CP-MAS ^{13}C -NMR, cross-polarization magic angle spinning carbon-13 nuclear magnetic resonance, spectroscopy (Tan, 2000; Hatcher et al., 1980). Humic

Chemical Composition of Humic Matter

matter samples in the solid and liquid state can be analyzed, whereas soil samples in the undisturbed state can also be used, provided the samples contain sufficient amounts of organic matter (Tan et al., 1992). The results of NMR analysis yield humic acid spectra that can be divided into several regions:

1. Aliphatic C region at 0 - 105 ppm chemical shift
2. A polysaccharide sub-region can often be distinguished from 65 to 105 ppm
3. Aromatic C region at 105 - 165 ppm chemical shift
4. Carboxyl C region at 165 - 185 ppm chemical shift

The intensity of the signals (peaks) for aliphatic, aromatic, and carboxyl C may vary depending on origin (differences in soils) and type of humic matter analyzed. These peak resolutions form the basis for calculating the group composition percentages by integration of the spectra. An example of such data on group composition is provided in Table 5.5. Generally, fulvic acids exhibit spectra dominated by strong signals for aliphatic C, which yield by integration high percentages of aliphatic carbon. The signals for aromatic C are often manifested as very weak peaks, which then translate into low percentages of aromatic carbon in fulvic acids. On the other hand, humic acids show spectra with weaker signals in the aliphatic, but with very strong signals in the aromatic regions. This gives by integration low percentages of aliphatic carbon, and high percentages of aromatic carbon, respectively. The data also show that the group composition of aquatic fulvic acid from the black water streams closely resembles that of soil fulvic acids, testifying that the aquatic fulvic acid from the streams of the southeastern United States are allochthonous in nature (Tan et al., 1990).

Aromaticity. - This is a measure of the relative amount of phenolic or aromatic structures in a humic molecule. Aromaticity is a function of the group compound percentages, providing information on how

Table 5.5 Average Group-Compound Composition of Humic and Fulvic Acids Extracted from Soils, from Black Water Streams in the Southeastern United States, and from Lignite

	Aliphatic C	Aromatic C	Carboxyl C	Aromaticity	Aliphaticity
	--%--				
Soils					
Humic acid	37.3	51.6	11.1	58.0	42.0
Fulvic acid	61.6	20.7	17.7	25.2	74.8
Black Water					
Fulvic acid	61.6	22.3	16.1	26.6	73.4
Geologic Deposit					
Lignite	63.3	30.6	6.1	32.6	67.4
Humic acid	27.2	63.8	9.0	70.1	29.9
Fulvic acid	45.7	34.9	19.4	43.4	56.6

Sources: Tan (2000); Tan et al. (1992).

much of the aromatic compounds relative to the aliphatic substances are present in the molecular structure of humic matter. Data on aromaticity or aliphaticity are not only of importance for assessing the origin, stability, and chemical behavior of humic substances, as indicated by Schnitzer et al. (1991), but they can also be used for distinguishing between types of humic substances.

The conventional method of determination of aromaticity is to analyze first by ^{13}C-NMR spectroscopy the percentages of aromatic, aliphatic, and carboxyl carbon, as explained above. However, Schnitzer et al. (1991) have reported that these percentages can also be measured by *x-ray diffraction analysis*. This is rather surprising, since this method is normally used to analyze crystalline compounds in soils and fails to yield satisfactory results with amorphous substances. More details about this issue will be given in Chapter 6. Nevertheless, the

Chemical Composition of Humic Matter

value for aromaticity can then be calculated using the weight percentages obtained by applying either one of the following formulas:

$$\text{Aromaticity} = \frac{\text{Aromatic C \%}}{\text{Total C\%}} \times 100\% \tag{5.11}$$

$$\text{Aromaticity} = \frac{\text{Aromatic C\%}}{\text{Aliphatic C \% + Aromatic C \%}} \times 100\% \tag{5.12}$$

Schnitzer et al. (1991) indicate that ^{13}C-NMR analysis determines both aromatic + phenolic C, hence aromatic + phenolic C should be used in equations (5.11) and (5.12), instead of aromatic C % alone. The contention that NMR spectroscopy analyzes both aromatic and phenolic C is indeed quite correct, but the term *phenolic* is also referring to a similar aromatic structure, hence using aromatic + phenolic C is in essence redundant.

Equation (5.11) will yield slightly lower aromaticity percentages than equation (5.12), but the differences are relatively small. The percentage aromicity for soil humic acid in Table 5.5 equals 58.0% as determined by equation (5.12). Using equation (5.11), it amounts to only 51.6%, and differences from 0 to 8% have also been detected in humic acids by Schnitzer et al (1991).

The data in Table 5.5 clearly show sharp differences in aromaticity between humic and fulvic acids. Humic acids appear to be twice as aromatic as fulvic acids, a fact supported by Schnitzer's data and his coworkers (1991), revealing average percentages for aromaticity in humic acids as high as 71.8% and 64.8%, as compared to 36.0% and 31.5% in fulvic acids, determined by x-ray diffraction and NMR analyses, respectively. Higher aromaticity in humic acids is considered by some to be the result of the humification process. During humification, it is believed that the amount of aromatic and alkyl carbons increases, whereas the O-alkyl carbon content decreases. This idea is based on the opinion that the carbohydrates (O-alkyl carbons) are broken down by the soil microbes, resulting in relatively increasing the aromatic and alkyl fractions (Chefetz et al., 2002).

5.1.7 Calculation of Formula Weights

Formula weights have attracted considerable attention since the beginning of humic acid science. As discussed in Chapter 1, it was perhaps Berzelius who presented the first formulas for his crenic and apocrenic acids. This was also the start for argument, which turned into a full-blown controversy on the existence of molecular weight formulas, especially as more formulas have been presented over the years as humic acid chemistry advances into a modern science. Unfortunately, the differences in opinion has polarized the scientific community, resulting in a division into two groups, with one group firmly believing in and the other highly critical about the existence of formula weights for humic substances. Regardless of the convictions on formula weights from the group of nonbelievers, several methods have been reported for the construction of such formulas by the group of believers. Steelink (1985) has also stated unequivocally "yes" to the question whether a reasonable structural formula can be created for humic acids. The most modern methods are reported by Schnitzer (1994) and Schulten (1994: 1996), who are using sophisticated methods, such as pyrolysis-field mass spectrometry and Curie-point pyrolysis-gas chromatography, to come up with formulas and graphical models of humic structures. Not only are these methods very complex, but they are also extremely expensive and out of reach for many similarly competent scientists. Less complex, but also qualified and easy to follow are the methods proposed by Steelink (1985) and Orlov (1985). Both have attempted using elemental compositions expressed in atomic percentages and atomic ratios for the calculation of formula weights of humic substances. However, in contrast to Steelink (1985), who considers only C, H, and O, Orlov (1985) includes N in devising chemical formulas for the humic substances. As indicated in Chapter 1, the latter author assumes the humic substance to be characterized also by a *minimum molecular weight* through conversion of the composition to one atom of N. A combination of Orlov's and Steelink's

Chemical Composition of Humic Matter

methods is presented below as an illustration of how to calculate an empirical formula for humic acid from the elemental composition and atomic ratios in Tables 5.2 and 5.3, respectively. The average values are used instead of rounding-up the figures to one digit:

			Average
H/C = 0.8 – 1.2	→		1.0
O/C = 0.4 – 0.6	→		0.5
N/C = 0.05 – 0.07	→		0.06

The average values for H/C and O/C are similar to those used by Steelink (1985) in his calculations. It could be coincidental, but such similarities can also be used to enforce opinion on the existence of a definite composition, hence supporting the idea of humic acids being discrete or real compounds in nature. Regardless of what the reasons are, the above composition ratios mean that for every carbon, there are one hydrogen, 0.5 oxygen and 0.06 nitrogen. Consequently, conforming to the combined theories of Orlov (1985) and Steelink (1985), the formulas for this humic acid can be written as follows:

0.06 N:	$C_1H_1O_{0.5}N_{0.06}$	
One N:	$C_{16.7}H_{16.7}O_{(16.7 \times 0.5)}N$ or $C_{17}H_{17}O_8N$ →	mol.wt. = 363
Two N:	$C_{34}H_{34}O_{16}N_2$	mol.wt. = 726
Three N:	$C_{51}H_{51}O_{24}N_3$	mol.wt. = 1089
Four N:	$C_{68}H_{68}O_{32}N_4$	mol.wt. = 1452
Five N:	$C_{85}H_{85}O_{40}N_5$	mol.wt. = 1815
Ten N:	$C_{170}H_{170}O_{80}N_{10}$	mol.wt. = 3630

The above formulation gives a humic acid with an elemental

composition of: 56.2% C, 4.7%H, 35.3% O, and 3.9% N. If now the humic acid above also contains the following amounts of functional groups (see Table 5.4):

			Average
COOH groups:	2.4 –5.4 meq/g	→	3.9 meq/g
Phenolic–OH groups:	2.8 –4.4 meq/g	→	3.6 meq/g

then the humic acid unit capable of containing 3.9 meq COOH and 3.6 meq phenolic–OH groups can be calculated as follows. Since COOH is monovalent, hence 3.9 meq = 3.9mmol/g or 3.9 x 363 mmol = 1415 mmol COOH/mole of humic acid with a formula weight of 363 g. This means that there are 1.4 moles of COOH per $C_{17}H_{17}O_8N$, or 10 x 1.4 = 14 COOH groups per $C_{170}H_{170}O_{80}N_{10}$. A similar calculation and reasoning conducted with 3.6 meq/g of the phenolic hydroxyl groups yield 1.306 mol phenolic–OH per mole of humic acid with a formula weight of 363 grams. Consequently 10 x 1.3 = 13 phenolic-OH groups are present per $C_{170}H_{170}O_{80}N_{10}$. The conclusion can be made that humic acid with a formula weight of $C_{170}H_{170}O_{80}N_{10}$ is capable of accommodating both 14 COOH and 13 phenolic–OH groups.

The validity of the method above for formulation of molecular weights is supported by Schnitzer (1994) and Schulten (1994). From an elemental composition of 66.8% C, 6.0% H, 26.0% O, and 1.30% N, these authors arrive at a formula of $C_{308}H_{328}O_{90}N_5$. Considering that the atomic ratios used are H/C = 1.065, O/C = 0.292, and N/C = 0.0162, the formula weight suggested above is quite correct. However, judging from the elemental composition, Schnitzer and Schulten's compound is not representative for humic acid. The carbon content is extremely high, and the oxygen content too low for humic acids. They are not even close in range by comparison with data generally considered characteristic for humic acids by a multitude of authors, as presented in Table 5.1. The nitrogen content of 1.30% is also too low, though perhaps some kind of humic acids may be low in nitrogen, such as

Chemical Composition of Humic Matter

humic acids from geologic deposits. However, together with the extremely high C content, this low value for N gives a C/N ratio of 66.8/1.30 = 51.4, which is far too high for humic acids, and infringes on the basic concepts of humification. It is not clear and very confusing why the authors above didn't use the composition of humic acids reported in the same paper for the udic borolls, haplaquods, or haplaquolls. The data from either of these soils or from their average values are more related to humic acids present in or extracted from soils than those of *"proposed"* humic acids. The use of a hypothetical humic acid is very unfortunate, since it may aggravate the issue of artifacts.

5.2 MOLECULAR STRUCTURES

In close relation to molecular formulas is the formulation of a molecular structure for humic substances. This is a more complex and very difficult problem. Many attempts have been conducted and a monograph was even published by the International Humic Substances Society (IHSS) in 1989 under the title: *Humic Substances II. In Search For Structure* (Hayes et al., 1989). However, for a detailed review of the study of molecular structure and its basic consideration reference is made to Stevenson (1994) and Orlov (1985).

All of the theories suggested on molecular structures are based on aromatic constituents or phenols obtained by a variety of degradation analyses of humic substances. This is especially exemplified by the IHSS monograph, which has a lot of information on degradation techniques but which falls short in structural concepts and designs characterizing humic substances. Oxidative degradation, hydrolytic degradation, reductive cleavage and thermal degradation techniques always yield phenol or benzoic-like units, e.g., methoxy-benzaldehydes, methoxybenzenecarboxylic acid, hydroxybenzene, dihydroxybenzene and the like. These compounds are considered by many people to be broken down from the large plant biopolymers, such as lignin. Recently, a modern and very sophisticated method has been used, employing thermochemolysis in the presence of tetramethylammonium hydroxide (TMAH) and pyrolysis-gas chromatography-mass spectrometry (Py-GC/MS) for the detection of structural units in humic matter (Chefetz et al, 2002). The authors claim that these techniques will provide structural information on building blocks of natural macromolecules in soils without interference from conventional degradation procedures. The TMAH method is said to be highly selective in cleaving ester and certain ester linkages in the humic molecule. The method also avoids decarboxylation and produces the methyl esters of carboxylic acids for easy analysis by gas chromatography. Most importantly, the authors indicate they can detect lignin-derived phenols and benzenes, protein-derived pyridine, and polysaccharide-derived acetic acids and aldehydes by analysis with the Py-GC/MS method. Although this could be true, it is not irrefutably proven yet that these small phenol units originate from lignin, or pyridine units from protein. It is a matter of believing and trusting only, since as discussed in Chapter 4, phenol or benzene-like compounds are derived from two sources in soils, from lignin and from synthesis by soil micro-organisms. Nevertheless, regardless of their origin, the building blocks of humic matter detected by all the degradation, thermal and pyrolysis methods appear to be mostly small phenol monomers. These small units are then rearranged into a structural unit representing humic acid. However, two exceptions to this concept are available that use lignin monomers instead as the basic units.

Since a great number of methods and opinions are available in the construction of molecular structures of humic substances, many structural designs have been presented. Some of the structural models are relatively simple, whereas others are very complex and remarkably stunning. Many of the older structures, such as presented by Fuchs and Dragunov (1958), will not be shown in this book, since they have been discussed in almost all the other books covering humic acid structures. For these and several other humic acid structures of historical value, reference is made to the books of Stevenson and Orlov, mentioned above. In this section, the various schemes will be sorted out according to the different concepts on humification, and only the most pertinent examples will be used as illustration.

Chemical Composition of Humic Matter

5.2.1 Structures Based on the Ligno-Protein Concept

Lignin Monomer Concept of Flaig. - Structural designs of humic acid based on lignin monomer units have been proposed by Flaig (1967) and Steelink (1985). Flaig has studied the disintegration of lignin into the simpler monomers, and their further disintegration into smaller phenol units. Most of his ideas on humic acid structures, based on using the phenol units, have been covered in detail in several of the author's publications and also reproduced by other authors (Flaig, 1975; Flaig et al., 1975; 1959; Stevenson, 1994; Orlov, 1985). However, one aspect that has apparently escaped attention is Flaig's idea of using a lignin monomer for the molecular structure of humic acid. In this case, the author believes that carbohydrates may also play a role, hence he proposes the incorporation of a carbohydrate molecule in the formulation of a humic structure. A revised version of the structural design is shown in Figure 5.1 with coniferyl alcohol as the main building block. Flaig's structural design is composed only of a phenyl propane unit and a carbohydrate group. However, since protein is now considered an essential component of the humic acid structure, a revision is made by the present author to also include a protein or peptide unit in Flaig's structural model. Without protein, Flaig's structure reflects only a precursor of humic acid. A similar structural design for terrestrial humic acid has been suggested by Tan (2000) in his book entitled *Principles of Soil Chemistry*.

Lignin Tetramer Concept of Steelink. - A second example of how a lignin monomer can be used in forming a molecular structure of humic acid is proposed by Steelink (1985). This author has calculated with the atomic ratios of C/H = 1.0 and O/C = 0.5 the smallest formula for humic acid with a phenyl propane backbone to be $C_{18}H_{18}O_9$. To accommodate 1.5 moles of COOH groups per mole of humic acid, Steelink reasons that 2 [$C_{18}H_{18}O_9$] are needed to contain 3 moles of COOH. The formula has then to be changed into $C_{36}H_{36}O_{18}$, which in his opinion translates into a phenyl proprane tetramer. However, one of the three monomers exhibits a terminal COOH groups, suggesting it to be a hydroxyphenyl peruvic acid, and not a phenyl propane mono-

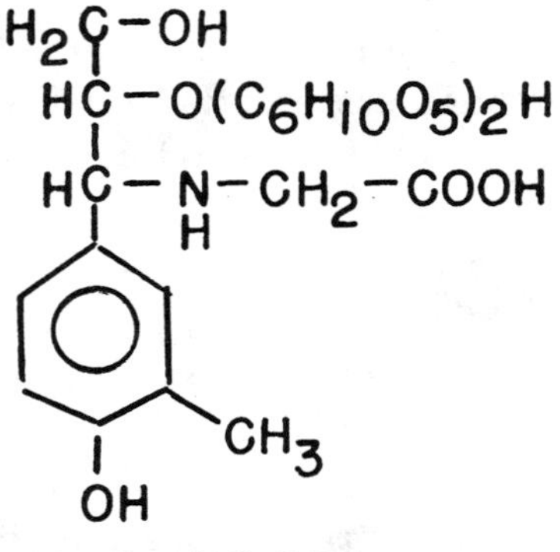

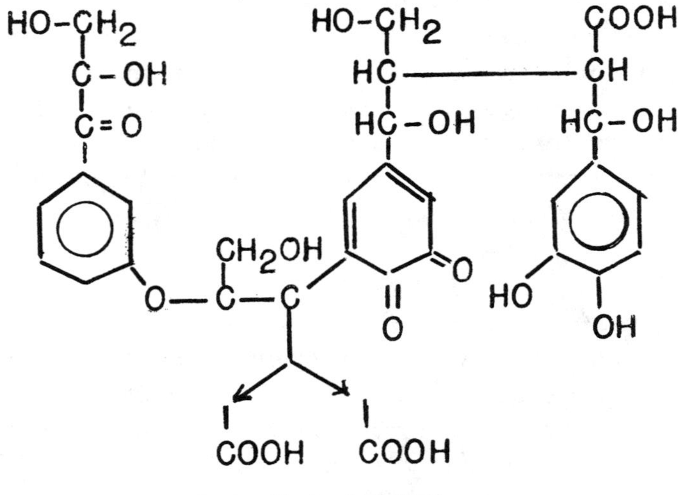

Figure 5.1 The author's version of humic acid structures according to the ligno-protein concept adapted from Flaig (1967) and Steelink (1985).

Chemical Composition of Humic Matter

mer, as can be noticed in Figure 5.1. Next to this is a quinone, which also does not qualify to be called a phenyl propane. In addition, lack of participation of a peptide unit only reflects the structure of a precursor of humic acid. Steeling's crooked and twisted structural design of the monomers has been corrected in Figure 5.1 to straight lines of -C-C-C-. This is the more common way to represent lignin monomer molecular structures.

5.2.2 Structures Based on the Phenol-Protein Concept

Phenol Oligomer Concept of Schulten. - Most of the structural designs for humic substances are in this category of the phenol-protein concept, and all of them show complex networks of phenol units linked together by other carbon units. For instance, a tangled web of phenolic units snaking around from the bottom to the top of a whole page has been presented for a structural concept of humic acid by Schulten (1994) and Schnitzer (1994). Schulten (1994) claims to arrive at such design from an extensive literature search, long-term investigations by Py-GC/MS, Py-FIMS, ^{13}C-NMR, chemical, oxidative, and reductive degradation, colloid chemical, and electron microscopic studies on soil humic substances. According to Chefetz et al. (2002), these complex instrumentations are only capable of detecting small organic compounds, presumably building blocks of humic substances. Although Schulten's structural concept can be considered as the most advanced design, the complexity and the lack of reasonable explanations, other than the complex analyses, for the rationale of a network of linkages make most scientists cautious about using it.

Phenol Polymer Concept of Schnitzer and Orlov. - Less complex structural designs that are more easy to understand have been presented by Schnitzer (1978), and Orlov (1985). Orlov's structural model, presented in Figure 5.2, underscores better than Schnitzer's the concept of phenol-protein linkages in the humic acid structure. Structural models of aquatic humic substances proposed by the U.S. Geological Survey Staff (1989) and Steinberg and Muenster (1985) are

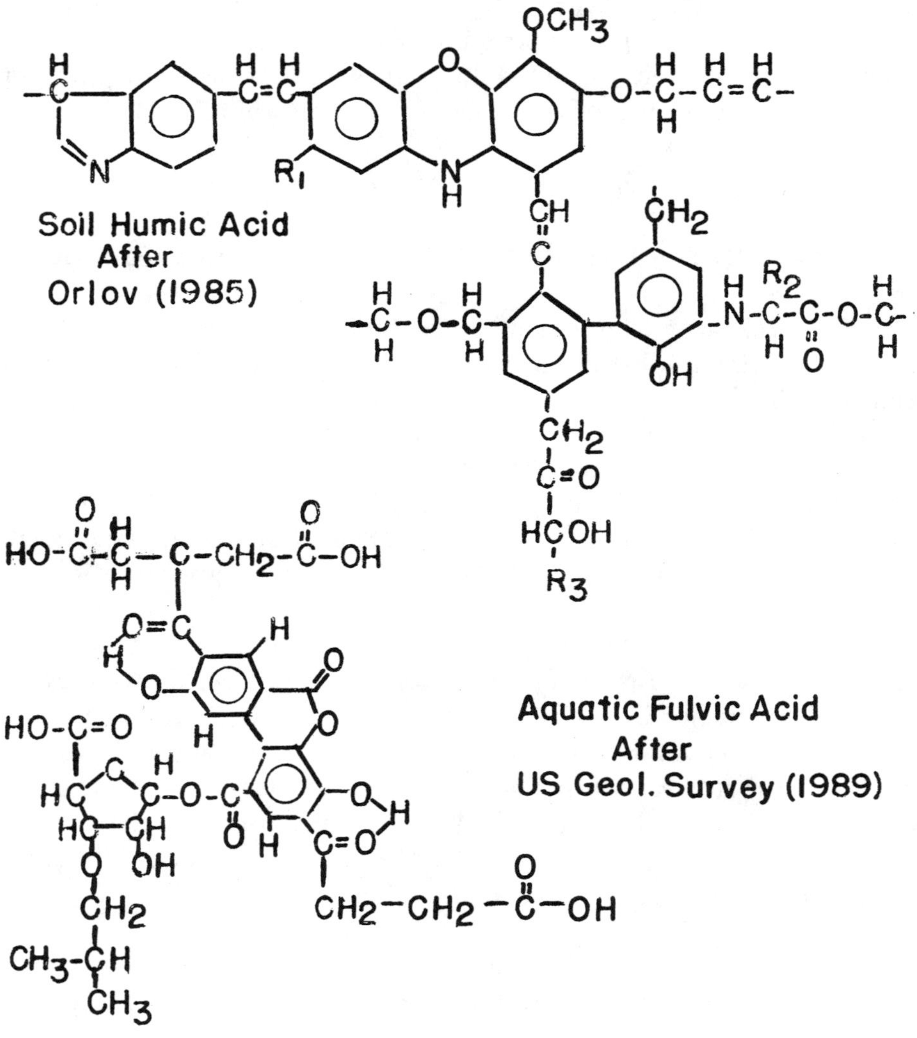

Figure 5.2 Structural models according to the phenol-protein theory adapted from Orlov (1985) and the U.S. Geological Survey (1989), respectively. The peripheral part in Orlov's model has been deleted for brevity.

Chemical Composition of Humic Matter

also in the same category, because of the large number of phenol units making up the structure. The example suggested by the U.S. Geological Survey, also shown in Figure 5.2, is supposed to be a structural model of an aquatic fulvic acid from the Suwannee River in South Georgia, USA. The phenol units making up the structure support the contention made earlier that these fulvic acids are allochthonous in nature. The four terminal carboxyl groups underscore the high acidity of fulvic acid in general, and also support the observation of fulvic acids containing high amounts of COOH groups as discussed above.

Phenol Dimer Concept of Stevenson. - The most simple design is the structure based on the dimer concept of Stevenson (1994). This author assumes the linkage of two phenol units to which a protein is attached as the basic or smallest unit of a humic molecule (Figure 5.3).

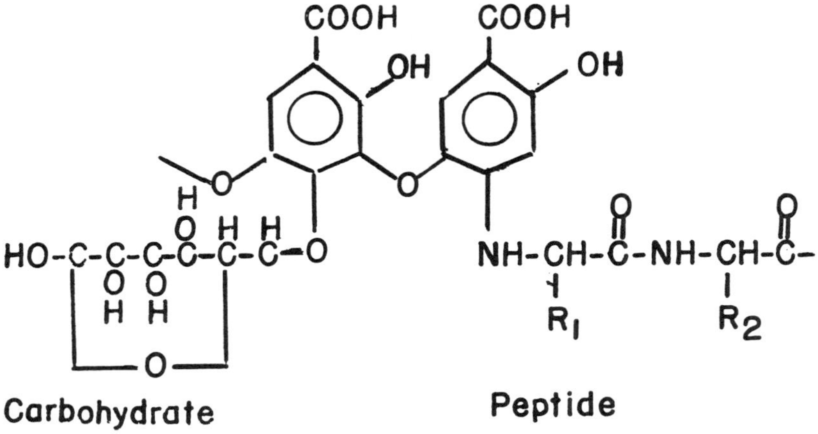

Figure 5.3 A revised structural model of humic acid based on the dimer concept of Stevenson (1994). The revision is made by the current author by including a carbohydrate molecular unit.

He also believes that the dimer is an excellent unit for explaining the chemical reactions characteristic of humic substances. Other structural designs based on linkages of phenol–protein units are available in the literature and for interested people reference is made to the respective publications listed in the reference lists of Tan (2000), Stevenson (1994), and Orlov (1985).

5.2.3 Structures Based on the Sugar-Amine Condensation Concept

These types of structures are characteristic for autochthonous aquatic humic substances. As explained in Chapter 4, the key components for an aquatic humic acid structure are sugar and amine or amino acid, and no lignin derivatives are required. In aquatic environments, plant materials, needed for formation of humic matter, usually do not contain lignin. Especially in marine environments, kelp, seaweed, and plankton, the major sources for the synthesis of marine humic acid, are composed mostly of carbohydrates. Therefore, these types of humic acids, if not affected by terrestrial material, exhibit molecular structures composed of carbohydrate and protein units only. As discussed earlier, sugar-amine complexes, called glucosylamine, are produced in the Maillard reaction. Condensation or polymerization of the glucosylamine units is considered to form humic substances. A structural model of such a humic acid is illustrated in Figure 5.4, where two or more glucosylamine groups are linked together through oxygen bonds.

5.3 Computer Modeling of Humic Acid Structures

With the recent advances in computer technology, attempts have been made to design computer models of molecular structures for humic substances. Though several methods are available, because of their similarities, only three of the methods will be discussed below.

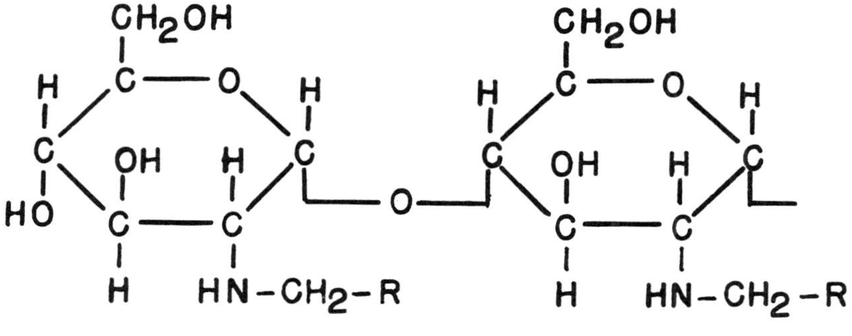

Chemical Composition of Humic Matter

Figure 5.4 A structural design showing the linkage of two glucosylamine molecules forming humic acid according to the sugar amine condensation theory.

One of the best methods is perhaps that reported by Schulten (1996) in relation to his formulation of the composition of humic acid monomers, trimers, decamers, and pentadecamers. The author claims to have developed a three-dimensional structure of a humic acid decamer from a combination of data obtained by wet-chemical, geochemical, biochemical, spectroscopic, and agricultural data and analytical pyrolysis. His wonderful colored 3-D pictures of a humic acid molecule, developed by computer modeling assisted by additional data obtained by pyrolysis-field ionization spectrometry (Py-FIMS) and Curie-point pyrolysis-gas chromatography/mass spectrometry, is, of course, quite impressive. A black and white reproduction of this structure is shown in Figure 5.5.

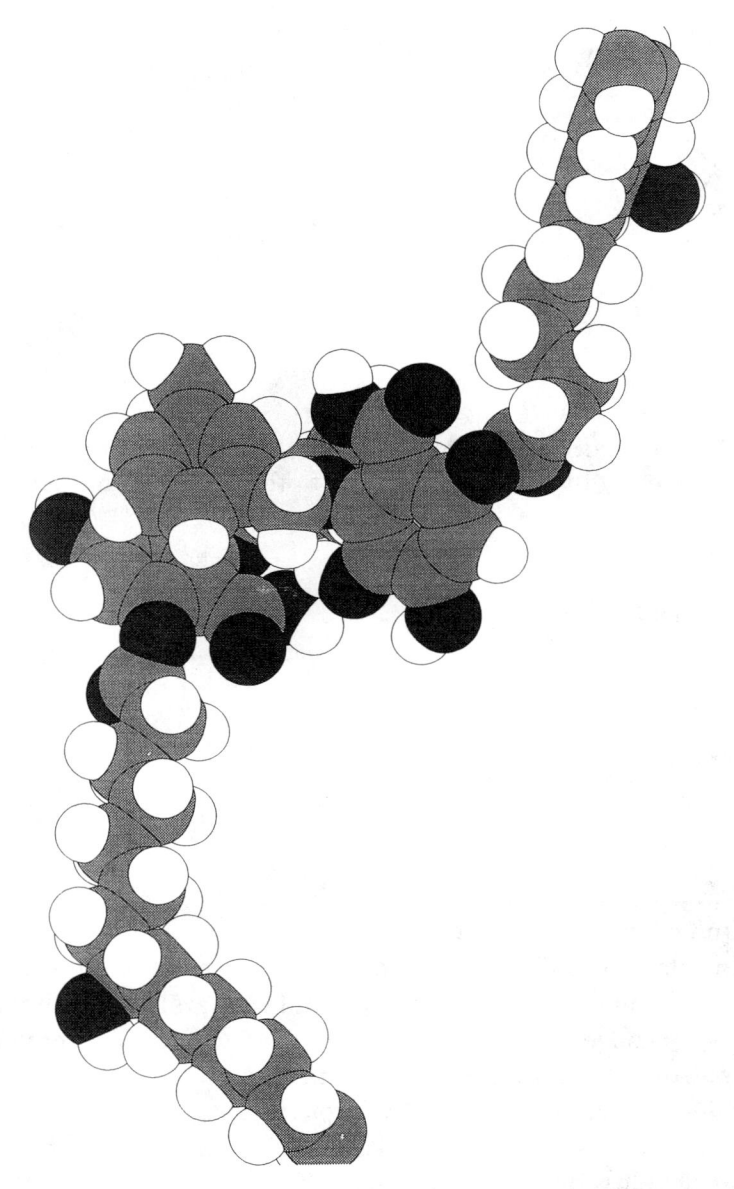

Figure 5.5 A black and white three-dimensional computer model of a hypothetical humic acid decamer structure: C = shaded-gray, H = white, O = black, and N = not shown. Reproduced with permission from Schulten (1996), copyright (1996) American Chemical Society, Washington, DC.

Chemical Composition of Humic Matter

Designing a molecular structure is a very complicated process, hence to simplify the problem a free 'translation' of Schulten's discussion is given below so that it can reach a wider audience of readers. The basis of Schulten's work is a hypothetical humic monomer, characterized by the formula $C_{308}H_{335}O_{90}N_5$, as discussed earlier. He has revised his formula by adding CH_2 to come up with a formula of $C_{315}H_{349}O_{90}N_5$, which is converted into the decamer $C_{3,150}H_{3,482}O_{896}N_{50}$. The latter has an elemental composition of 67.10% C, 6.22% H, 25.43% O, and 1.24% N, which hardly represents the composition generally noticed for humic acids by a variety of authors. The ratio of C/N = 67.10/1.24 = 54.1 indicates more likely the presence of *'raw'* organic matter, or litter, instead of a humified product.

Schulten (2002) justifies his selections above by stating that to simplify the process, it is necessary to use a humic molecule composed of a relatively low number of atoms, giving a corresponding mass below 6000 g mol^{-1}. Using HyperChem software, a molecular structure is then drafted by moving two of these molecules so that a phenolic-OH of one molecule is aligned and in close contact with a furan-O of the other molecule. Intermolecular hydrogen bonds and van der Waals forces are assumed to be linking the two molecules together into a dimer. By adding another molecule to the dimer a trimer can be formed and so on.

In his latest efforts, Schulten (2002, 2001) has used in addition to the 'Hyper-Chem,' a 'ChemPlus version 3.1' program for computing molecular properties, referring apparently to calculation of amount of energy in the formation of humic acid polymers. A HA-dimer is allegedly constructed, similarly as discussed above, by moving two geometrically optimized molecules of humic acid monomers along the inertial x-axis toward each other. When the distance between the two monomers is as close as 0.25 to 0.8 nm, a phenolic-hydroxyl group of one HA is placed in contact with a furan-oxygen of the other HA molecule. An energy gain of 6.37 kJ/mol has been calculated by Schulten for the formation of this HA-dimer. A HA trimer can then be formed in a similar way by adding a monomer to the HA dimer or by trapping a monomer in the dimer structure. Tetramers are formed by again (1) trapping a monomer in a trimer HA structure or by (2) linkage of four HA monomers through covalent bonding. *Trapping* is explained as a process of placement of a molecule in a structural void, and is considered an important mechanism for adding peptides, sugars, and other xenobiotics to the humic molecule. Linkage is assumed by intramolecular hydrogen and coordination covalent bonds, the latter resulting in complex formation.

The total energy of the covalently bonded HA tetramer is three times higher than that of the 'trapped' tetramer, and nine times higher than that of the HA monomer. These energy differences may indicate that humic acid polymers are more stable, hence will likely be more resistant to decomposition, than the humic acid monomers.

Schnitzer (2000) has also used a somewhat similar procedure for computer modeling of molecular structures. A two-dimensional structure is drawn first by hand, linking aliphatic chains covalently with isolated n-alkyl aromatic rings. Oxygen-containing compounds included in the structure are carboxyls, phenolic and alcoholic hydroxyls, esters, ethers and ketones. The N content is represented by nitriles and N-containing heterocyclic compounds. The resulting carbon skeleton, presumably of a humic acid monomer, has an elemental composition of 66.8% C, 6.0% H, 26.0% O, and 1.3% N, with a molecular mass of 5540 Da. This is not much different from the monomeric compound previously used by Schulten. This two-dimensional structure is converted into a three-dimensional model with the aid of *'Hyper Chem software.'*

Though all these efforts are indeed brilliantly executed and state-of-the-art modeling of humic acid structures, it is, however, very unfortunate that compounds are used, which hardly qualify to be called humic acid. Elemental compositions, calculated from extracted soil humic acids, are available in the literature and have even been presented by Schulten (1994) and Schnitzer (1994). They can be applied as easily as fake figures, and the use of such data is preferable to using data of fictitious materials. The application of fictional figures may create more problems rather than alleviating the issue of molecular structures of humic acids. It would increase the danger of making the critics 'die-hard' or 'hardcore' nonbelievers on the existence of a humic acid composition and structure.

Another version of computer modeling is presented by Hatcher et al. (1994). A *Signature Computer Program* is applied to construct a

Chemical Composition of Humic Matter

3D chemical structure by introducing the C, H, O, N weight percentages and selected NMR data obtained from an andosol. By asking the program to use benzene, naphthalene, chrysene, carbonyl and carboxyl units, the authors confess that they have obtained a very confusing equation for computing the amount of each unit required to match the analytical data. However, they claim to be able to redefine their equation by applying an *isomer generator* provided by the Signature program and manage to draw a 3D structure of the humic

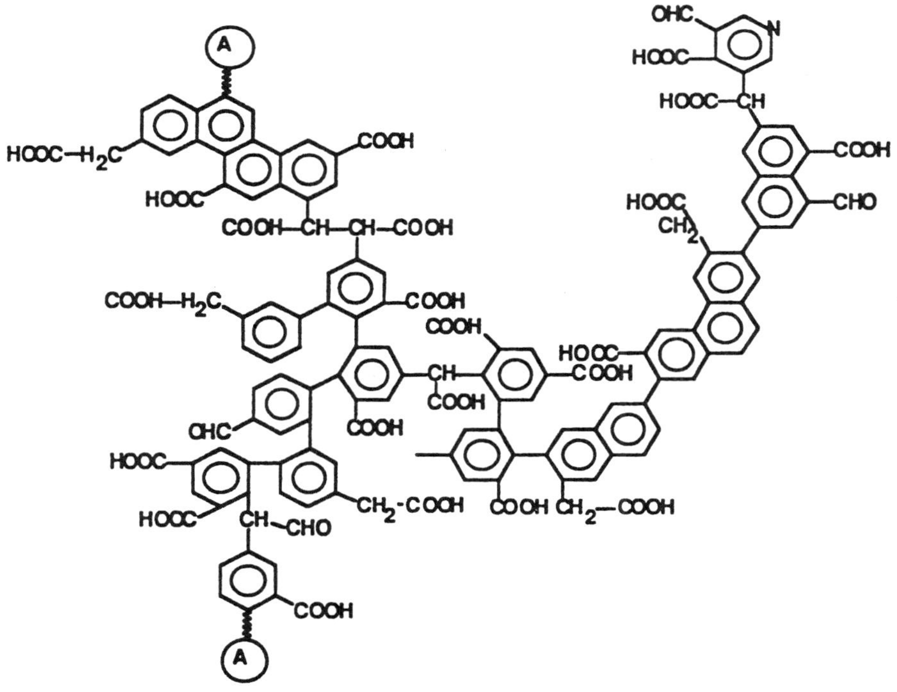

Figure 5.6 A two-dimensional structure of humic acid from andosol. Reproduced with permission from Hatcher et al. (1994), copyright (1994) Elsevier Science B.V., Amsterdam.

acid from the andosol. A two-dimensional reproduction is presented as an illustration in Figure 5.6. As can be noticed, the structure is composed only of benzene or phenol units. The functional groups are represented by carboxyl groups and no phenolic-OH groups have been considered. Polysaccharides, and peptides or amino acids, currently also considered important structural constituents of humic acids, have not been taken into account, though %N has been fed into the program. To consider this structure as representing a humic molecule is a matter of conjecture. A basic question also arises whether this is the structure of a humic acid monomer or a humic acid polymer? The authors, apparently, try to justify all these by declaring that their pure aromatic concept may deviate from other theories.

CHAPTER 6

CHARACTERIZATION OF HUMIC SUBSTANCES

6.1 CHEMICAL CHARACTERIZATION

Part of the chemical characterization of humic matter has been discussed in the preceding chapter on chemical composition. It was deemed necessary to cover it in a separate chapter, because of the many issues or topics, making it too long to include them in one chapter. In addition to the characteristics discussed earlier, humic substances also exhibit molecular weights and very distinctive spectroscopic features. Many scientists have tried using ultraviolet-visible, infrared, and NMR spectroscopy in the identification of various humic substances with results that are surprisingly reproducible (Orlov, 1985) for materials considered by many to be fake or operational compounds. These remaining characteristics will be discussed more in detail below.

6.2 MOLECULAR WEIGHTS

The topic of molecular weights is closely related to the issue of elemental composition and chemical formulas as discussed in Chapter 5. The possession of a formula composition implies the presence of a molecular weight, which is a basic physical property of humic substances of profound importance for their chemical activity. As discussed earlier elemental composition, chemical formulas and molecular weights are controversial issues in humic acid science. The use of molecular weights in characterization of humic substances also encounters many other problems, because of their polydispersive nature. They possess, therefore, a wide spread in molecular weights, causing many authors to consider humic substances to be very heterogeneous compounds (Felbeck, 1965). With homogeneous macromolecules, all particles have the same molecular weights. It is a well-known fact that molecular weights of humic substances can vary from as low as one thousand for fulvic acids to as much as several thousands for humic acids.

Physically, molecular weights can be expressed into: (1) number average, (2) weight average, and (3) z-average molecular weights. These types of molecular weights will be explained in more detail below.

6.2.1 Number-Average Molecular Weight, M_n.

This is formulated as follows:

$$M_n = (\Sigma nM)/n$$

where n = number of component molecules and M = molecular weight of component molecules. The methods used to determine M_n are osmometry, diffusion, and isothermal and cryoscopic distillation. Osmometry is considered the best method, but it appears not to be applicable to analysis of molecular weights >200,000.

6.2.2 Weight Average Molecular Weight, M_w.

This is defined as: $M_w = (\Sigma nM^2)/(\Sigma nM)$

which is usually measured using viscosity analysis and gel filtration. Of the two, gel filtration is the simplest method.

6.2.3 Z-Average Molecular Weight, M_z.

This is defined as:

$$M_z = (\Sigma nM^3)/(\Sigma nM^2)$$

This is normally measured by the sedimentation method and creates many problems in humic compounds due to their negative charges balanced by cations creating a diffuse double-layer system. Because of the latter, the molecules tend to repel each other, offsetting the sedimentation process. Intermolecular repulsion yields high-diffusion and low-sedimentation coefficients owing to faster sedimentation of the larger molecules than the counterions, resulting in an electrostatic drag. In addition, the polydisperse nature makes it difficult to achieve well-defined sediment boundaries with humic substances.

For a heterogeneous or polydisperse system: $M_n < M_w < M_z$, but for a homogeneous or monodisperse system: $M_n = M_w = M_z$.

6.2.4 Characterization by Molecular Weight

For the study of humic substances, it is common to use M_w because of its simple determination by filtration. Values reported for average molecular weights of humic matter may vary from 1000 to 30,000. Flaig and Beutelspacher (1951) state molecular weights of >100,000, and values of 2 million have been reported occasionally.

Apparently any number within these ranges can be obtained, depending on the filtration procedures employed, with fulvic acids usually exhibiting the lower, and humic acids the higher molecular weight values. Ultrafiltration by Lobartini et al. (1997) with an amicon cell, employing a membrane with a 10,000 daltons exclusion limit at the start, also indicates that humic acid would yield molecular weight fractions as imposed by any exclusion limits used in the analysis. However, the elemental composition, infrared spectra, and electron micrographs show that these different fractions contain essentially the same components, suggesting a composition more homogeneous in nature than previously expected.

The methods of filtration and gel chromatography are in fact measuring molecular weight ranges, rather than the weight average molecular weights or the mean values. By means of gel filtration using gels with a series of exclusion limits, a range of molecular weight values from 2,600 to 1,360,000 has been reported (Cameron et al., 1972). However, Stevenson (1994) is of the opinion that the most abundant part of the molecular weight distribution is around 100,000 and assumes that the highest value recorded of 1,360,000 is caused by formation of aggregates or attributed to an extended molecular weight tail. He believes that the upper weight average molecular weight of humic acid is approximately 200,000 daltons and the lower limits are perhaps in the range of 50,000 to 70,000.

In addition to filtration techniques, molecular weights of humic acids can also be determined by a variety of other methods. However, the values obtained may vary widely from one to another method used. This is evident from the data reported by Stevenson (1994) summarizing the information from the literature. Molecular weights of humic acids may vary from 36,000, to 25,000 and 1,390, as determined by viscosimetry, freezing point, and x-ray diffraction techniques, respectively. Molecular weight values of > 20,000 and in the range of 24,000 - 230,000, and 53,000 - 100,000 have also been obtained by electron microscopy, equilibrium sedimentation and sedimentation - diffusion analytical procedures, respectively. Methods by x-ray diffraction are questionable, since these methods are applicable only for analysis of crystalline materials. The author above has also cited small-angle x-ray scattering analysis as being capable of

Characterization of Humic Substances

detecting molecular weights of humic acids between 200,000 and 1,000,000.

6.2.5 Relationship Between Molecular Weight and Size or Shape

Molecular Size

From results of filtration analysis using sephadex gels with different exclusion limits, Tan and McCreery (1975) note that the degree of polymerization and the sizes of the molecules isolated affect molecular weights of humic matter. A summary of the data listed in Table 6.1 demonstrates the relation between the size of the molecule

Table 6.1 Molecular Weights and Size (in Å and nm) of Humic Acids Obtained by Sephadex Gel Filtration

Molecular weight	Molecular volume Å	Radius Å	Radius nm
30,000	23,622	17.8	1.78
5,000	3,937	9.8	0.98
1,500	1,181	6.6	0.66
1,000	787	5.7	0.57

Source: Tan and McCreery (1975); Tan (2000).

and molecular weight. By assuming that the humic molecules are spherical in shape, the larger the size of the molecule of the humic compound isolated, the larger will be the numerical value of the average molecular weight of humic acid.

Molecular Shape and Frictional Ratio

Particle shape can be determined by calculating the so-called *frictional ratio*, which is defined as f/f_o, in which f = frictional coefficient and f_o = frictional coefficient of an unsolvated sphere of the same mass (Cameron et al., 1972; Richie and Posner, 1982). These coefficients are calculated using the following equations:

$$f = \frac{RTs}{M} \tag{6.1}$$

where R = gas constant, T = absolute temperature, s = sedimentation coefficient, and M = molecular weight.

$$f_o = 6\pi\eta\,(3Mv/4\pi N)^{1/3} \tag{6.2}$$

where η = viscosity, v = partial specific volume of colloid, and N = Avogadro's number.

The values of f/f_o or frictional ratios are unity (equal to one) as reported by Flaig and Beutelspacher (1968). This is the reason for considering the humic molecules to be spherical or globular in shape. The ratio will exceed unity for shapes differing from spheres or when an interaction takes place between the humic molecule and the solvent. However, more recent observations indicate that the frictional ratios may increase with molecular weight as can be noticed from the data listed in Table 6.2. High values for f/f_o of 1.4 - 2.4 are exhibited by humic acids with molecular weights between 20,000 and 1,360,000, whereas low values of 1.14 and 1.28 are displayed by humic acids with low molecular weights of 2,600 and 4,400, respectively. Considering

Characterization of Humic Substances

standard errors and variations, these low f/f_o values can be taken as approaching unity, hence may perhaps indicate that the humic molecules are spherical in shape. Judging from the data in the table, it can be expected for certain that this is true for humic molecules with

Table 6.2 Relation between Frictional Coefficients and Molecular Weights of Humic Acids

Molecular weight	f/f_o
2,600	1.14
4,400	1.28
12,800	1.41
20,400	1.46
20,400 (pH 11)	1.39
23,800	1.52
23,900 (pH 7.0)	1.36
83,000	1.96
108,300 (pH 11)	1.42
125,900 (pH 9.0)	1.44
127,000	2.18
412,000	2.12
1,360,000	2.41

Sources: Cameron et al. (1972); Ritchie and Posner (1982).

molecular weights < 2,000. At the higher molecular weights, the humic acid molecules are believed to have shapes in the form of *random coils* (Cameron et al., 1972). They are conceived to be negatively charged branched threads that coil and wind randomly with respect to time and space. Coil density is envisioned to increase with branching, yielding shapes of the more compact spherical types than the linear types. The solvent is trapped within the internal regions but can move freely in the peripheral areas. From their studies with surface pressure and viscosity measurements, Ghosh and Schnitzer (1980) believe that humic and fulvic acids behave like rigid *spherocolloids* at high sample concentration, at low pH or in the presence of sufficient amounts of neutral electrolytes. At low sample concentrations, they are flexible linear colloids.

6.3 Ultraviolet and Visible Light Spectrophotometry

The color of humic substances is a physical property that has attracted the attention of many scientists who have attempted using it for characterization of humic substances (Flaig et al., 1975; Tan and Van Schuylenborgh, 1961; Schnitzer, 1971, Tan and Giddens, 1972; Kumada, 1987). In Germany, especially, color properties of humic substances have been investigated by a number of scientists, who are of the opinion that the intensity of light absorption was characteristic for the type and molecular weight of humic substances. The absorbancy

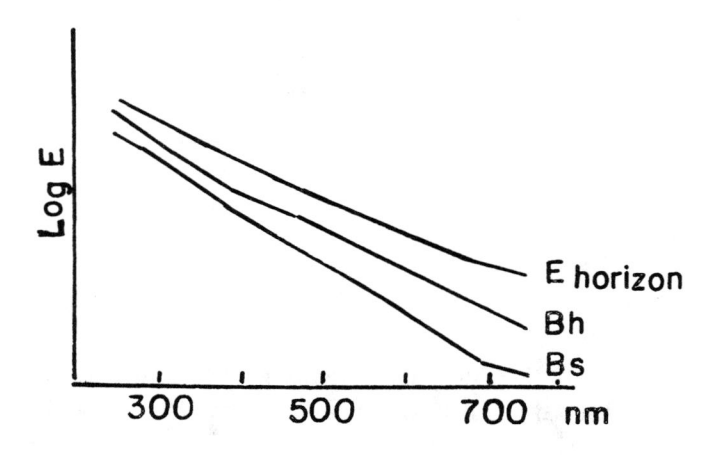

Figure 6.1 Visible light absorption of humic and fulvic acids of a spodosol in tropical region soils (Tan and Van Schuylenborgh, 1961).

or extinction of humic matter is recorded at various wavelengths from 300 to 800 nm. By plotting the logarithm of the absorbencies against the wavelengths, a straight line is usually obtained (Figure 6.1). The slope of such a line has been used for differentiation of humic substances, and its importance as a humification index has been discussed in Chapter 4. Fulvic acids are noticed to yield spectra with steep slopes, in contrast to humic acids. As explained earlier, the slope of the spectral curve can be expressed as a ratio or quotient of the absorbencies at two arbitrarily selected wavelengths. Many people chose the absorbency or extinction values at 400 and 600 nm, and the formula of the ratio, designated as E_4/E_6 or $Q_{4/6}$, called *color ratio*, is given earlier as equation (4.9). Other scientists opt to use extinction values at 465 and 650 nm and the ratio is formulated as D_4/D_6, where D stands for *optical density*. Orlov (1985) is even of the opinion that the coefficient of extinction, E, can be used for characterization of humic substances.

This color ratio is used as an index for the rate of light absorption in the visible range. A high color ratio, 7 - 8 or higher, corresponds to curves with steep slopes and is usually observed for fulvic acids or humic acids of relatively low molecular weights. On the other hand, a low color ratio, 3 - 5, corresponds to curves that are less steep. These curves are exhibited by humic acids and other related compounds with high molecular weights. The data in Table 6.3 show some E_4/E_6 ratios of humic substances extracted from temperate region soils. It can be noticed that humic acids with high molecular weights (m.w. > 30,000) have lower E_4/E_6 values (4.32 - 4.45) than humic acids with lower molecular weights (m.w. = 15,000). The lower molecular weight humic acids exhibit E_4/E_6 values of 5.47 - 5.49. This is supported by data from the literature, which in general show humic acids to be characterized by E_4/E_6 ratios between 3.3 - 5.0 in contrast to fulvic acids whose E_4/E_6 ratios are between 6.0 - 8.0. The values of D_4/D_6, as reported by Orlov (1985), seem also to agree by showing a range of 4.1 - 4.8 for humic acids as compared to a range of 9.0 - 17.7 for fulvic acids. The corresponding E values are higher for humic acids (0.061 - 0.104) than for fulvic acids (0.010 - 0.016). These observations are not in conformity with Orlov's assumption that the E value is related to the

Table 6.3 Color Ratios, E_4/E_6, of Humic Substances Extracted from Temperate Region Soils

Soil	Humic substance	E_4/E_6
Ultisols (Cecil soil)[a]	Humic acid, m.w. > 30,000	4.32
Ultisols (Greenville soil)	Humic acid, m.w. > 30,000	4.45
Ultisols (Cecil soil)	Humic acid, m.w. = 15,000	5.49
Ultisols (Greenville soil)	Humic acid, m.w. = 15,000	5.47
Alfisols[b]	Humic acid	3.5
Andosols[c]	Humic acid	3.4
Aridisols[b]	Humic acid	4.3
Mollisols (Chernozem)[b]	Humic acid	3.3
Mollisols (Chestnut soil)[b]	Humic acid	3.9
Spodosols[b]	Humic acid	5.0
Ultisols (Cecil soil)[d]	Fulvic acid	8.0
Unknown[b]	Fulvic acids	6.0 - 8.0

Sources: [a] Tan (2000); [b] Schnitzer and Khan (1972) and Kononova (1966); [c] Kumada (1987); [d] Tan and Giddens (1973).

molecular weight of humic acid, which he formulated as follows:

$$E = \frac{\epsilon}{MW \cdot 100} \tag{6.3}$$

$$MW = \frac{\epsilon}{E \cdot 100} \tag{6.4}$$

where MW = molecular weight, ϵ = molar coefficient of absorption, and E = measured extinction coefficient. Equation (6.4) indicates that the value of MW increases when E decreases (at constant ϵ). Similarly the value for MW decreases when E increases. Fulvic acids are noticed to be substantially lower in E values than their humic acid counterparts, hence should exhibit higher molecular weight values if equations (6.3) and (6.4) are valid assumptions.

Since UV-visible light spectra of humic compounds are generally featureless straight lines, Salfeld (1975) suggests modifying the analysis by measuring the absorbencies at intervals of 10 nm in the range of 230 to 700 nm. The difference between two adjacent absorbencies (= ΔE) is considered to reflect the slope of the curve. By plotting the logarithms of ΔE against the wavelengths, a curve is obtained with several peaks, called the *derivative spectrum*. Another more simple variation to express the inclination of the spectral lines is the method of Kumada (1978), who uses Δlog k values, which have been defined and discussed in Chapter 4.

Perhaps, it is also important to mention that humic compounds can also be characterized by fluorescence spectra. By using fluorescence excitation spectroscopy, Ghosh and Schnitzer (1980b) show both fulvic and humic acids to yield spectral curves with distinctive bands at 465 nm. Fulvic acid appears to distinguish itself from humic acids by displaying an additional band at 360 nm.

Finally, mention should also be made briefly that colorimetric analysis of humic acid solutions obeys the Bouger-Lambert-Beer law (Orlov, 1985; Tan, 1995), hence provides applicabilities for measurements of the concentrations of humic substances. This law is usually formulated as follows:

$$\log (I_o/I_t) = D = \epsilon l c \qquad (6.5)$$

where I_o = intensity of incident light, I_t = intensity of transmitted light, ϵ = extinction coefficient, l = thickness of sample, and c = concentration.

By using a sample holder of 1 cm thickness, l = 1.0, hence the law above indicates that the optical density or absorbance is directly proportional to ϵc, in other words to concentrations. Conformity to Lambert and Beer's law gives a linear regression if optical density or absorbance is plotted against concentrations (Tan, 1995). However, the few colorimetric procedures presented in the literature for a rapid quantitative determination of humic acid have been accepted only with mixed blessings. The method proposed by Holmgren and Holzhey (1984), using 2-amino-2-methyl propanol buffer, is apparently based on measurement of color related to the amount of Fe and Al chelated by the humic substances.

6.4 INFRARED SPECTROSCOPY

This method is another important tool and relatively simple for use in the characterization of humic substances. Infrared spectroscopy has been used extensively in the past to characterize humic substances, although some doubt exists about the significance of the infrared spectra. Spectroscopic methods in general are deemed by MacCarthy and Rice (1985) severely limited in the study of humic substances. However, of the several spectroscopic methods available, e.g., UV-visible, spectrofluorometry, and electron spin resonance spectroscopy, it is the opinion of the authors above that infrared spectroscopy is by far the most useful. Reservations for infrared analysis are perhaps caused in part by the complexity of the infrared spectra of humic preparations. Humic substances are mixtures of polyelectrolytic molecules and their spectra reflect the responses of the many different molecular species. The use of poorly prepared humic samples and the publication of poorly resolved spectra have aggravated the problem immensely. In spite of these issues, infrared analysis has proven to be very useful. It is deemed to be very valuable in the identification of (1) functional groups and their structural arrangements in the humic molecule, and (2) organic and inorganic impurities. Several typical vibrations of C-H and oxygen-containing functional groups absorb light in the infrared region, yielding peaks,

Characterization of Humic Substances

called *absorption bands*, characterizing the spectrum. For people interested in the basic principles of infrared vibration properties and analytical procedures, reference is made to Tan (2000; 1995), Stevenson (1994), MacCarthy and Rice (1985), and Schnitzer (1965).

6.4.1 Infrared Spectra of Humic Matter

Though several procedures are available in infrared analysis of humic substances, the most commonly applied method is the pressed KBr pellet technique (Tan, 1995). The humic matter is mixed with KBr and pressed into a transparent pellet, which is scanned from 4000 to 600 cm^{-1}. Sometimes scanning is continued to 400 cm^{-1}, but frequently the characteristic infrared bands are located mostly within 4000 to 600 cm^{-1} (Table 6.4). The spectrum is often divided into two regions, a *group frequency region* (4000 - 1300 cm^{-1}) and a *fingerprint region* (1300 - 650 cm^{-1}). In the group frequency region, the principal bands may be assigned to vibration units that consist of only two atoms to a molecule. In the fingerprint region, single bond stretching and bending vibrations of polyatomic systems are major features. Molecules similar in structure may absorb similarly in the group frequency region, but will show differences in absorption in the fingerprint region.

Notwithstanding the many arguments on the usefulness of infrared analysis of humic substances, the method is capable of detecting and distinguishing between the different types of humic substances and organic compounds in general. Examples of spectra are given in Figure 6.2 as illustrations.

Fulvic Acid

As can be noticed in Figure 6.2, the spectrum of fulvic acid has very different infrared absorption features than humic acids or the other substances, hence can be used as a fingerprint for identification purposes. The fulvic acid spectrum has a strong absorption band at 3400 cm^{-1}, a weak band between 2980 and 2920 cm^{-1}, a *shoulder* at 1720 cm^{-1} followed by a strong band at 1650 cm^{-1}, and a strong band

Table 6.4 Infrared Absorption Bands of Functional Groups in Humic Matter

Wavenumber cm^{-1}	Wavelength μm	Proposed assignment
3400 - 3300	3.94 - 3.03	O-H and N-H stretch
3380	2.950	Hydrogen bonded OH
2985	3.35	CH_3 and CH_2 stretch
2940 - 2900	3.40 - 3.44	Aliphatic C-H stretch
1725 - 1720	5.79 - 5.81	C=O stretch of COOH groups
1650 - 1630	6.00 - 6.10	C=O stretch (amide I), aromatic C=C, hydrogen bonded C=O, double bond conjugated with carbonyl and COO$^-$ vibrations
1650 - 1613	6.00 - 6.19	COO$^-$ symmetrical stretch
1460	6.85	Aliphatic C-H, CC-H_3
1440	6.95	C-H stretch of methyl groups
1435	6.97	C-H bending
1400	7.14	COO$^-$ antisymmetrical stretch
1390	7.20	Salts of COOH
1280 - 1230	7.80 - 8.10	C-O stretch, aromatic C-O, C-O ester linkage, phenolic C-OH
1170 - 950	8.50 - 10.5	C-C, C-OH, C-O-C typical of glucosidic linkages, Si-O impurities, C-O stretch of polysaccharides
1035	9.67	O-CH_3 vibrations
840	11.9	Aromatic C-H vibrations

Sources: Tan (2000, 1995); Stevenson (1982, 1994); Mortenson et al. (1965).

at 1000 cm^{-1}. These bands are attributed to vibrations of OH, aliphatic C-H, carbonyl (C=O) followed by carboxyls in COO$^-$ form, and ethyl, vinyl -CH-CH_2, aromatic aldehyde, amine and SH groups, respective-

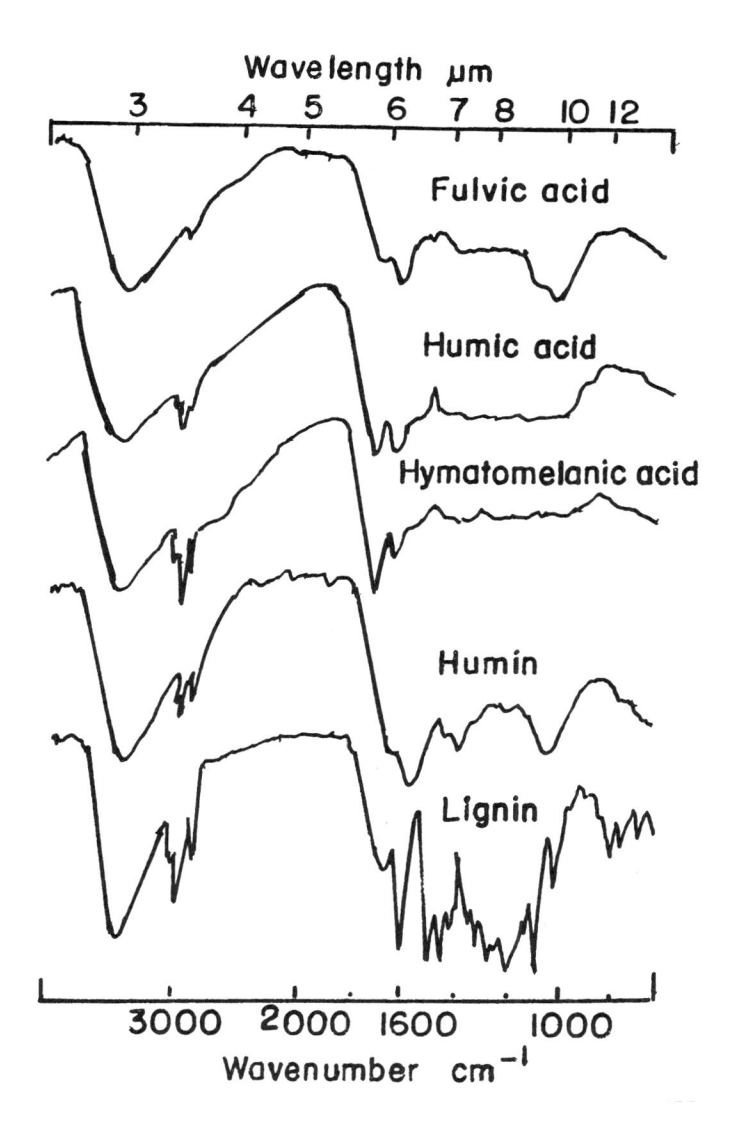

Figure 6.2 Characteristic spectra of major humic substances and lignin.

ly. This infrared spectrum shows close similarities to the infrared spectrum of polysaccharides (Tan and Clark, 1968).

Humic Acid

In contrast to fulvic acid, the humic acid spectrum is characterized by a strong aliphatic C-H absorption band between 2980 and 2920 cm^{-1} and two strong absorption bands for carbonyls and carboxyls in COO$^-$ at 1720 and 1650 cm^{-1}, respectively. In addition, the humic acid spectrum lacks the strong band at 1000 cm^{-1}. This feature frequently distinguishes it from fulvic acid. The presence of a band at 1000 cm^{-1} in a humic acid spectrum is ordinarily associated to impurities with SiO_2. Such an impurity can be removed by washing the humic acid specimen with a dilute HCl-HF mixture.

Some humic acids, especially those extracted from ultisols, may not exhibit the two bands at 1720 and 1650 cm^{-1}, respectively, but may have spectra featuring only the strong band at 1650 cm^{-1}.

Hymatomelanic Acid

The infrared spectrum of hymatomelanic acid has very strong absorption bands between 2980 -2920 cm^{-1}, and at 1750 cm^{-1}, attributed to aliphatic C-H and C=O stretching vibrations, respectively. It has been discovered by Clark and Tan (1969) that hymatomelanic acid is an ester compound formed from humic acid and polysaccharide. This is supported by subsequent investigations by Tan and McCreery (1970) and Tan (1975) that also provide evidence indicating the C-H group, belonging to the polysaccharides, is esterified to the carboxyl group of the humic acid molecule.

Humin

The infrared spectrum of humin closely resembles that of fulvic

Characterization of Humic Substances 185

acid. However, a stronger aliphatic C–H absorption between 2980 – 2920 cm^{-1} distinguishes it from the spectrum of fulvic acid. Such close similarities in infrared features of humin and fulvic acid are rather surprising due to the concept that humin is a condensed form of humic acids.

Lignin

Lignin has an infrared spectrum that distinguishes it clearly from humic and fulvic acid. Humic matter is believed to be a decomposition product of lignoid or lignin-like compounds.

6.4.2 Classification of Infrared Spectra

Some authors try to group humic matter spectra into several different types. Stevenson has cited Kumada (1987) to have classified infrared spectra into types A, B, R_p and P. However, these four symbols were used by Kumada (1978) in his book *Chemistry of Soil Organic Matter* for distinguishing types of humic acids by Δlog k values obtained from colorimetric analyses. This has been explained earlier several times. The infrared data are supplied by Kumada as additional characteristics for the four types of humic acids and not for the purpose of classification of infrared spectra. More real perhaps is the idea of Stevenson and Goh (1971), who have attempted to distinguish the infrared spectra of humic substances into types I, II, and III. *Type I* spectra are the spectra of humic acids, with the absorption bands at 1720 and 1650 cm^{-1}, considered as being equal in intensity. *Type II* spectra are typical for fulvic acids with strong absorption at 1720 cm^{-1} and weak absorption at 1640 cm^{-1}. The strong band at 1720 cm^{-1} is attributed by Stevenson (1994) to the occurrence of more COOH groups in fulvic acids than in humic acids. The author also believes that this band at 1720 cm^{-1} will progressively disappear with an increase in color intensity. These features deviate from those presented for fulvic acids in Figure 6.2, where the band at 1720 cm^{-1} is very weak, and the band at 1640 cm^{-1} is the strongest and the most prominent band. Most spectra of fulvic acids are of this nature. A strong intensity band at 1640 cm^{-1} conforms more to the presence of large amounts of carboxyl groups, since this is the absorption band caused by vibrations of carboxyls in COO$^-$ form. *Type III* spectra have infrared features similar to type I, but show in addition strong bands between 2900 and 2840 cm^{-1}, indicative for more aliphatic C-H compounds.

The infrared spectra reported above by Stevenson are not much different from those presented in Figure 6.2, and by other authors. Selected spectra, provided for comparison in Figures 6.3 and 6.4, indicate that the three fulvic acid spectra shown do not differ dramatically from each other. The fulvic acid spectrum from MacCarthy and Rice (1994) exhibits a weak shoulder and a strong band at 1720 and 1650 cm^{-1}, respectively, similarly as those reported for the fulvic acid above in Figure 6.2. This is supported by the fulvic acid spectrum of Kemp and Mudrochova (1975). The humic acid spectra are also comparable and show little variations from one to the other. The 'hump' or broad band near 2900 cm^{-1} in Stevenson's type I spectrum is apparently caused by an error in recording or improper analysis. Usually, humic acid spectra exhibit in this region a series of sharp bands as evidenced by the spectra from Bedrock et al., Kemp and Mudrochova, and Tan. The humic acid spectrum from Kemp and Mudrochova has a strong band at 1000 cm^{-1}. This is usually attributed to impurities by chelated silica, which should have been removed by washing the sample in a dilute HCl-HF solution prior to infrared analysis. Humic acid spectra are commonly reported not to have a band at 1000 cm^{-1}.

The humic substances above have been extracted from different soils in different regions and have been analyzed with different models of infrared spectrophotometers. The only thing in common is that they were extracted with the same NaOH procedures. Nevertheless, the similarities of the spectra are very apparent. Such reproducibility in infrared spectra, regardless of the different sources, tends to confute the concept of humic substances being fake compounds or artifacts.

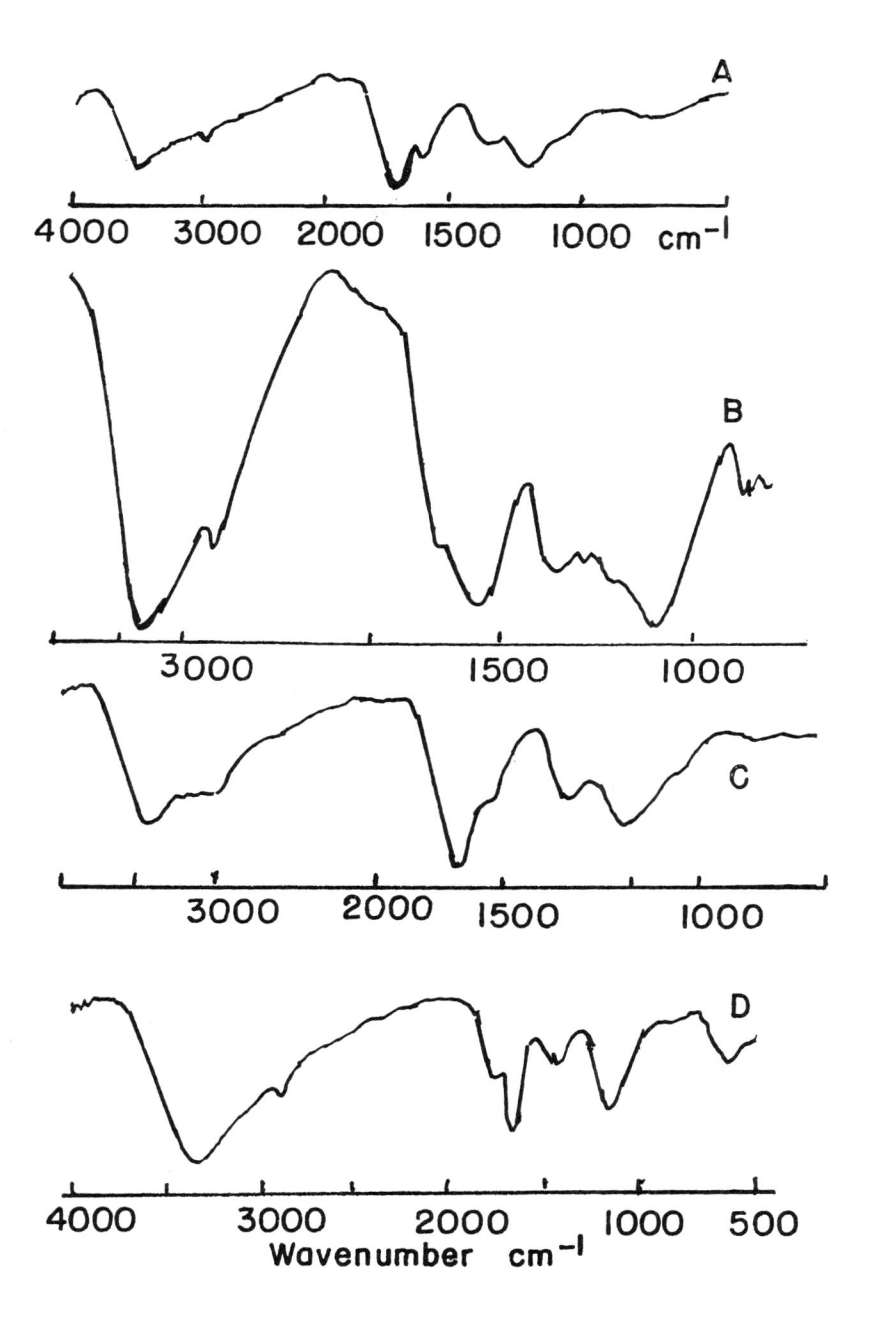

Figure 6.3 Infrared spectra of fulvic acids from different sources as recorded by (A) Schnitzer (1975); (B) MacCarthy and Rice (1985); (C) Stevenson (1994), type II; and (D) Kemp and Mudrochova (1975).

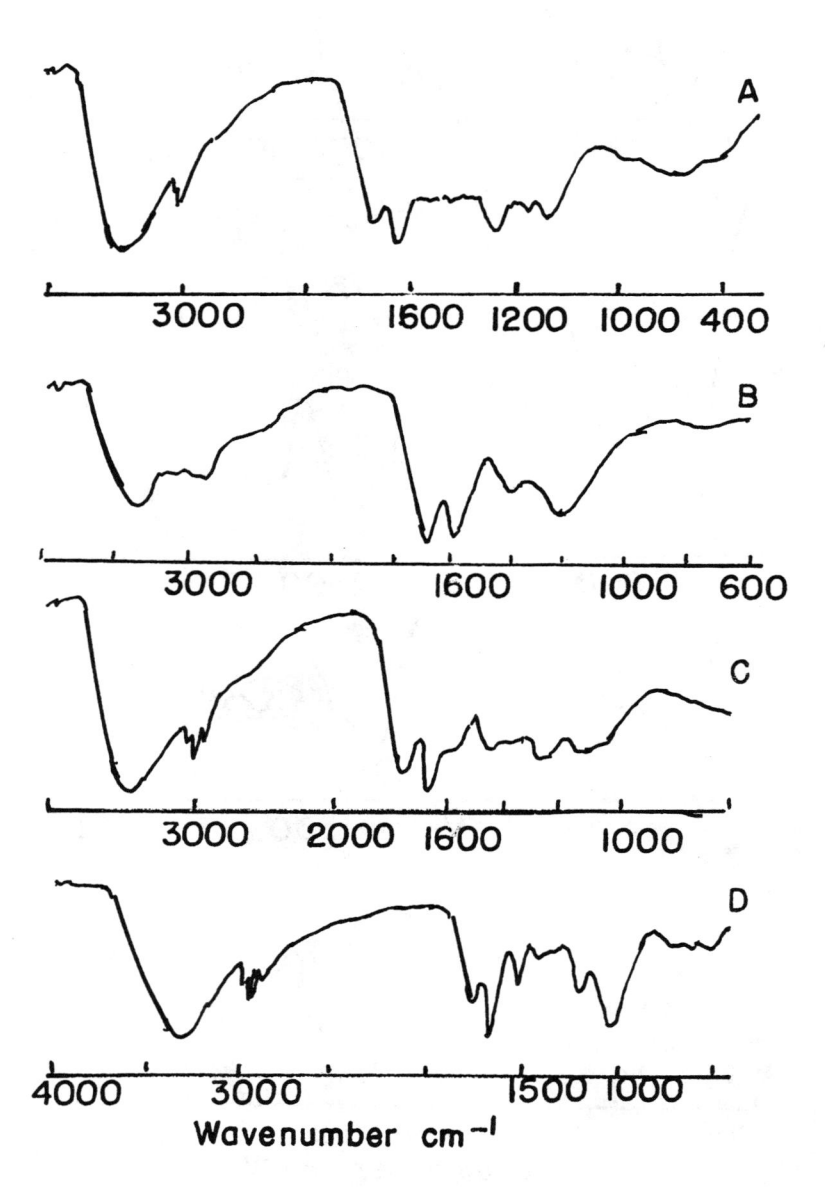

Figure 6.4 Infrared spectra of humic acids from different sources as recorded by (A) Bedrock et al. (1994); (B) Stevenson (1994), type I; (C) Tan (1976); and (D) Kemp and Mudrochova (1975).

Characterization of Humic Substances

6.5 NUCLEAR MAGNETIC RESONANCE SPECTROSCOPY

Magnetic resonance spectroscopy in general makes use of magnetic radiation and can be distinguished into two categories: (1) electron spin resonance (ESR) or electron paramagnetic resonance (EPR) and (2) nuclear magnetic resonance (NMR) spectroscopy. Electron spin resonance analyzes electron spin resonance of large free radicals in large polymers of soil organic compounds, and NMR analysis was used in the beginning for determination of proton resonance in relatively small organic compounds. With the rapid development of the technique, today NMR has been expanded for use with ^{31}P, ^{19}F, ^{13}C, ^{15}N, and ^{113}Cd (Pfeffer and Gerasimowicz, 1989). For the study of organic carbon in humic substances, ^{13}C NMR spectroscopy is the preferred method. Nitrogen-15 and ^{31}P NMR are suitable for analysis of organic N and organic P, respectively (Thorn et al., 1996; Lobartini et al., 1998; 1989).

6.5.1 Electron Paramagnetic Resonance

In the study of humic acids, EPR or ESR spectroscopy has been used occasionally and for several reasons it has been abandoned since the 1980's. The method is rather complicated and requires a lot of calculations of spectroscopic splitting factors, called g-values. In principle, a solid ground sample placed in a quartz tube is analyzed by an ESR spectrometer. The magnetic field at the sample is calibrated relative to DPPH (diphenylpicrylhydrazyl) or another suitable standard, and the spin concentration of the sample is determined by comparison with known concentrations of DPPH diluted with powdered KCl. The differences in field between maxima in the derivative signals are considered as the line widths (G). The spectroscopic splitting factors (g-values) are calculated using the following equation derived from values of the magnetic field (H) at which resonance occurs for the sample:

$$H_2/H_1 = g_s/g_r \tag{6.6}$$

where H_1=resonance of sample, H_2 = resonance of reference (standard), g_s = g-value of sample, and g_r = g-value of reference (DPPH has a g_r = 2.0036).

The ESR spectra of humic substances consist of single lines with hyperfine splitting with g-values ranging from 2.0031 to 2.0045 and line widths from 2.0 to 3.6 G (Ghosh and Schnitzer, 1980; Riffaldi and Schnitzer, 1972). Examples of ESR spectra from a spodosol and soil humic substances are shown in Figure 6.5. The peak in the spodosol spectrum is identified as the organic radical in the humic acid molecule (Steelink and Tollin, 1967). The spectrum of humic acid supports this observation.

Steelink (1964) and Steelink and Tollin (1967) were perhaps the first who tried to apply ESR methods to show paramagnetism in humic acids owing to the presence of semiquinones and hydroxy-quinones. This was followed later by Riffaldi and Schnitzer (1972), Senesi and Schnitzer (1977) and Ghosh and Schnitzer (1980), who confirmed by ESR analyses the presence of semiquinone radicals in humic acids.

More recently, ESR spectroscopy finds application in the study of metal chelation by humic substances for the determination of speciation of metals, forming inner sphere complexes difficult to analyze by any other methods. An example of an ESR spectrum of an Fe-fulvic acid complex is shown in Figure 6.5. Three main signals or resonances have been reported in the ESR spectra of naturally occurring Fe-fulvic acid complexes (Senesi et al., 1977) :

(1) an antisymmetrical resonance at g = 4.1 with an average line width = 125 G

(2) an isotropic resonance at g = 2.0028 - 2.0043 and a line width = 4.3-6.4 G

(3) a resonance at g = 2, composed of poorly defined, broad, lines.

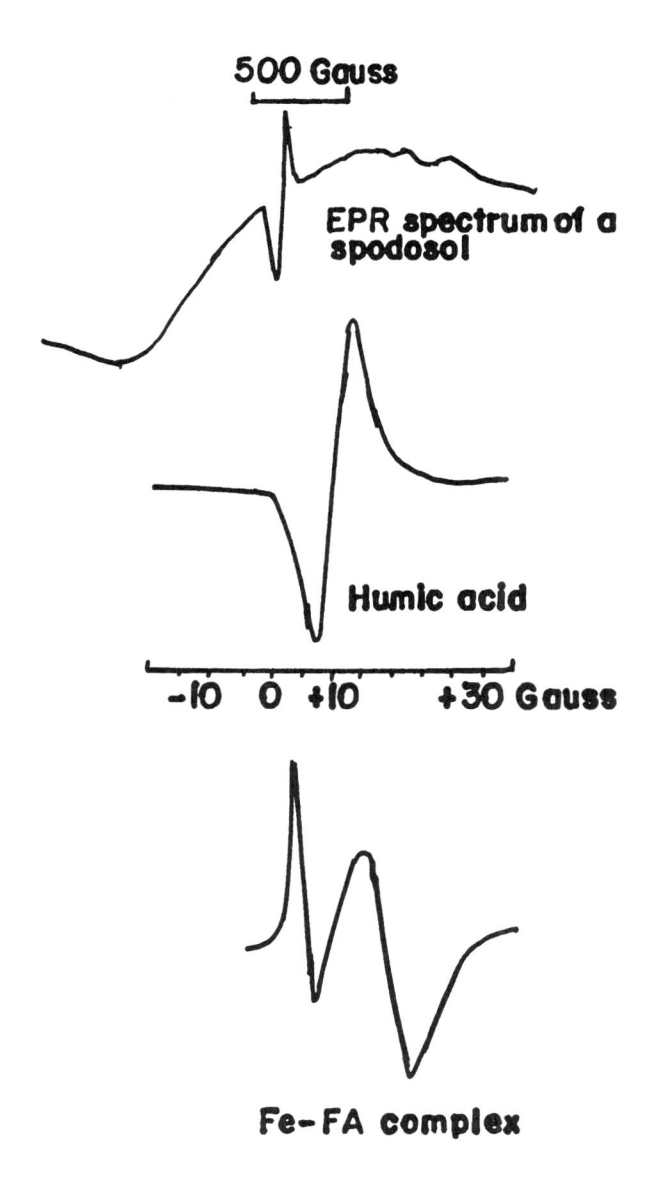

Figure 6.5 Electron spin resonance spectra of a spodosol (top), humic acid (middle), and Fe-fulvic acid complex (bottom). (Steelink and Tollin, 1967; Schnitzer, 1986).

From these studies, Senesi and coworkers believe that Fe is attracted to fulvic acid as Fe^{3+} by two possible mechanisms: (1) by tetrahedral and/or octahedral coordination bonds, and (2) by adsorption on external fulvic acid surfaces.

6.5.2 Carbon-13 Nuclear Magnetic Resonance

Of more importance, apparently, is ^{13}C NMR spectroscopy, which has attracted substantially more attention than ESR, and now is even heralded as the most important method in the study of humic acids. For the basics of NMR spectroscopy in general and its application in agriculture and humic acids reference is made to Wilson (1981), Wershaw (1985), Pfeffer and Gerasimowicz (1989); and Bortiatynski et al. (1996). In this section, only the subjects necessary for a proper understanding of an NMR spectrum will be briefly discussed, such as the chemical shift, δ.

Spectroscopic analysis involves interaction of radiation with matter, and that part that is absorbed or emitted by the sample is detected as a function of wavelength or frequency, which is defined as the spectrum. The frequencies in NMR analysis are usually in the range of 100 to 600 MHz, hence associated to relatively closely spaced transitions between energy levels. These levels depend on the different magnetic states of the nucleus and are proportional to the applied magnetic field as can be noticed from the following equation (Pfeffer and Gerasimowicz, 1989):

$$\Delta E = h\gamma B \tag{6.7}$$

where ΔE = difference in energy states, h = Planck's constant, γ = magnetogyric ratio characteristic of each nucleus, and B = applied magnetic field.

In addition to the applied magnetic field, a small localized

Characterization of Humic Substances

magnetic field is present around the nucleus, produced by the electronic currents of the atoms. Because of this, each nucleus within a molecule absorbs energy at a slightly different frequency or resonance position. The separation of these resonance frequencies from a standard (chosen) reference is called the chemical shift, δ, which is usually calculated using the following equation:

$$\delta = \frac{\nu_s - \nu_r}{\nu_r} 10^6 \qquad (6.8)$$

where δ = chemical shift in ppm, ν_s = resonance frequency of sample peak, ν_r = resonance frequency of standard reference peak.

In the NMR spectrum each signal (peak) is identified by a value of δ. The standard reference in ^{13}C NMR and ^{1}H NMR is tetramethylsilane (TMS), which has a chemical shift, δ = 0.0 ppm.

In its early development, NMR was capable of analyzing only the hydrogen atoms, or protons, of the humic molecules. The usefulness of such analysis is questioned by several scientists, because of lack of success in obtaining spectra of humic substances by ^{1}H NMR analysis. In the early days, the main problem was that ^{13}C NMR analysis required the sample to be dissolved in a suitable solvent. The solvents frequently used at that time were CCl_4 (chloroform) and $CDCl_3$. Humic acid, however, is usually not soluble in these reagents, and must first be methylated or broken down into smaller molecules by degradation procedures. An aqueous medium (water), needed for ^{1}H NMR, is unsuitable for ^{13}C NMR. Solid samples at that time cannot be used since they interfere with magnetic interaction. Another solvent is D_2O, which finds application with analysis of fulvic acids. Today, the use of NaOD has apparently eliminated this obstacle. In addition to the foregoing problems, difficulties also arise from the use of radio waves in NMR analysis, which are low energy forms of electromagnetic radiation. The level of energy is considered very small, but still large enough to affect the nuclear spin of the atoms in the poorly defined complex polymers of humic acid molecules. This makes an analysis by NMR very time consuming, and even with today's powerful modern machines one analysis may take several hours of scanning time.

Both liquid and solid samples can now be used. Soil samples can also be used in the undisturbed state, provided the sample contains sufficient amounts of organic carbon (Tan, 2000; Tan et al., 1992). With the very powerful instruments available today, *cross-polarization magic angle spinning carbon-13 NMR* - CP-MAS ^{13}C NMR - spectroscopy produces better spectra with solid than with liquid samples of humic acids. Magic angle spinning is a technique by which a sample is rapidly rotated at an angle of 54.7°, known as the *magic angle*, in order to decrease line broadening of the spectrum.

The analysis with ^{13}C NMR spectroscopy is capable of measuring the distribution of C in the various types of compounds and this information can be used in structural analysis and in differentiating the different types of humic matter. A ^{13}C NMR spectrum of humic matter can usually be divided into several regions (Hatcher et al., 1980), and the various peaks in the spectrum assigned to specific carbon functional groups. Peak heights and peak areas are used for quantitative measurements by the integration method. This has been discussed earlier in Chapter 5, and to support the capability of ^{13}C NMR analysis as contended above, the following spectra are provided as illustrations in Figure 6.6. As can be noticed, the spectrum of lignite-humic acid has a strong signal between 105 and 165 ppm chemical shift for aromatic carbon, and a weak signal between 165 and 185 ppm for carboxyl carbon. In contrast, the spectrum of fulvic acid shows a weak signal for aromatic carbon, but a substantially strong signal for carboxyl carbon. This is supported by the other two spectra from Suwannee River aquatic humic substances, where the spectrum of humic acid is also dominated by strong signals between 105 and 165 ppm. On the other hand, the aquatic fulvic acid spectrum shows again a weak signal in the aromatic region, but has a very strong peak in the carboxyl region at 177 ppm chemical shift. These observations agree with the chemical data indicating the fulvic acids contain larger amounts of carboxyl groups than humic acids. Additional evidence for these differences between fulvic and humic acids are provided by other spectra shown in Figure 6.7, where the

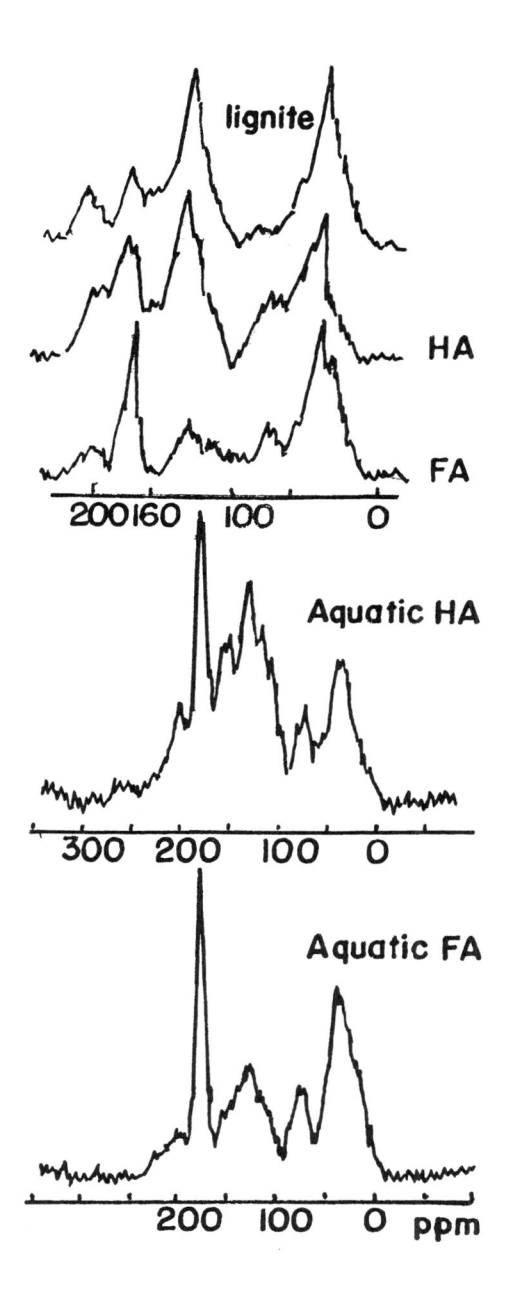

Figure 6.6 Solid state CP-MAS ^{13}C NMR spectra of humic and fulvic acid extracted form lignite, and liquid state ^{13}C NMR spectra of aquatic humic substances from the Suwannee River (Tan et al., 1992; Thorn et al., (1989).

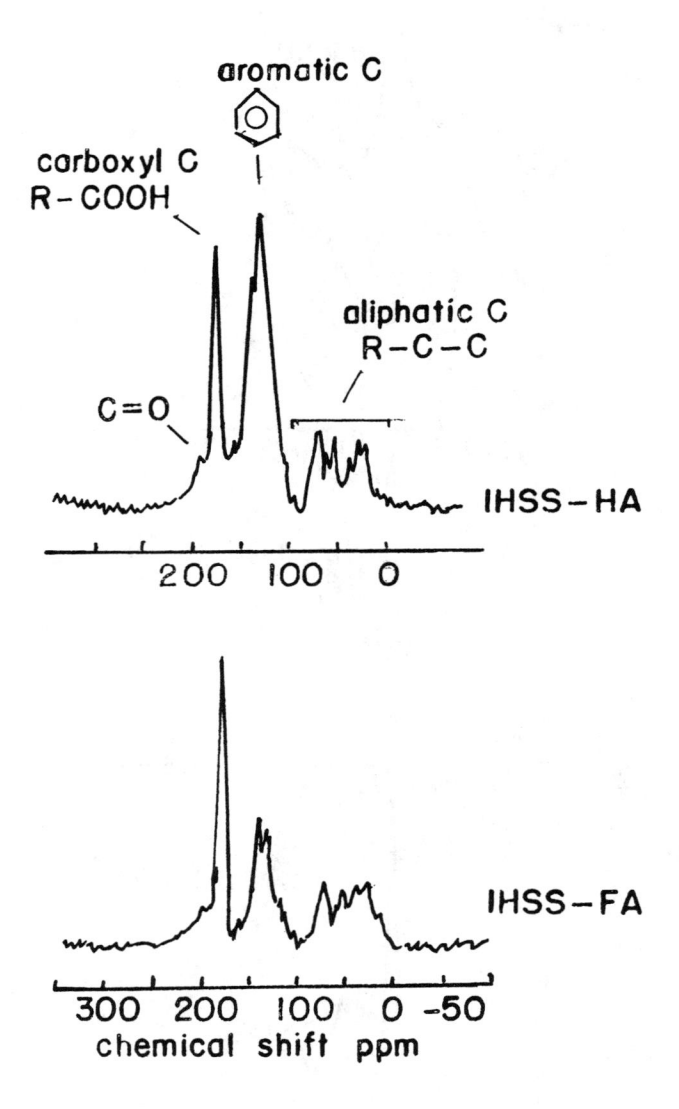

Figure 6.7 Differences in aromaticity between standard humic and fulvic acids from the International Humic Substances Society (IHSS) as revealed by liquid state ^{13}C NMR spectroscopy (adapted from Thorn, 1989; Thorn et al., 1989).

Characterization of Humic Substances

peaks in the aromatic region of the spectrum of soil fulvic acid are small in intensity relative to those in the spectrum of soil humic acid. Both samples are supposed to be standard or reference samples from the International Humic Substances Society.

The data in Figures 6.6 and 6.7 also tend to suggest that these spectral features are reproducible, as was the case with infrared spectra as discussed earlier. The aquatic fulvic and humic spectra reported by Thorn et al. (1989) for the Suwannee River samples (Figure 6.8) are in perfect agreement with those produced by Mills et al. (1996) from samples collected 5 years later from the same river in January 1994 (Figure 6.6). Apparently different handling and techniques also have little influence on reproducibility of spectra as long as the humic substances have been obtained by a similar extraction procedure. The standard IHSS fulvic and humic acid spectra (Figure 6.7) from Thorn et al. (1989) were produced by dissolving in D_2O plus NaOD prior to ^{13}C NMR analysis, whereas those from Thorn et al. (1996) were dissolved in DMSO, dimethylsulfoxide, an organic extracting reagent for humic substances. Nevertheless, the soil fulvic and humic acid spectra shown in Figure 6.7 are in agreement with those in Figure 6.6, except for the strong DMSO peaks in the latter.

6.5.3 Nitrogen-15 Nuclear Magnetic Resonance

As can be noticed from the discussion above, ^{13}C NMR analysis can only determine qualitatively and quantitatively several major types of carbons, e.g., aliphatic C, polysaccharide C, aromatic C, carboxyl C, and carbonyl C. This method is not capable, however, of measuring the nitrogen content in the nitrogenous constituents of the humic molecule. As discussed in Chapter 4, these constituents are believed to be composed of amino acids, peptides or protein, amino sugars, and HUN, surviving microbial and chemical degradation due to incorporation in the humic structure by the humification process. The peptides may be linked to the central core of the humic molecule by H-bonding. The conventional way to analyze N compounds in soil and humic acid is by using standard hydrolysis and chromatographic techniques. However, the method is believed to be suspect due to the inherent analytical

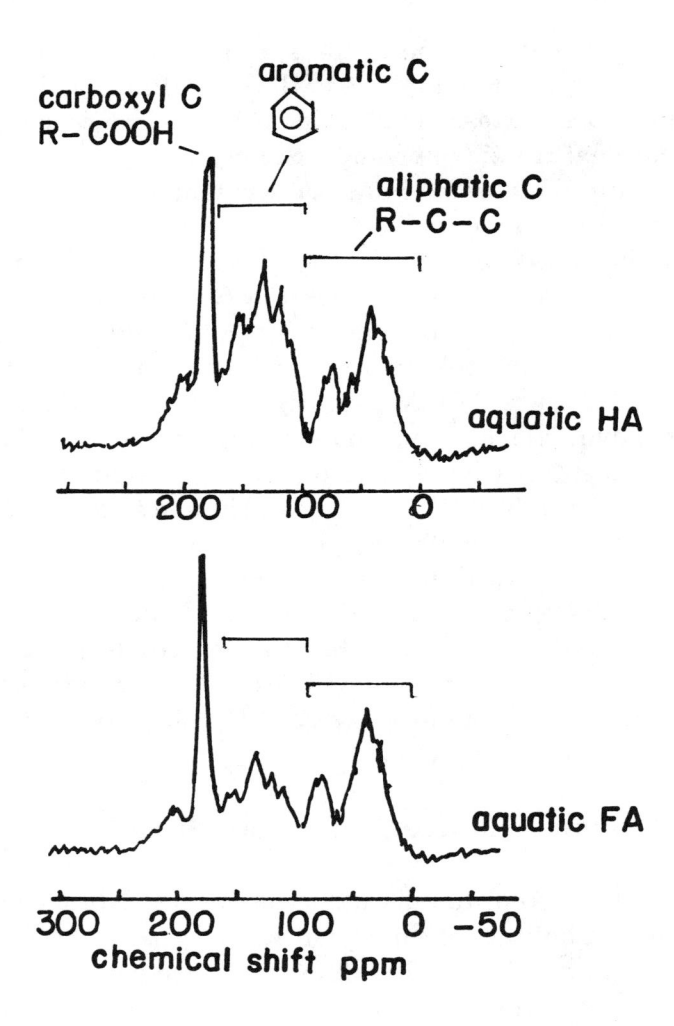

Figure 6.8 Differences in aromaticity between aquatic humic and fulvic acids from the Suwannee River as determined by liquid ^{13}C NMR spectroscopy (adapted from Mills et al., 1996).

problems, such as incomplete extraction and losses during hydrolysis (Stevenson, 1994). Using the old methods only 30 to 50% of the total N in soils can be recovered, mostly as soluble amino acids and amino sugars.

With the recent advancement in NMR spectroscopy, many scientists are now turning their attention to this technique for the analysis of N substances in humic acids. It is a nondestructive method and still in its early stages of development. In this respect, two isotopes are available for application in NMR spectroscopy, e.g., ^{14}N and ^{15}N. Nitrogen-14 is the most abundant isotope in organic substances, but ^{14}N NMR analysis has been reported to yield broad line spectra or spectra with poor resolutions. On the other hand, ^{15}N, with only 0.37% abundance, is less available in nature, but ^{15}N NMR analysis produces spectra with higher resolutions, though it is still considered 50 times less sensitive than ^{13}C NMR (Bortiatynski et al. (1996). Most of the analyses with ^{15}N NMR spectroscopy have so far been confined to ^{15}N labeling studies. Using ^{15}N-labeled soil samples or melanoidins the method is reported to have detected the presence of amine and pyrrole-type N (Stevenson, 1994; Bortiatynski et al., 1996). Recent investigations using ^{15}N-labeled aniline indicate that aniline undergoes nucleophilic reactions with the carbonyl groups of humic acids (Thorn et al., 1996).

On the other hand, very little is known yet on ^{15}N NMR spectroscopy of the distribution of N in the various functional groups in the humic molecule. Several scientists are of the opinion that ^{15}N NMR can be applied to analysis of N, as is the case of ^{13}C NMR for the analysis of C. Similarly as with ^{13}C NMR, solid state CP-MAS ^{15}N NMR spectroscopy is also noted to work better than the liquid state. Bortiatynski et al. (1996) are convinced that the solid state spectra are also quantitative for measuring N in the various types of compounds in the humic molecule. However, assignments of ^{15}N NMR signals relative to the standard nitromethane (δ = 0 ppm) are given by Bortiatynski and coworkers mostly in the negative region of the chemical shift. The main signal with the strongest intensity is at -257 ppm, identified by the authors to be caused by secondary amides. Less intensive signals are noticed between -285 and -325 for NH_2 derivatives, and between -325 and -350 ppm for aliphatic amines, as

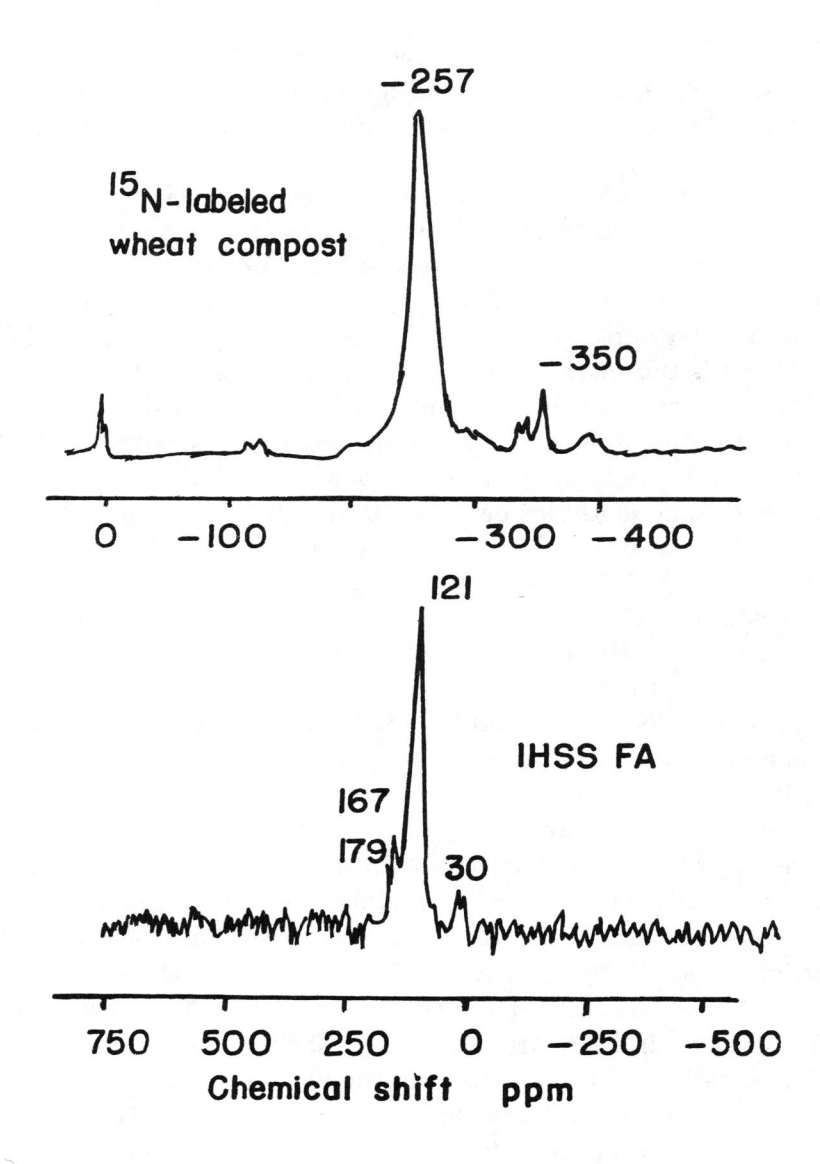

Figure 6.9 A comparison between solid state ^{15}N NMR spectra of compost from ^{15}N-labeled wheat, and 'reference' IHSS-fulvic acid, showing naturally occurring N at contrasting chemical shifts (Bortiatynski et al., 1996; Thorn et al., 1996).

Characterization of Humic Substances

illustrated in Figure 6.9. This in contrast to Thorn et al. (1996), who show spectra of reference IHSS-fulvic acid with ^{15}N signals in the positive region of chemical shifts (Figure 6.9, bottom). The dominant signal at +121 ppm in the reference fulvic acid sample is believed to be caused by amide-nitrogen, whereas the peak at 30 ppm shows the presence of free amino nitrogen, including amino acids and amino sugars. Indoles, pyrroles, imides and lactam nitrogens are suspected to be the reasons for the signals between 135 and 185 ppm, with maxima at 156, 167, and 179 ppm. The ^{15}N NMR spectrum of humic acid (not shown) has only one dominant peak at 120 ppm, and differs from that of fulvic acid by the absence of signals between 135 and 185 ppm.

6.5.4 Phosphorus-31 Nuclear Magnetic Resonance

Another NMR method that has recently also attracted research attention is ^{31}P NMR spectroscopy for the characterization of P in various substances in soils. However, since the P content in humic substances is usually very low, it has not been applied to analyze the distribution of P in the humic molecule. The method has mainly been used in the past to determine the form of P extracted from soils. Several authors claim to have detected with ^{31}P NMR analysis the presence of orthophosphate ions, orthophosphate diesters, phosphate esters, and humophosphates in soil extracts. The latter is a chelated form of P, in which $H_2PO_4^-$, HPO_4^{-2}, and/or PO_4^{-3} ions are chelated by the carboxyl and phenolic-OH groups of the humic acid molecule. Some of the results indicate that most of the organic P in soils is present in the form of phosphate esters, whereas smaller amounts exist as alkylphosphonates (Ogner, 1983; Newman and Tate; 1980; Glonek et al., 1970). Recently ^{31}P NMR spectroscopy has been applied to identify forms of P chelated by humic acid from apatite, $AlPO_4$, and $FePO_4$ minerals (Lobartini et al., 1994; 1998; 1989). A ^{31}P NMR spectrum of the dissolution product resulting from the reaction between apatite and humic acid (at pH 9.0) is shown in Figure 6.10. As can be noticed the main signals indicate the presence of orthophosphate diesters at −0.81ppm, orthophosphate ions at 2.69 ppm, and humophosphates at 4.04, and 18.31 ppm chemical shifts. The signals at 4.04 and 18.31 ppm

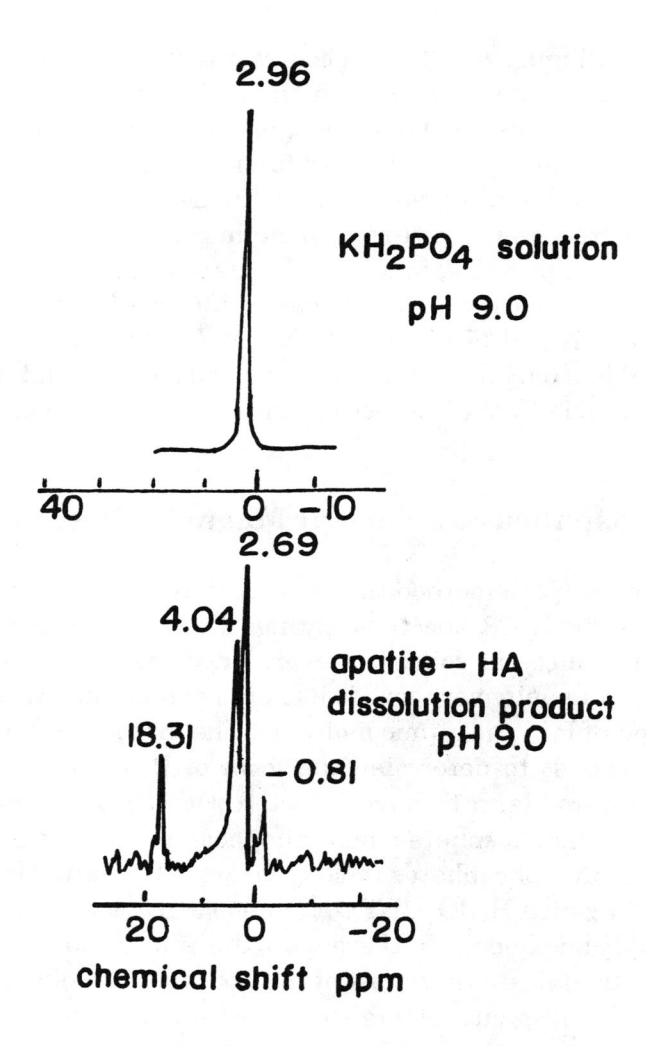

Figure 6.10 A ^{31}P NMR spectrum of pure $H_2PO_4^-$ and a dissolution product from the reaction between apatite mineral and humic acid at pH 9.0 (Lobartini et al., 1994).

Characterization of Humic Substances

disappear when the reactions between apatite and humic acids have been carried out at pH 5.0. At lower pH values the functional groups, COOH and phenolic-OH, are protonated, preventing the interaction of P ions with humic acid (Lobartini et al., 1994; Tan, 1986).

6.6 ELECTRON MICROSCOPY OF HUMIC MATTER

The use of electron microscopy in the study of humic compounds has received considerable research attention. Two types of electron microscopy are available for this purpose, transmission electron microscopy (TEM), and scanning electron microscopy (SEM). The advantage of TEM is that it can reach high magnification without distorting the resolution. In contrast SEM can only be used at lower magnification for good resolution, but it can provide a three-dimensional picture that is not possible with TEM. Sample preparation is also less complex in SEM than in TEM analysis.

6.6.1 Transmission Electron Microscopy

Flaig and Beutelspacher (1951; 1954) are perhaps among the first who have tried electron microscopy for analyzing the shape and size of humic particles. Employing a transmission electron microscope, the authors above notice that humic acids exist as very small spherical particles on the order of 10 to 15 nm in diameter. The spheres are frequently joined together in racemic chains. Since then the investigations carried out using TEM have yielded mixed results on the macromolecular structure of humic substances. Visser (1963) shows a structureless film of humic acid by TEM, whereas Dudas and Pawluk (1970) claim that humic acids from chernozemic soils are composed of tiny spherical particles united into spongy aggregates. Although Khan (1971) has also presented electron micrographs showing loose spongy structures for humic acids, he believes that on the basis of viscosity measurements the humic particles are in fact nonspherical. More recently, three types of fulvic acid particles have been reported from TEM investigations by Schnitzer and Kodama (1975): (1) small spheroids (1.5 - 2.0 nm), (2) aggregates of spheroids (20 - 30 nm), and (3) an amorphous material perforated by voids (50 - 110 nm). From electron diffraction analysis Schnitzer and Kodama (1975) indicate that fulvic acid (pH 2.0) exhibits a crystalline structure. Since the spherical nature of the humic particle has also been mentioned in the Russian literature (Orlov, 1985), many European scientists tend to believe that at low concentrations humic acids are tiny balls. These balls coalesce at higher concentrations into sheets with a granular surface as frequently noticed by TEM. Using a replica technique, Stevenson and Schnitzer (1982) show a sequential process of coalescence of fulvic acid spheroids into aggregates with a chainlike structure in samples examined by TEM from dilute to more concentrated areas. In dilute aqueous solutions the humic substances exist as flat, multibranched, stretched fibers or filaments, 20 to 100 nm wide, whereas at the higher concentrations, these chainlike or fibrous-like structures seem to be converted into perforated sheets. Six major types of humic structures have been recognized by the authors above, e.g., small spheroids, flattened aggregates of spheroids, linear, chainlike assemblies of aggregates, flattened filaments, and perforated sheets.

6.6.2 Scanning Electron Microscopy

This method has been used extensively in the study of humic substances. Especially in Canada, Schnitzer and coworkers have applied it in studying the texture and fabric of humic and fulvic acids. As indicated above, scanning electron microscopy has the advantage over TEM of being capable to furnish a three-dimensional picture with a resolution depth of 5 to 10 µm. In addition, humic particle surfaces and orientation can also be shown. However, the method used is complicated, and the preparation of samples for SEM is very tedious, and time consuming, hence scaring many people away from using it. Chen and Schnitzer (1976) use a freon-liquid-N preparation technique adapted from the preparation of animal tissue for surface scanning electron microscopy. However, the inclusion of freon gas treatment and preparing and drying samples first on glass slides or mica sheets for later transfer onto SEM specimen stubs are unnecessary. To make it simpler and faster Tan (1985) suggests placing a drop of humic solution directly on a SEM specimen stub. Ultrarapid freezing in liquid-N and appropriate drying of the frozen specimen under a high vacuum (6.5×10^{-10} MPa) for 24 hours are required for producing the proper micrographs. Failure to do so, like slow freezing in a refrigerator and slow drying in a desiccator, will yield poor resolutions or pictures showing only featureless, massive, structures of humic substances (Tan, 1985).

The modified method proposed by Tan (1985) produces similar tissue-like structures for fulvic acid and humic acid as with the freon-liquid-N method on mica strips. Except for better 3-D depth, the structures revealed by SEM are also not much different from those produced by TEM. As can be noticed in Figure 6.11, fulvic acid at pH 3.0 exhibits a fiber-like structure. The fulvic acid fibers ranged in thickness from 0.2 to 5 µm, and some soils possess fulvic acids with thick fibers, whereas other soils contain fulvic acids with thin fibers. The humic acid (pH 7.0) structure is almost similar but tends to be more like shredded sheets in appearance. The observations above for slight differences in fulvic acid and humic acid structure can also be noticed for aquatic humic substances. The micrographs in Figure 6.12 show aquatic fulvic acid to be characterized by thin fibers, whereas the humic acid counterpart has a structure composed of relatively slightly thicker fibers.

Chen and Schnitzer (1976) believe that the differences in humic structure are the effect of pH. Protonated fulvic acid (pH 2-3) is reported by the authors to have an open structure formed from elongated fibers with rounded tips. This structure is noted to change into a spongelike structure at pH 7.0, and at pH 9.0 it becomes a homogeneous sheet in which grains are visible. However, the effect of pH on humic acid structure cannot be supported or reproduced by the current author. Perforated sheet structures have been noticed by the present author in humic acids at pH 3.0, as can be seen in Figure 6.13. The humic acid solution was prepared and purified by thorough leaching through a H-saturated cation exchanger column according to

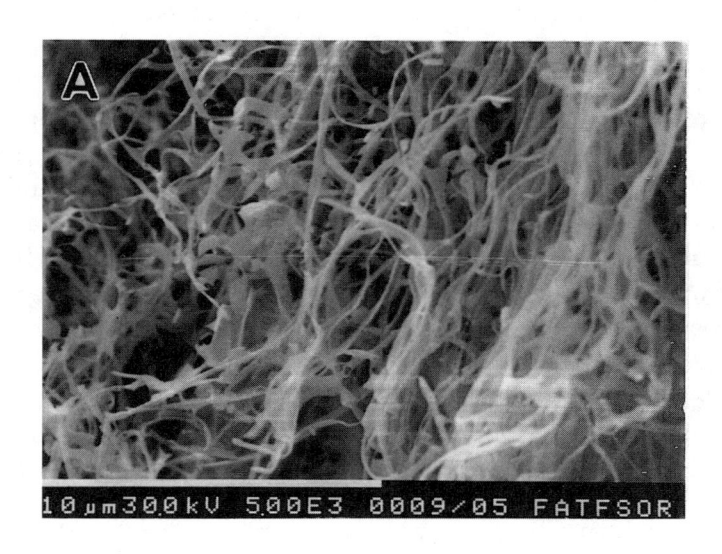

Figure 6.11 Scanning electron micrographs of (A) fulvic acid, and (B) humic acid, extracted from a Tifton soil (ultisols) in Georgia, USA, showing the characteristic fiber structures. Magnification: white bar = 10 μm.

Characterization of Humic Substances 207

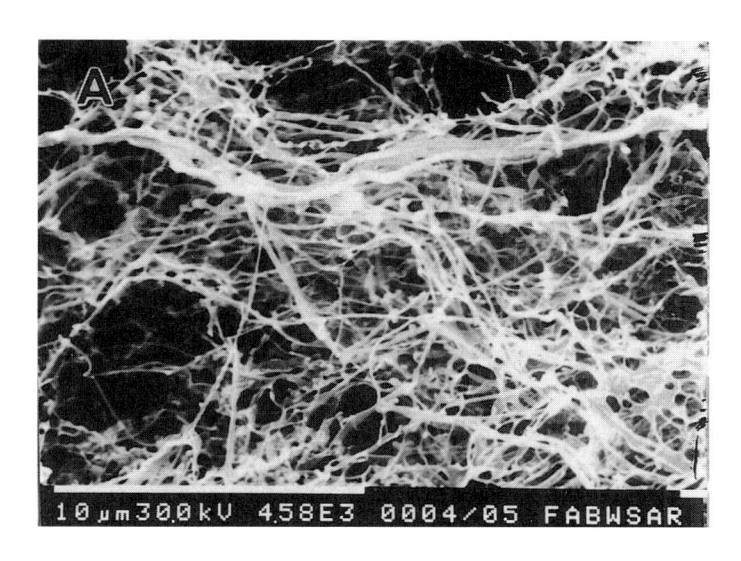

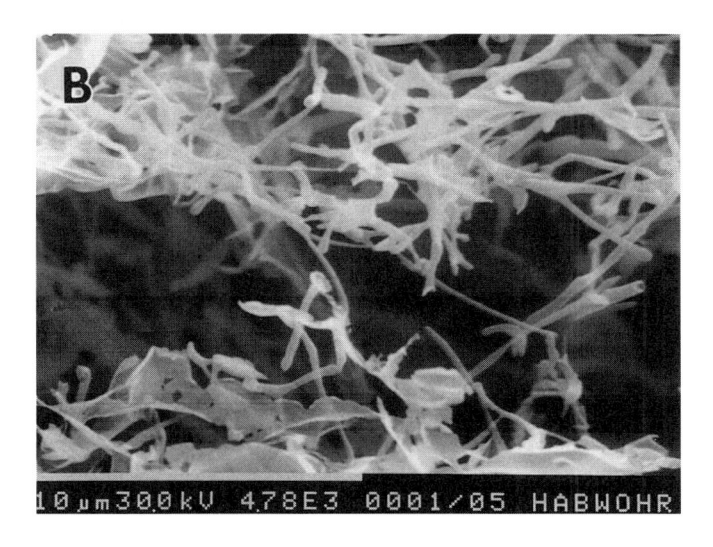

Figure 6.12 Scanning electron micrographs of aquatic (A) fulvic acid, and (B) humic acid, extracted from black water of the Satilla and Ohopee rivers, respectively, in southeast Georgia, USA, showing similar fibrous-like structures. Magnification: white bar = 10 μm.

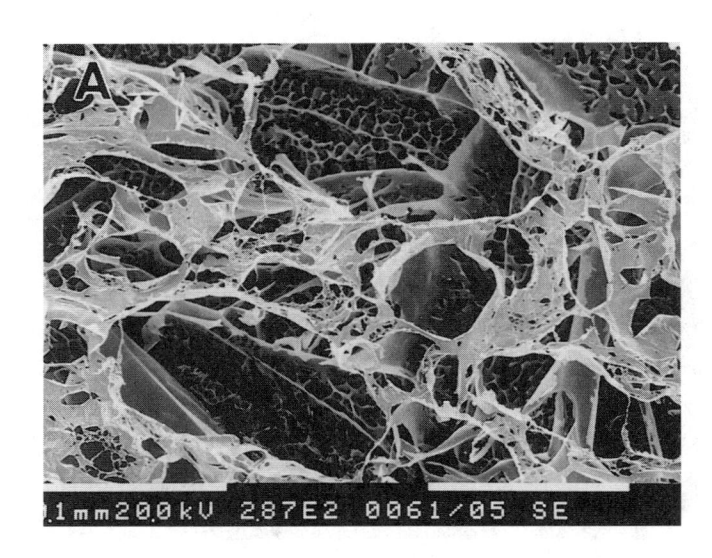

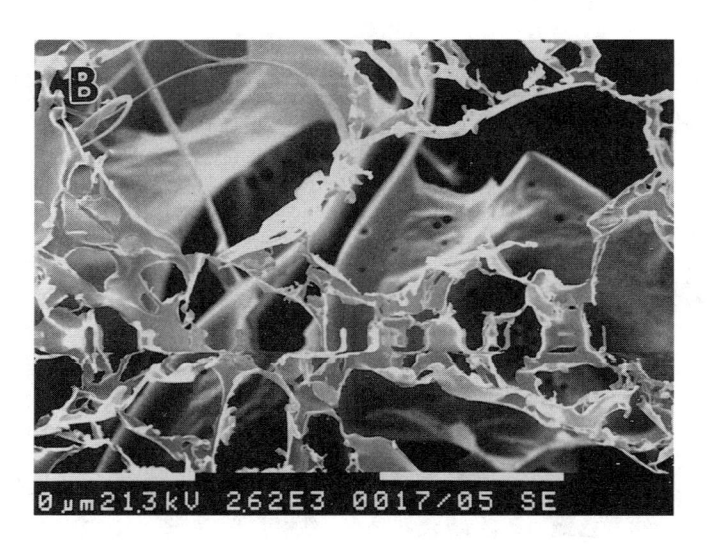

Figure 6.13 Scanning electron micrographs of humic acids (pH 3.0) extracted from (A) a Cecil soil (ultisols) in Georgia, USA, (magnification: white bar = 0.1 mm), and (B) an andosol in Indonesia (magnification: white bar = 10 µm), showing perforated sheet structures.

Characterization of Humic Substances

the method of Lakatos et al. (1977) as described earlier. The purified humic acid solution flowing from the cation exchange column is highly protonated and has a pH = 3.0 (Tan, 1996), and is used directly for SEM analysis. From the discussion above, it appears that in general only two major types of structures can be distinguished, e.g., (1) stretched fibers, and (2) perforated shredded sheets, as illustrated by the examples in Figures 6.11 and 6.12, and Figure 6.13, respectively.

The other types of structures recognized by Schnitzer and coworkers in Canada can be considered intergrades between the two major types above. Though soil pH has some effect, the current author believes that the concentration has perhaps a more dominant role in affecting structural changes. With increased humification, more humic substances are being produced, and in a more concentrated (crowded) condition several of the humic fibers are bundled together and forced to fuse or coalesce into thin, flat strands, which eventually become perforated or shredded sheets with a continued concentration increase of humic substances.

The fiber- or tissue-like structures are apparently characteristic of the humic substances, regardless of origin or method of determination. As indicated earlier, both TEM and SEM yield similar structures, suggesting reproducibility of results. The humic substances used for TEM and SEM analyses by the Canadian scientists have been extracted from spodosols, acid soils of the humid temperate and cool regions of North America. On the other hand, the humic substances used for SEM by the current author originate from ultisols, acid soils of the humid warm temperate to subtropical region of the southeastern United States. Yet the spodosol humic substances exhibit similar fibrous structures as the ultisol humic compounds. By comparison no differences have also been noticed for the structures of aquatic humic matter. The question is raised now whether artifacts could have produced such reproducible structures, or whether all these suggest the possibility for the presence of real humic and fulvic acids in nature.

CHAPTER 7

ELECTROCHEMICAL PROPERTIES OF HUMIC MATTER

7.1 ORIGIN AND TYPES OF ELECTRIC CHARGES

Humic and fulvic acids are considered amphoteric compounds, but Stevenson (1994) assumes them to be weak acids. However, it is well known that humic substances can, in fact, react with both bases and acids, hence carry both positive and negative charges. These properties and behavior are regarded as distinctive characteristics of amphoteric substances. The negative charges are usually studied more intensively and consequently are better known than the positive charges. All these charges are developed by the ionization or dissociation of various functional groups.

7.1.1. Negative Charges

The negative charges are attributed to dissociation of protons from the functional groups in the humic molecule. The two most

Electrochemical Properties of Humic Matter

important functional groups in this respect are the carboxyl and phenolic-OH groups. In general, these two functional groups control the electrochemical behavior of humic matter and are the main reasons for adsorption, cation exchange, complex and chelation reactions. The carboxyl, COOH, groups start to dissociate their protons at pH 3.0 (Posner, 1964) and the humic molecule becomes negatively charged (Figure 7.1). At pH < 3.0, the charge is very small, or even zero. At pH 9.0, the phenolic-OH groups also dissociate their protons, and the humic molecule attains a high negative charge.

Since the development of the negative charge is pH dependent, this charge is called pH dependent charge or variable charge (Tan 1998). At low pH, the charge is expected to be low, whereas at high pH, the negative charge is high, which corresponds to low cation exchange capacity (CEC) at low pH and high cation exchange capacity at high pH. According to the theory of CEC, the negative charge will eventually reach a maximum value at pH 8.2. This will be explained further in Section 7.4 on cation exchange capacity.

The Significance of the Henderson-Hasselbalch equation

Generally, the ionization of amphoteric compounds can be studied by using the concept of pK values. By assuming that the dissociation of humic acid (HA) proceeds as follows:

$$HA \rightleftarrows H^+ + A^- \tag{7.1}$$

then, the ionization constant K of the reaction above is given by:

$$K = \frac{(H^+)(A^-)}{(HA)} \tag{7.2}$$

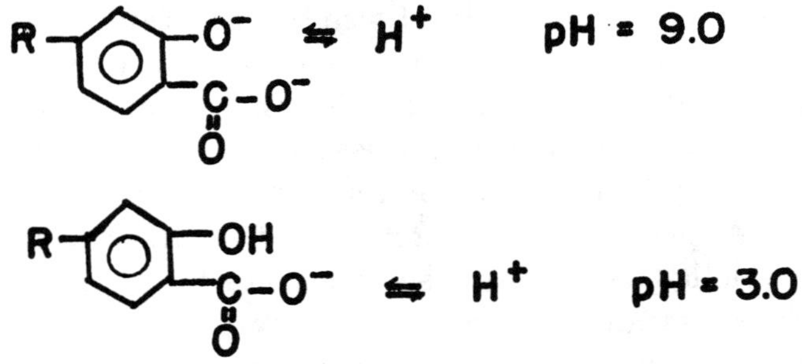

Figure 7.1 Development of variable charges in a humic molecule by dissociation of protons from carboxyl groups at pH 3.0, and from phenolic-OH groups at pH 9.0.

By converting into $-\log$, equation (7.2) changes into:

$$-\log K = -\log \frac{(H^+)(A^-)}{(HA)} \tag{7.3}$$

$$pK = -\log (H^+) + [-\log \frac{(A^-)}{(HA)}] \tag{7.4}$$

$$pK = pH - \log \frac{(A^-)}{(HA)} \tag{7.5}$$

Electrochemical Properties of Humic Matter

Equation (7.5) is the famous Henderson–Hasselbalch equation. It describes the ionization process of amphoteric compounds, hence applies to ionization of humic acids. When ionization has proceeded to the point where the concentration or activity of (A⁻) = (HA), the equation changes into:

$$pK = pH \qquad (7.6)$$

This pK is often referred to as pK_a or *ionization constant*. In titration analysis, the condition, defined by equation (7.6), usually occurs at half-neutralization. The pK_a is considered to be of intrinsic value and should apply to all the acidic or COOH groups in the humic molecule.

Use of pK_a in Determining Negative Charges

In soil chemistry, the magnitude of the ionization constant K or the pK_a value is used as an indication for the degree of ionization. As can be noticed from equation (7.2), the higher the value of the ionization constant K, the larger will be the value of (H⁺)(A⁻) and the smaller the amount of (HA). This means that at high K values (or low pK_a values), large amounts of (HA) are ionized into H⁺ and A⁻ ions. Ionization is less at low K or high pK values. In pure chemistry, substances characterized by high ionization constants (or low pK_as) are called strong acids, in contrast to those with low ionization constants (or high pK_as), which are considered weak acids.

Conforming to the Henderson-Hasselbalch concept, ionization amounts to only 50% at pH = pK_a. Stevenson (1994) assumes that at one pH unit above the pK_a, the acidic groups of the humic molecule will be 90% ionized, whereas at two pH units above the pK_a, the acidic groups are estimated to be 99% ionized. In contrast, at one pH unit below the pK_a, the functional group is only 10% ionized, whereas at two pH units below the pK_a, ionization amounts only to 1%. Because the degree of ionization determines the level of negative charges created, the present author believes that the ionization constant K_a or pK_a can also be used for indicating the extent of variable negative charges created at higher or lower pH values. Consequently, substantial amounts of negative charges are expected to be present at the pK_a, which will increase in magnitude and reach maximum values at two pH units above the pK_a. At two pH units below the pK_a, the humic molecule is practically noncharged or does not carry any substantial charges at all.

The Issue of COOH Groups

The level or degree of electronegative charges is not affected only by the degree of ionization of the active functional groups, but it also depends on the concentration or relative distribution of these groups in the humic molecule. The larger the concentrations of the functional groups, the higher will be the negative charges of the humic molecule. The relative distribution of these functional groups is noticed to vary widely from soils to soils, and a considerable variation is also present for humic matter within similar soil groups. As discussed in Chapter 5, the opinion is that fulvic acids are generally higher in carboxyl group contents than humic acids. Schnitzer (1977) has reported even more dramatically larger differences in carboxyl group contents between fulvic and humic acids than shown in Table 5.4 of Chapter 5. The carboxyl contents in fulvic acids are shown by Schnitzer to range between 5.20 and 11.20 me/g as compared to a range of 1.50 and 5.70 me/g for humic acids, extracted from soils over the world. However, the above is contradicted by the studies conducted by Tsutsuki and Kuwatsuka (1978), involving a large number of humic acids, extracted also from a wide variety of soils. Their results indicate that the COOH content increases whereas the phenolic-OH group content decreases during the humification process. This suggests that humic acid, the product of advanced humification, would be higher in COOH content than fulvic acid, the substance formed at the start of humification. This is in sharp contrast with Stevenson's (1994) theory on diagenetic transformation of humic acid into fulvic acid as discussed earlier. The controversial revelations above make the issue of COOH content very confusing and leave us wondering whom to believe. However, all these

Electrochemical Properties of Humic Matter 215

do not indicate that Schnitzer's or Stevenson's data are incorrect, they only mean that one has to use caution in accepting the facts on COOH content in humic substances. The latter finds its origin in the considerable difficulties encountered in the analysis of functional groups, where the exact measurement for accounting the acidic groups is subject to many errors.

Both carboxyl and phenolic-OH groups generally contribute to development of negative charges, but the opinion exists that the carboxyl groups are the most important in the formation of negative charges. This is perhaps true and can be explained by applying the Henderson-Hasselbalch concept. If we can assume that the COOH groups dissociate their protons at pH 3.0 as postulated by Posner (1964), and at this condition pH = pK_a, then 99% ionization will be reached at pH 5.0, a 'normal' pH value in most acidic soils generally productive for agricultural operations, especially forestry. In contrast, the phenolic-OH groups will be dissociating their protons at pH 9.0, a pH value seldom occurring in agricultural soils. If the assumption is made again that at this condition pH = pK_a, then 99% ionization will be reached at pH 11.0, a pH value too high to be agriculturally productive in even the best aridisols. A possibility is that the pH value at 9.0, as postulated by Posner (1964) for the dissociation of phenolic-OH groups, is far too high and valid only for laboratory conditions, but not valid for natural soil condition. Chelation and complex reactions are noticed to take place at pH 4.0 to 8.0 in natural soil environments. Apparently, more research has to be conducted to confirm or revise the exact pH for the dissociation of especially phenolic-OH groups in natural soils.

The Significance of Total Acidity in Negative Charges

As explained in Chapter 5, the sum of the carboxyl and phenolic-OH groups is defined as *Total Acidity*, hence this property should also reflect the level of negative charges of humic substances. A high total acidity value is then indicative for the presence of high negative charges. A low total acidity value, in turn, points to the presence of low negative charges. Since fulvic acids exhibit higher total acidity values than humic acids, they are expected to be higher in negative charges than humic acids. However, this does not necessarily mean that fulvic acid has higher chemical activity than humic acid. Results of studies on chelation and complexation analyses indicate that metal chelation by humic acids appears to be more effective than that by fulvic acids. The amounts of metals chelated by humic acids are always higher than those chelated by fulvic acids (Tan, 1978a and b; Lobartini, 1994). Most people assume this to be caused by the differences in sizes and complexity between the two humic substances (Stevenson, 1994). The substantially larger molecules and the more complex structures of humic acids are accepted to be the reasons for more binding sites and higher binding capacity in contrast to fulvic acids, which are smaller and less complex. In this respect, the following hypothesis is added by the current author for further contemplation. In the preceding sections above, fulvic acids have been described as possessing higher COOH contents than humic acids. Carboxyl groups, in general, exhibit their chemical activities through their acidic (H^+) reactions only. They are effective in cation exchange reactions, but they display little or no chelation, although some complex reactions may be present (Tan, 1986). Acetic acids and formic acids are compounds in this category, since their acidic characteristics are attributed to the presence of only COOH groups in their molecules. On the other hand, humic acids exhibit acidic characteristic attributed to the presence of COOH groups and especially substantial amounts of phenolic-OH groups. Because of these groups, humic acids have the advantage over fulvic acids, by being able to exert both an acidic (H^+) reaction and a strong or large interaction effect. The interactions can be in the form of electrostatic attraction, complex formation or chelation, and water bridging, as illustrated in Figure 7.2. By virtue of the higher phenolic-OH group content, chelation is then substantially higher by humic acids than fulvic acids. Hence, the lower content of phenolic-OH groups in fulvic acids (see Chapter 5) is perhaps an additional reason for their lower chelation capacities. In summary, the conclusion can be drawn that a high total acidity, generated by high COOH and low phenolic-OH group contents, will be less effective in chelation and complexation reactions than a total acidity caused by the presence of lower carboxyl contents but in combination with high amounts of phenolic-OH groups.

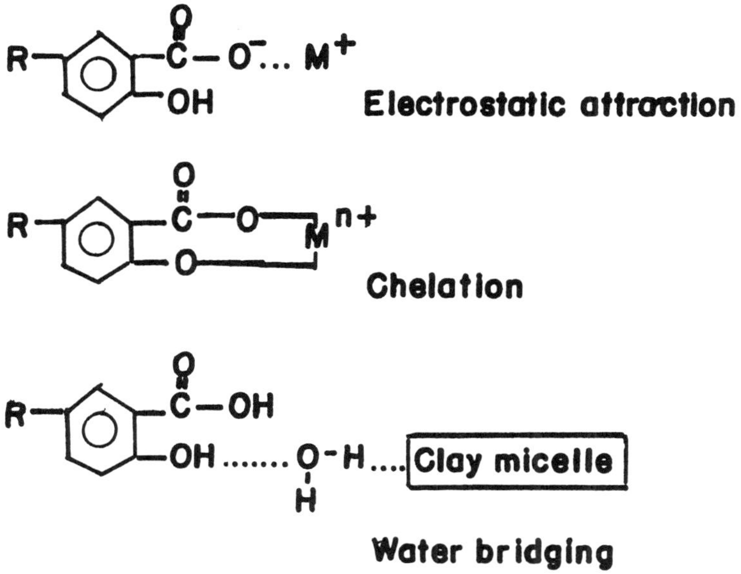

Figure 7.2 Adsorption or electrostatic attraction by humic acid (top), complex or chelation reaction (middle), and water bridging or coadsorption (bottom). M^{n+} = cation with charge n^+, and R = remainder of the humic acid molecule.

7.1.2 Positive Charges

The positive charges are caused by the presence of amino groups. Protonation of amino groups will create positive charges (Tan, 2000). By comparison with the oxygen-containing functional groups, the concentration of amino groups in humic substances is often believed to be relatively small. This is perhaps one of the reasons why the positive charges of humic substances are considered to be only of minor importance. However, the N contents of humic matter are substantial and do not confirm the opinions above. Considerable amounts of NH_2 groups must be present especially to account for the substantially high contents of N in humic acids. It is perhaps the inability of today's techniques in determining NH_2 groups in humic substances that have created a misconception of low amino group contents. Even though, the nitrous oxide method, a standard method for analysis of free NH_2 groups in proteins, shows 30% of the humic-N to be present as amino groups, the analysis is subject to many errors due to interference by lignin and phenolic groups in the humic molecule (Stevenson, 1994). Other scientists have also shown mixed results in detecting measurable amounts of amino groups in humic substances (Sowden, 1957; Sowden and Parker, 1953). Because of the uncertainty in getting reliable results, the issues of NH_2 group contents and positive charges in humic matter are usually ignored.

In clay mineralogy, it is noted that positive charges can also be created on mineral surfaces by protonation of exposed OH groups. Not only can protons be dissociated from these OH groups, but the latter can also adsorb and gain protons (Tan, 1998). This process of protonation is important only in a strongly acidic condition. The reactions for dissociation and protonation of exposed OH groups in clay mineralogy can be summarized as follows:

Alkaline medium: $-Al-OH + OH^- \rightleftarrows -Al-O^- + H_2O$ (7.7)
Octahedron

Acid medium: $-Al-OH + H^+ \rightleftarrows -Al-OHH^+$ (7.8)
Octahedron

Humic substances are known to contain substantial amounts of OH groups, though, of course not associated as octahedral-Al-OH groups. They are in fact present in the aromatic core, as phenolic-OH groups, as well as on the aliphatic C-chain of the humic molecule, as alcoholic-OH groups (see Chapter 5), and most of them, if not all, are located in exposed positions. Since they also react as weak acids, it is perhaps conceivable that these OH groups can also behave similarly as in reactions (7.7) and (7.8). Phenolic-OH groups have been thought by some to dissociate their protons also in an alkaline medium, which is considered one of the reasons for the development of variable negative charges in the humic molecule. However, how they behave in an acidic medium is another question. It has been speculated earlier that at two pH units below the pK_a, the phenolic-OH group is practically nondissociated, hence this group is essentially neutral. Though positive charges are developed on clay minerals at pH values below their ZPC, it is still a very big question why at pH values below the *'isoelectric point'* of the phenolic-OH group above, the acidic condition can induce protonation of phenolic-OHs. Such a positive charge may also reduce the negative charge developed by the carboxyl group, creating another issue for the possibility of the humic molecule becoming a 'zwitter ion.' The latter has been established for amino acids, whereas clay minerals are known to be negatively charged on planar surfaces but positively charged on broken edge surfaces. No direct information is available to refute or support all these assumptions with humic substances, though their cation exchange and complex reactions seem to point to these directions by decreasing substantially with a decrease in soil pH.

The Significance of pK_a and pK_b

The difficulty with protonation of amino groups is that the process can only occur in an acidic condition when soil pH is below the pK_a value of humic acids. The rules in basic soil chemistry indicate that amino groups will be protonated, hence carry positive charges, in acid soils or when pH < pK_a, a condition for providing the required large amounts of H^+ ions. The amino groups are neutral or carry no charges in basic soils or when pH > pK_a. The reaction of the amino group is in fact governed by a constant called pK_b, which is related to the pK_a as explained below. Protonation of an amino group can be illustrated by the following reaction:

$$R-NH_2 + H_2O \rightleftharpoons R-NH_3^+ + OH^- \qquad (7.9)$$

The equilibrium constant K of the reaction above is:

$$K_b = \frac{(R-NH_3^+)(OH^-)}{(R-NH_2)(H_2O)} \quad (7.10)$$

At standard conditions, the activity of water is unity, hence:

$$K_b = \frac{(R-NH_3^+)(OH^-)}{(R-NH_2)} \quad (7.11)$$

Multiplying by $-\log$ gives:

$$-\log K_b = -\log \frac{(R-NH_3^+)(OH^-)}{(R-NH_2)} \quad (7.12)$$

$$-\log K_b = -\log(OH^-) - \log \frac{(R-NH_3^+)}{(R-NH_2)} \quad (7.13)$$

$$pK_b = pOH - \log \frac{(R-NH_3^+)}{(R-NH_2)} \quad (7.14)$$

When the activity of $(R-NH_3^+) = (R-NH_2)$:

$$pK_b = pOH \quad (7.15)$$

Electrochemical Properties of Humic Matter

Since pH + pOH = 14, and at half neutralization pK_a = pH, hence:

$$pK_a = 14 - pK_b \tag{7.16}$$

In contrast to the concept of ionization of acidic groups as explained earlier, protonation of amino groups is now expected to be lower at pH values above the pK_a as defined now by equation (7.14), and higher at pH values below the pK_a. When pH = pK_a, conforming to the Henderson–Hasselbalch concept only 50% of the amino groups are protonated. However, at one pH unit below the pK_a, protonation of amino groups amounts to 90%, whereas at one pH unit above the pK_a only 10% of the amino groups are protonated (Stevenson, 1994). High positive charges are therefore expected to be present when pH < pK_a, whereas no positive charges or only low positive charges are present at pH > pK_a.

7.2 SURFACE CHARGE DENSITY

In soil chemistry the negative charges created by soil colloids are theoretically point charges. However, for practical reasons these charges are considered evenly distributed over the colloidal surface. The magnitude of these charges is then usually expressed in terms of amount of charges per unit area. The latter is called *surface charge density*, σ_s, which can be formulated as follows:

$$\sigma_s = e/S \tag{7.17}$$

in which σ_s = surface charge density in esu/mμ^2 (1 mμ^2 = 100 Å^2), e = number of charges per unit formula, and S = specific surface (Fripiat, 1965).

However, since the total charges on colloidal surfaces are in fact the contributions of permanent and variable charges, the following relationship exists:

$$\sigma_s = \sigma_p + \sigma_v \qquad (7.18)$$

where σ_s = surface charge density in esu/cm² (esu = electrostatic unit and 1 esu = 300 volts), σ_p = surface charge density due to permanent charges, and σ_v = surface charge density due to variable charges. The value of σ_p is constant, but the value of σ_v is variable. Since permanent charges in humic matter are usually very small and can be neglected, the following relationship is assumed to be valid for humic substances:

$$\sigma_s = \sigma_v \qquad (7.19)$$

in which σ_v customary can be calculated using the *Gouy-Chapman* equation as follows:

$$\sigma_v = \sqrt{\frac{2\eta\epsilon kT}{\pi}} \; \sinh \frac{ze\psi}{2kT} \qquad (7.20)$$

in which σ_v = variable surface charge density in esu/cm², η = electrolyte concentration in numbers of ion/cm³, ϵ or D = dielectric constant of the medium, k = Boltzmann constant in erg/ion degree, T = absolute temperature in degrees Kelvin, π = a constant = 3.14, z = valence, e= electron charge in esu, and ψ = surface potential in statvolt.

The unit esu/cm² for surface charge density can be changed into meq/cm² by taking into consideration that 1 coulomb = 3 x10⁹ esu, and 1 Faraday = 96500 coulombs/g.eq.

However, not much information is available yet on the

application of the surface charge density equation in humic matter.

7.3 ELECTRIC DOUBLE LAYER

The concept of double layers is always discussed in relation to charged clay surfaces and no information is available that it also pertains to charged surfaces of humic matter. The present author cannot find any reason why double layers cannot also exist at the surfaces of humic substances. Both clays and humic matter are colloids that are negatively charged. As stated in an earlier section, the negative charges of humic substances are even substantially higher than those of clays. Hence, the surfaces of humic substances will also attract counterions in the same way as the clays. These counterions are attracted similarly by negative charges, and it makes no difference whether the negative charges come from the clay or humic matter surfaces. The issue lies perhaps more in the fact that not much research has been conducted on double layers in humic acids, which is also the case with surface charge densities as stated above.

Because of the presence of electronegative charges, the colloid surface in general can attract cations. These positively charged counterions are held at or near the colloid surface, hence the negatively charged surface is screened or covered by an equivalent cloud or swarm of counterions. This is nature's way of maintaining electroneutrality in the soil's ecosystem. Together the negatively charged surface and the swarm of counterions in the liquid phase are called the *electric double layer*. Theoretically, the negative charge is a localized point charge within the solid surface, as indicated earlier, but customarily this charge is considered to be distributed uniformly over the colloidal surface. The distribution zone of the counterions in the liquid phase varies according to the theories existing on electric double layers. At the state of present knowledge four theories are available in the literature, e.g., (1) Helmholtz, (2) diffuse double layer theory of Gouy and Chapman, (3) Stern double layer theory, and (4) triple layer theory of Yates, Levine and Healy. Since these theories are well covered in the literature, for those interested reference is made to Tan (1998) and other basic soil chemistry textbooks.

7.3.1 Fused Double Layer

In the existing theories on electric double layers, the concepts presuppose that two particles in suspension approaching each other will repel each other because the outer zones of their double layers are equally positive in charges. Such a repulsion prevents the colloidal suspension from flocculating and the suspension is called stable. Flocculation by interparticle attraction can only occur when the double layers are suppressed to very thin layers by for example increasing the concentration of counterions. The thin double layers then decrease the interparticle distance between the approaching particles making a close approach possible. If the interparticle distance decreases to ≤ 20 Å, the theories assume that the *van der Waals attraction* becomes larger than the repulsive forces, and this results in flocculation of the particles.

The present author is of the opinion that the presence of electric double layers surrounding individual colloidal particles is only possible in very dilute condition or very thin soil suspensions, containing only very small amounts of particles. This condition allows the particles to remain in suspension as true individual particles, each exhibiting electric double layers. The thick double layers separate them from each other by considerable distances. In natural conditions, even minor puddling of soils causes dispersion of relatively large amounts of organic and inorganic colloids. These particles, each surrounded by their counterion clouds, are close to each other. However, the double layers are in fact not repelling the particles, but two double layers, confronting each other, are more likely to fuse together to become just one layer. This fused double layer is shared by the two adjacent particles in question. The negative surface of one particle is unable to distinguish whether the counterions belong to its own or to the neighbor's surface. Neither can the counterions. Squeezed between two adjacent surfaces, they are unable to distinguish to which charged surface they actually belong. This conforms to the concept of cation exchange, which dictates that for example Na^+ ions from one surface

Electrochemical Properties of Humic Matter

can freely exchange for Na^+ ions from the other surface. For more details and further implications of the fused double layer concept the interested reader is referred to Tan (2000).

7.4 CHEMICAL REACTIONS AND INTERACTIONS

Because of the presence of electrical charges and electrochemical properties as discussed in the preceding sections, a number of reactions and interactions can take place. At low soil pH, the humic molecule is expected to exhibit positive charges of importance in phosphate fixation and other types of interactions with anionic substances. At a pH range common in most natural and agricultural soils, the humic substances are more likely negatively charged and are capable of adsorption or attracting cations, which leads to cation exchange reactions. When both the carboxyl and phenolic-OH groups are completely ionized or dissociated, humic matter is able to undergo complex and chelation reactions with metal ions or other soil constituents, both xenobiotics and natural compounds (Figure 7.2). These reactions play an important role in soil fertility, plant nutrition, and detoxification of soils, and in enhancing environmental quality as will be discussed in more detail in Chapter 8. Both adsorption and complex reactions can also take place by a water and metal bridging reaction. This is the process by which two negatively charged soil constituents can attract each other. The interaction between humic acid and clay, made possible by water or metal bridging, is also called *coadsorption*. It is reported to also play an important role in adsorption of phosphate ions. Water or any of the metal ions, Ca^{2+}, Al^{3+}, Fe^{3+}, Fe^{2+}, and Mn^{2+} can serve as a bridge between the organic ligand (humic substance) and the clay micelle. Sodium, Na^+, formed by fusing of two opposing electric double layers, was explained earlier (Tan, 2000) to play an important role in interparticle attraction and repulsion.

Each of these reactions will be discussed in more detail in the following sections.

7.4.1 ADSORPTION

The electrochemical properties discussed in the preceding sections find many practical applications in soils. Besides the beneficial effect of flocculation on soil conditions and plant growth, they are why soils develop the capacity to adsorb gas, liquid, and solid constituents. Cation exchange reactions, interactions between clay and organic compounds, including complex reactions and chelation between metal ions and inorganic and organic colloids are additional implications of the electrochemical behavior of soil colloids. The latter reactions are more pronounced in humic substances than in clay minerals. Not only are adsorption and cation exchange exhibited much more by humic substances, but complex and chelation reactions are within the active chemical domain of humic matter.

In contrast to the above, the rate of a true chemical reaction increases as temperature is increased, as formulated by the *Law of Van't Hoff*. Therefore, these differences can be used to distinguish an adsorption process from a true chemical reaction, although a similar equilibrium can be reached in the latter.

Recently, the tendency exists to refer specific adsorption to complexation of solutes by inner-sphere surfaces of clay minerals and nonspecific adsorption for complexation of solutes by outer-sphere surfaces of clays (Sposito, 1989; Zachara and Westall, 1998). If a solute or ion does not form a complex with the charged surface of clay, it is believed to be adsorbed in the *diffuse-ion swarm*. This issue will be addressed in more detail in the next section.

A whole lot is known about the concept of adsorption by inorganic soil constituents or clay minerals, but not much research data are available on adsorption by humic acids and the like. The theory of adsorption in soils is more concerned with the type of concentrating material at the solid-liquid interfaces of clay minerals, as manifested by the counterions in double layer positions. This type of adsorption is often distinguished into *positive* and *negative adsorption*. Positive adsorption is defined as the concentration of solutes on the clay mineral surfaces. It is also referred to as *specific adsorption*. The solute usually decreases the surface tension. On the other hand, negative adsorption is the concentration of the solvent on the clay surface, and the solute is then concentrated in the bulk solution. In this case, surface tension is increased. Since clay minerals are usually negatively charged in ordinary soil conditions, cationic counterions are subject to positive adsorption, whereas anions will be mostly affected by negative adsorption. For more details on the subject reference is made to Tan (1998) and Gortner (1949).

Since humic substances obey similar rules in the development of electrical charges as the clay minerals, it is perhaps fair to expect that they will also exhibit the same two types of adsorption processes as do the clay minerals. The following information provides additional support in this aspect. The soil pH and pK_a of organic adsorbates have been reported to affect the extent of negative and positive adsorption of these organics by negatively charged clay minerals (Frissel, 1961; Bailey and White, 1970). In general, it is noted that negative adsorption is dominant at soil pH > 4.0, whereas the organics are positively sorbed at soil pH ≤ 4.0. Therefore, negative adsorption of organic substances seems to occur first until the pH in the soil approaches the pK_a value of the adsorbates, after which (or below which) positive adsorption takes place and increases as the soil pH decreases. According to White and Bailey (1970) positive adsorption starts when the soil pH is approximately 1.0 to 1.5 pH units higher than the dissociation constant of the organic compounds.

In view of the discussions above, it is perhaps clear that a soil pH of 4.0 is above the pK_a of humic matter, hence the humic substances are by rule mostly negatively charged, causing, in their role as the adsorbates, their repulsion by the also negatively charged clay surfaces. The use of a limit of soil pH = 3.0 is perhaps better, instead of 4.0, since this corresponds with the start of dissociation of COOH groups as explained before. Therefore, in considering now humic matter as the adsorbent, its negative charge is attracting cations by positive adsorption as expected at soil pH > 3.0, and at the same time causing negative adsorption or repulsion of anions. However, the negative charge of humic acids will decrease with a decrease of soil pH, and the charge will become positive if soil pH decreases below 3.0, or the dissociation constant, pK_a, of COOH group in humic matter. This is then the condition where positive adsorption of anions can become of significance.

Adsorption Characteristics

Adsorption reactions are defined as reversible and equilibrium reactions (Gortner, 1949). Sometimes an adsorption process results in chemical changes of the adsorbed material. The changes are of such a nature that desorption is inhibited; hence the process is neither reversible nor in equilibrium. This type of adsorption is called *pseudoadsorption*.

Another important characteristic is that adsorption generally decreases as temperature increases; in other words, adsorption is less at elevated temperatures. This is caused by increased kinetic energies of the molecules at higher temperatures, interfering with the concentrating process. To illustrate this issue, the results of adsorption of fulvic acids by a Cecil soil, a Typic Hapludult, in Georgia, USA, are provided in Figure 7.3 (Tan et al., 1975). The isotherms for adsorption at 25°, 35° and 50° C show adsorption of fulvic acids by the Cecil soil to decrease with increased temperature.

In contrast to the above, as mentioned earlier, the rate of a true chemical reaction increases as temperature is increased, as formulated by the *Law of Van't Hoff*. Therefore, these differences can be used to distinguish an adsorption process from a true chemical reaction, although a similar equilibrium can be reached in the latter.

As indicated earlier, the tendency exists to refer specific adsorption to complexation of solutes by inner-sphere surfaces of clay minerals and nonspecific adsorption for complexation of solutes by outer-sphere surfaces of clays (Sposito, 1989; Zachara and Westall, 1998). If a solute or ion does not form a complex with the charged surface of clay, it is believed to be adsorbed in the *diffuse-ion swarm*. The formidable statistics, accompanying these new developments, have convinced many scientists to *jump eagerly onto the band wagon*. However, to a large number of other scientists, they only result in making the subject more complex and very confusing. Questions are often raised about the inner- and outer-space surfaces in clay minerals and especially in organic compounds (Tan, 2003). Many also wonder what the difference is between a diffuse ion swarm and ions 'complexed' by outer-sphere surfaces. Complexation of ions by outer-sphere surfaces is defined as nonspecific adsorption attributed to elec-

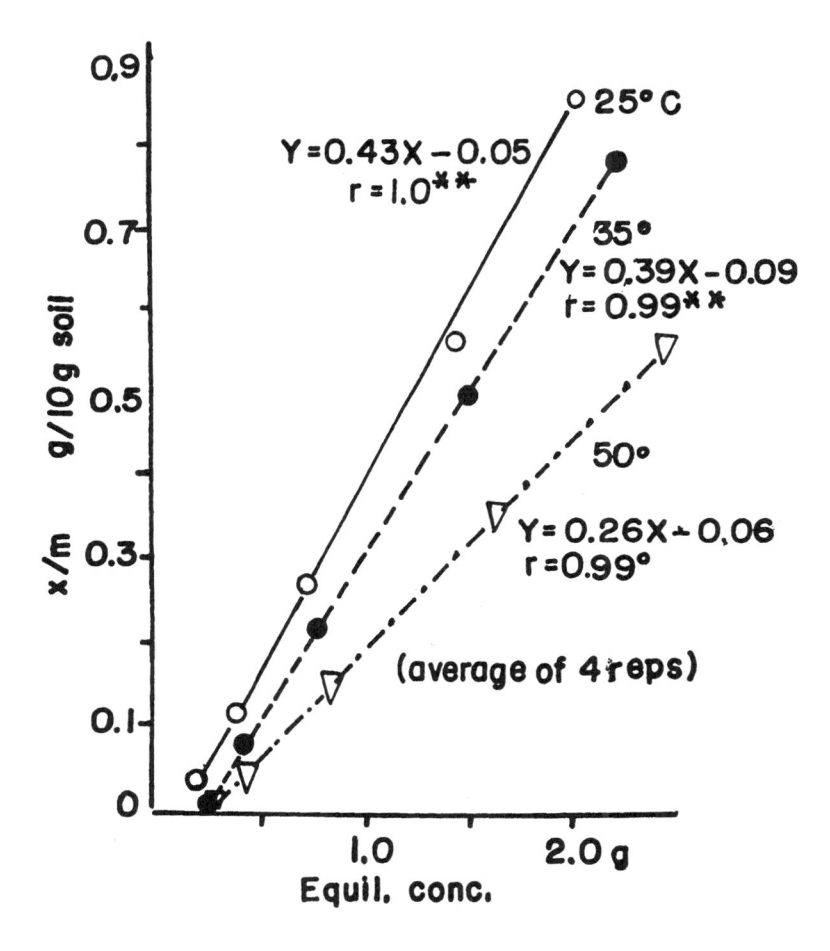

Figure 7.3 Adsorption of fulvic acid, extracted from broiler litter, by a Cecil topsoil at 25°, 35°, and 50° C, respectively (Tan, Mudgal, and Leonard, 1975)

trostatic attraction. But, this is also the definition of the diffuse-ion swarm. Complexation of ions by innersphere surfaces makes the confusion worse, because in the triple-layer theory adsorption in innerspheres is limited to adsorption of potential determining ions only, creating the so-called 'effective surface.' The charge is usually reversed, since the effective surface carries the charge of the adsorbed potential determining ions. All the unanswered questions above find their origin perhaps in regarding adsorption as similar to complex reactions. In basic chemistry, complex reactions are usually considered to occur only with certain cations, and in particular with the transition metals, Al, Fe, Mn, Cu and Zn, binding organic compounds. These reactions, yielding the so-called metal-organo complexes, are to be viewed as rather different from the adsorption of cations in a double-layer region of clay surfaces. The complexed ion usually assumes a central position and the coordination number of the metal determines the number of organic molecules complexed (Murmann, 1964; Mellor, 1964). Unless another definition is available, the concept of complexation in basic soil chemistry differs from that of adsorption in inner- and outer-sphere surfaces as discussed above.

Adsorption Models

Several models are available for describing adsorption processes in soils, some are very simple, and others are very complex. Though most of them have been developed for inorganic compounds, in view of the presence of similar electrochemical properties, there is no reason why the models cannot also apply to organic compounds, such as humic substances. Since adsorption is an equilibrium reaction, fundamental principles of soil chemistry, such as the Law of Mass Action or the Law of Equilibrium, have been applied for interpretation of the process, which is considered as the *scientific approach*. Apparently this method has yielded mixed results because of the extreme difficulties obtained when attempts were made to extend it by involving the double layer concept. In contrast, another group of methods tries to explain adsorption by just accepting the facts obtained without relating them to any basic chemical principle. This second group is called the empirical method, which includes the Freundlich and Langmuir equation models. Since the latter two are well-established models and closely related to one another, only the Langmuir model will be provided below as an example:

Electrochemical Properties of Humic Matter

$$\frac{x}{m} = \frac{k_1 C}{1 + k_2 C} \tag{7.21}$$

where x = amount adsorbed, m = amount of adsorbents, k_1 and k_2 = constants, and C = concentration in equilibrium solution.

At low concentrations, the value of $k_2 C$ becomes so low compared to the factor 1 that it can be neglected, and equation (7.21) reverts to the Freundlich model: $x/m = k_1 C^{1/n}$, in which $1/n = 1$. The Freundlich equation suggests adsorption of solutes to increase indefinitely, whereas the Langmuir indicates that at high values of C, adsorption reaches a maximum. The latter corresponds more to soil conditions where the capacities for adsorption and ion exchange are noted to become saturated.

Another method of describing adsorption processes is by the identification of shape and curvature of adsorption isotherms. In this respect four basic types of adsorption models have been recognized, e.g., S, L, C and H-type isotherms (Weber, 1970; Giles et al., 1960). The S- and L-type adsorption curves are considered to predict similar processes as the Langmuir isotherm (Choudry, 1983). A detailed discussion on these adsorption isotherms and other classical adsorption models, e.g., Brunauer, Emmett, and Teller (BET) and Gibbs, is provided by Tan (1998).

Recently, several scientists have regarded adsorption as identical to cation exchange reactions. Impressive names have been used to re-distinguish adsorption, e.g., surface complexation nonelectrostatic model (SC-NEM), surface complexation-electric double-layer model (SC-EDL), mechanistic, and semiempirical approach (Zachara and Westall, 1998). In this new approach, adsorption in inner- and outer-sphere surfaces is redefined as a complex reaction, forming a stable molecular unit when an aqueous species reacts with a surface functional group. For more details on the merits of these redefined concepts on adsorption, reference is made to Tan (2003), since modeling of adsorption processes is more the subject of soil chemistry than the science of humic matter.

Forces and Mechanisms of Adsorption

Forces responsible for adsorption reactions include (1) physical forces, (2) chemical forces, (3) hydrogen bonding, (4) hydrophobic bonding, (5) electrostatic bonding, (6) coordination reactions, and (7) ligand exchange. Such a listing, compiled from several textbooks and journal articles, is subject to many arguments. Since the issue has been examined in detail earlier (Tan, 1998, 2003), this section will discuss the material related to the forces and adsorption mechanisms exerted by humic substances only. In this respect, the role of humic matter as an adsorbent should be clearly distinguished from that as an adsorbate.

As an adsorbent, humic substances are attracting solutes or cations in a similar fashion as negatively charged clay minerals. As an adsorbate, the humic compounds are then attracted by the clay minerals. Not much research has been conducted on these adsorption reactions by humic acids, and much of the discussions below are based on conjecture or scientific reasoning.

Physical Forces. - A major physical force is the van der Waals force that is active at close distance among all types of molecules. It is additive in nature, hence will increase in force with an increase in size of the compounds or an increase in molecular weight, such as humic acid. Van der Waals forces decreases rapidly with interparticle distance. It is also believed to be especially of importance with nonpolar organic compounds and neutral organic substances.

Electrostatic Bonding. - This type of reaction occurs between two molecules that are opposite in charges. The negatively charged humic molecule will attract cations and these ions create the so-called counter-ion atmosphere of the double layer. Since these counterions can be easily exchanged by other similarly charged ions, they are considered also exchangeable cations. Therefore, an adsorption process may develop into the process of an exchange reaction, when an exchange takes place. The new theories consider adsorption identical

Electrochemical Properties of Humic Matter

to exchange reactions, as pointed out earlier. However, it must be realized that adsorption ≠ exchange reaction, since the counterions in the double layer can remain in the adsorbed condition. This type of reaction is indeed closely related to exchange reactions, but then it is also related to adsorption by H-bonding and ligand exchange.

Electrostatic bonding is the reason for (1) adsorption of water, (2) adsorption of cations, leading to cation exchange reactions, and (3) adsorption of organic compounds. The latter may not only develop into complex reactions, but also into a reaction involving an exchange of the adsorbed organic ligand for an inorganic cation, called *ligand exchange*.

Chemical Forces. - The following discussion is limited only to protonation, since many other forces can also be considered chemical forces. Protonation can occur at the humic molecule surface, and in the solution phase, as well as in the hydration shell of cations. It is of importance for adsorption of anions and organic compounds that are basic in nature. Ammonia, NH_3, is noted to be chemisorbed by clays in the form of NH_4^+, which is considered a protonated form of NH_3 (Mortland et al., 1963). The protons are provided by the dissociation of adsorbed water or by water in the hydration shell of cations. Protonation of basic organic compounds has been noted to occur by clay minerals saturated with H and/or Al. For a more in-depth discussion reference is made to (Tan, 1998).

Hydrogen Bonding. - The bond by which a hydrogen atom acts as the connecting linkage is called a hydrogen bond. As such it is perhaps clear that hydrogen bonding is very closely related to protonation. However, protonation involves a full charge transfer from the electron donor (base) to the electron acceptor (acid), whereas hydrogen bonding is a partial charge transfer only (Hadzi et al., 1968). Hydrogen bonding is a very important adsorption force for humic substances, because of the existence of functional groups containing hydrogen in their molecules, e.g., N-H, -NH_2, -OH, and COOH groups. Hydrogen bonding then takes place between these functional groups and the oxygen on clay mineral surfaces.

Hydrophobic Bonding. – This is the type of bonding associated with adsorption of nonpolar molecules, where the latter compete with water molecules adsorbed on the adsorption sites. In the process, the adsorbed water is expelled by or exchanged for the nonpolar molecule, which is the reason for calling the process hydrophobic. Polysaccharides, for example, are adsorbed in this way by clay minerals, and the expulsion of water, especially from the intermicellar clay surfaces, reduces swelling.

Coordination Reaction and Complex Formation. – The reaction involves coordinate covalent bonding, in which the ligand donates electron pairs to the metal ion. The latter is usually a metal in the series of *transition metals*. The ligand, therefore, fits the definition of a *Lewis base* and the metal is then the *Lewis acid*. The compound formed is called a coordination compound, complex compound, or an organo-metal complex. Coordination compounds or organo-metal complexes are substances containing the metal as a central atom, surrounded by a cluster of organic ligands. The total number of ligands that can be complexed corresponds to the coordination number of the metal.

The terms inner-sphere and outer-sphere complex reactions have been used recently to indicate adsorption processes by these types of complex reactions. As discussed previously, inner-sphere complex reactions are referred to as adsorption processes. It is perhaps a very special type of adsorption that does not obey the rules of reversibility and desorption. The desorption process is often only partially possible, and is also reported to occur only with other transitional metals at a very specific pH value (Lindsay, 1974; Lindsay and Norvell, 1969). For example, Na^+ cannot be exchanged for Fe^{3+} and cannot assume the central position in the complex compound. It is up to the reader to accept the inner-sphere theory as an adsorption process.

Ligand Exchange. – This process entails the replacement of a ligand by an adsorbate molecule. The adsorbate can be an inorganic ion or an organic molecule, but in either case it must have a stronger chelation

Electrochemical Properties of Humic Matter

capacity than the ligand to be replaced.

7.4.2 CATION EXCHANGE CAPACITY

The cation exchange capacity (CEC) of soils is defined as the capacity of soils to adsorb and exchange cations. The basic concepts on CEC have been provided elsewhere in detail (Tan, 1998). Scientifically, CEC is related to the surface area and surface charge of clay minerals or humic substances. This relationship is expressed by the following equation, which serves as a statistical model for CEC:

$$CEC = S \times \sigma \qquad (7.22)$$

in which S = specific surface (cm^2/g), and σ = surface charge density (meq/cm^2). The surface charge density can be calculated by using equation (7.17) or (7.20). It is usually expressed in esu/cm^2, but as explained before this unit can be converted into meq/cm^2. However, it is common practice in the determination of CEC to analyze all exchangeable cations, and the CEC is then:

$$CEC = \sum meq \text{ exchangeable cations per 100 g soil} \qquad (7.23)$$

The magnitude of the CEC of humic substances is considered related to their total acidity value. As discussed in Chapter 5 and other sections of this book, the average figures for total acidity vary from 6.0 to 8.9 meq/g for soil humic acids, and from 10 to 12.3 meq/g for fulvic acids.

In the past, two types of CEC for humic substances have been distinguished (1) a *measured CEC*, determined by exchange with any suitable cation, and (2) *potential CEC*, which is defined as the sum of the measured CEC and the CEC attributed to blocked sites (Schnitzer, 1965). These blocked sites will be exposed by liming or when the humic substances are extracted from the soil, and their CEC is often larger than that of the measured CEC (Piccolo and Stevenson, 1981). However, a more recent concept recognizes a total CEC defined as:

$$CEC_t = CEC_v + CEC_p \qquad (7.24)$$

in which the subscripts t = total, v = variable, and p = permanent (Tan, 1995; 2000).

Humic substances are believed to contain a very small amount of permanent charges, responsible for the development of CEC_p. These charges are estimated to amount to only 10% of the total negative charges, whereas the remaining 90% are attributed to the variable charges (Brady, 1990; Tan, 1992). The CEC_t is formerly called CEC_m in which m = maximum (Mehlich, 1960). In the case of clay minerals, the latter is measured at pH 8.2, and no further increases in CEC values can be obtained by analysis at higher pH values. If this is true, then it can be expected that the variable negative charge reaches a maximum value at pH 8.2. This is in disagreement with the formation of negative charges based on the theory of dissociation of functional groups in humic matter, as postulated by Posner (1964). As discussed earlier, Posner believes that the carboxyl groups dissociate their protons at pH 3.0, whereas the phenolic-OH groups are ionized at pH 9.0. Consequently, negative charges in humic matter are more likely to increase at pH values > 8.2, which is in contrast to clay minerals.

Several scientists, in fact, believe that contrary to clay minerals, humic substances do not possess limits in their CECs. Stevenson (1994) is also of the opinion that the exchange capacity of humic matter increases markedly with pH, which is attributed to increased ionization of COOH groups at high pH values. The latter differs sharply from Posner's findings above, and also deviates from basic rules in chemistry where COOH groups are listed as strong acidic groups with low pK_a values. Formic acid, a carboxylic acid with the formula of HCOOH, has a $pK_a = 3.62$, and peruvic acid with a formula of $CH_3COCOOH$ exhibits a $pK_a = 2.50$ (Conn and Stumpf, 1967), all testifying for a strong acidic behavior. It is the phenolic-OH group that

Electrochemical Properties of Humic Matter

is responsible for the increase in CEC of humic substances at high pH. This group is known to behave as a weak acid, and its half-dissociation point is apparently at $pK_a = 9.0$, whereas 99% dissociation was expected to be attained at pH 11.0, as discussed before. Consequently, the CEC of humic substances will correspondingly increase with the increase in negative charges due to increase dissociation of phenolic-OH groups at high pH. However, the present author cautions against using extremely high pH (12.0 - 13.0) values, because of the danger for humic substances to be broken down by hydrolysis.

Cation Exchange Reaction Models

In soil chemistry, a number of cation exchange reaction models have been developed mostly for application with clay minerals. Similarly as with adsorption models, there is no reason why the exchange reaction models cannot be applied to exchange reactions by humic substances. However, since modeling is more in the realm of soil chemistry rather than in the science of humic substances, the most common models only will be discussed as examples in these sections, e.g., the models based on the Law of Mass Action. The purpose is to show that the same statistical modeling process can be applied for exchange reactions with humic acid as well as those with clay minerals. After all, *"a statistician is a scientist who can draw a straight line from an unwarranted assumption to a foregone conclusion"* - *Yale Hirsch*. For details on the other models, such as the empirical equations, kinetic equations, the Donnan equation, and thermodynamic equations of cation exchange, reference is made to Tan (1998).

Mass Action Law equations. - Two of the most common models in this group that find extensive application are the Kerr's and Gapon equations. The two models are, in fact, closely related to each other as can be noticed from the following discussion.

Let us assume that a mono-divalent cation exchange reaction at equilibrium can be written as follows:

$$2Na^+ + Ca\text{-}HA \rightleftarrows (2Na)\text{-}HA + Ca^{2+} \tag{7.25}$$

in which HA = humic acid. According to the Law of Mass Action, the equilibrium constant of the reaction is then:

$$K_{eq} = K_{ex} = \frac{[Na^+]^2 (Ca^{2+})}{(Na^+)^2 [Ca^{2+}]} \tag{7.26}$$

The signs [ ] and ( ) denote adsorbed and free cations, respectively. Because this is an exchange reaction, K_{eq} can also be considered the *exchange constant*, K_{ex}. Equation (7.26) is called the *Kerr's equation*. By taking the square root, equation (7.26) changes into:

$$\sqrt{K_{ex}} = K^{'}_{ex} \frac{(\sqrt{Ca^{2+}})\,[Na^+]}{[\sqrt{Ca^{2+}}]\,(Na^+)} \tag{7.27}$$

This is known as the *Gapon equation*.

7.5 COMPLEX REACTION AND CHELATION

These are the outstanding properties of humic substances that have made them very conspicuous in soils, agricultural, pollution, and environmental issues. It is these reactions that have propelled humic substances to be regarded as one of the most active components in soils, not matched by other soil constituents.
The terms complex formation and chelation have often been used in soil science interchangeably. In the preceding sections above,

Electrochemical Properties of Humic Matter

it has been pointed out that complex formation has also been considered identical as adsorption processes in inner- and outer-sphere regions of clay minerals. Consequently, it is deemed necessarily to provide the common definitions for these processes as they appear in many textbooks. Owing to the nature of reaction and bonding process, a distinction must be made between adsorption, complex formation, and chelation. The difference from adsorption has been discussed earlier, and at present the focus will be on concepts and definitions differentiating complex formation from chelation.

Complex formation is the reaction of a metal ion with a compound through electron-pair sharing (Murmann, 1964; Mellor, 1964; Calvin, 1952). The resulting product is called a *metal coordination compound*, which can be in the form of cations, anions, or neutral molecules. The metal is the electron-pair acceptor, whereas the compound is the electron-pair donor. The latter is usually called a *ligand,* and may assume the form of an anion (HA^-) or a neutral molecule (NH_3). The metal ion serves as the central atom, and the ligands are coordinated around it in a first coordination sphere. The number of ligands bonded to the central atom in a definite geometry depends on the *coordination number* of the metal. Almost any metal atom can serve as an acceptor atom, including K^+, Na^+, Li^+, Ag^+, and Au^+. A long-known complex compound with a monovalent ion is potassium ferrocyanide. However, the most common metals capable of complexation reactions are the transition metals, e.g., Al, Fe, Cu, Mn, and Zn. Mention is also made in the literature of distinguishing two categories of metals on the basis of differences in forming coordinate covalent bonding (Stevenson, 1998). Category A includes metals capable of forming complexes with ligands containing oxygen as the electron-pair donor. Category B is a group of metals that preferentially coordinate with ligands containing N, P, and S. Not much is known about this preferential nature of metals for ligands and more information is required to accept or reject the contention above.

Chelation is the reaction when a ligand can bind the metal ion with two or more donor functional groups to form a ring structure. The compound formed with the characteristic heterocyclic ring is called a chelate (Greek chele = lobster claw), which in fact is referring to the pincer-like bonding of the metal (Figure 7.2: middle). If one ligand is involved in the formation of a chelate, the compound is called a *monodentate* chelate, and when two ligands form a chelate, the compound is called a *bidentate*, etc. For more details, reference is made to the references given above and to Tan (1998).

The reactions in Figure 7.4 are provided to illustrate and underscore the differences between adsorption, complexation and chelation between a metal (Zn) ion and humic acid. The complexation reaction appears to be similar to an adsorption reaction on paper. However, ad-

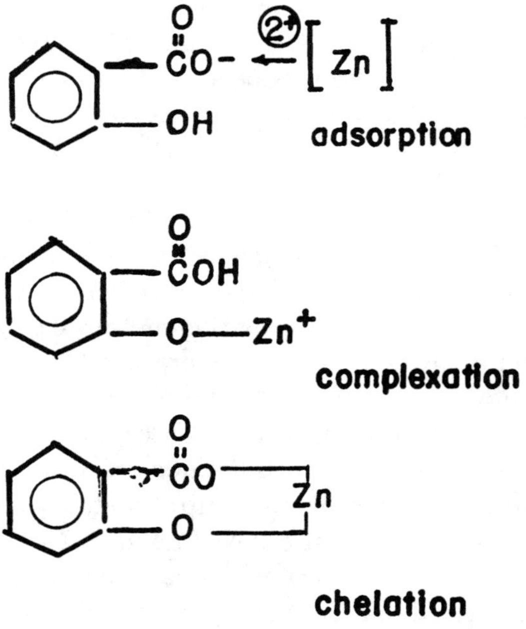

Figure 7.4 Adsorption, complex formation and chelation of Zn by humic acid.

Electrochemical Properties of Humic Matter

sorption is electrostatic attraction of Zn^{2+} by the negative charge of the humic acid surface, whereas complexation is a coordinate covalent bonding through electron pair sharing. In this case the Zn ion is bonded and becomes an integral part of the humic acid molecule, as indicated on the figure. The chelate as illustrated is an example of a monodentate.

7.5.1 The Significance of COOH Groups

Recently, the idea has surfaced that formation of metal complexes is attributed to the interaction of the metal with the COOH groups. From a study on photochemical and microbial processes, affecting metal-humic interactions, McKnight et al. (2001) have arrived at the conclusion that the carboxyl groups account for the majority of strong complexing sites in the humic molecules. These functional groups can form charge transfer complexes with metals such as Fe and Cu.

The COOH groups can be linked to the aromatic or to aliphatic C structure. Humic substances, which are highly aromatic, may contain large amounts of aromatic-COOH groups, hence according to the authors above will exhibit strong complexing sites for Cu. The COOH groups on the aliphatic sites are allegedly important for complexing especially Ca ions. Fulvic acids isolated from rivers with 'soft water' (low Ca and Mg contents) are believed to have stronger Cu-binding sites than fulvic acids extracted from rivers containing 'hard-water' (Breault et al., 1996; Leenheer et al., 1998).

Simple sorption, or simultaneous coprecipitation, and adsorption of humic substances on the Fe oxide surfaces have also been considered complex reactions by McKnight et al. (2001). Variations in the Fe complexation constant among carboxylic groups are believed to influence these reactions. To identify Fe complexation reactions as similar as precipitation or coprecipitation and adsorption is very confusing. First of all, a definition must be provided by the authors for their Fe complexation constant. In the literature, a stability constant for chelates is available, but no information can be found about an Fe complexation constant. Perhaps, the authors meant the *affinity* of the various carboxyl groups for Fe, which has a totally different meaning. Purdue (2001; 1988) has mentioned complexing capacities (CC) of humic acids, which are related to the total acidities (TA) of the humic compounds, and proposes to define it as follows:

$$CC = TA/z \qquad (7.28)$$

where z = valence of the cation. His reasoning in including the valence of the cation is that the electrostatic driving force for complexation of a cation by humic substances is gradually decreased as the negative charge of the metal-humic complex decreases to zero. The latter is usually reached when the humic compound becomes fully saturated with metal ions. Though it is a very sound concept, not much information is available and more research is required to test the suitability of using complexation capacity (CC) in complex reactions by humic substances.

Secondly, precipitation and 'simple' adsorption have been interpreted by the authors as coordination reactions. As far as the rules are in basic soil chemistry, precipitation takes place only with noncharged clay minerals or noncharged humic substances and no covalent bonding is involved. This process occurs normally at the ZPC or isoelectric point values. On the other hand, adsorption is defined in basic soil chemistry as the attraction of counterions on the surfaces of charged clay minerals and humic substances. In the strict sense of 'simple' adsorption, the process does not include electron-pair sharing, a requirement for complex reactions. It is commonly known by now that complex reactions involve covalent bonding through electron pair sharing. This capacity is exhibited especially by the phenolic-OH groups, since they behave usually as 'Lewis acids.' The carboxyl groups, on the other hand, are acidic in nature and behave as Brønsted-Lowry or as Arrhenius acids. Consequently, they are expected to react more through electrostatic attraction rather than through electron-pair sharing. The chemistry of the phenolic-OH and COOH groups in relation to complex reactions and chelation will be discussed in more detail in the following section.

7.5.2 The Significance of pK_a

The two most important functional groups in humic substances responsible for complex formation and chelation are the COOH and phenolic-OH groups. They will dissociate and become active at two different pK_as or soil pH. As indicated earlier, the carboxyl groups usually behave as strong acidic groups and as explained before exhibit pK_a values around 3.0, with a chance of 99% ionization at soil pH = 4 -5. Bonding by this group constitutes only adsorption and according to McKnight et al. (2001) also complex reaction. Chelation starts to become of importance when the soil pH increases to a value approaching the conditions for the phenolic-OH groups to ionize. As stated earlier at pH = 9.0 the pK_a of these functional groups is reached, and ionization is expected to be 50%. Therefore, chelation is predicted to take place at pH values above 4 to 5, and increases gradually, especially at pH = 11, when 99% ionization of the phenolic-OH is attained. Studies on the effect of pH on chelation seem to support the above. As a result of his investigations, Tan (1978) indicates that the amounts of Al chelated by humic acids increase from 0.226 meq/100g at pH = 7.0 to 0.445 meq/100g at pH = 11.5. The amounts of Fe chelated by humic acids are also reported to increase from 0.255 meq/100g at pH 7.0 to 0.443 meq/100g at pH = 11.5.

In conclusion, it can perhaps be stated that complex formation in humic substances occurs mostly at low soil pH, whereas chelation reaction is more dominant at higher pH values.

7.5.3 Stability Constants of Chelates

Humic matter is capable of forming soluble and insoluble complexes and chelates (Tan, 1978: Stevenson, 1978a,b). Some of the factors affecting this solubility are (1) type of humic substance, (2) types of metals, (3) dissociation of functional groups, and (4) saturation of binding sites.

Metal complexes of fulvic acids are in general more soluble than those of humic acids. This is perhaps attributed to the lower molecular weights and higher solubility of fulvic acids.

Several of the cations have the potential for enhancing polymerization of humic matter by linking the individual molecules together into chainlike structures. The metal-humic acid complex remains soluble when the metal/humic ligand ratio is low (Stevenson, 1994). The metal-humic acid complex becomes insoluble and precipitates as the metal bridges increase and the chainlike structure grows. The maximum chelating or complexing capacity of humic matter equals its total acidity, which is the amount of H^+ ions from both COOH and phenolic-OH groups. A total acidity of 1000 meq/100g corresponds to 90 mg of Al^{3+}/g of humic acid bonded by the functional groups. If this maximum capacity is satisfied, the humic ligand is saturated with the metal, and the complex or chelate becomes insoluble.

Dissociation of functional groups induces repulsion of charged groups, as indicated by the dissociation of carboxyl groups as follows:

$$-R-COOH \rightleftarrows -R-COO^- + H^+$$

As shown in the reaction, it gives to the humic molecule a stretched configuration. After reaction with a cation, the negative charge is neutralized and the stretched molecule collapses, thereby reducing solubility.

The problem of solubility of metal-humic acid complexes and chelates is usually expressed in terms of *stability constants*. These constants are derived as follows. Assume that the following complex or chelation reaction occurs:

$$M^{2+} + 2HA \rightleftarrows M\text{-}(2A) + 2H^+ \qquad (7.29)$$

in which M^{2+} = divalent cation, HA = humic acid, M-(2A) = bidentate metal chelate.

According to the Law of Mass Action, the equilibrium constant of equation (7.29) is then:

Electrochemical Properties of Humic Matter

$$K_{eq} = \frac{[M-(2A)](H^+)^2}{(M^{2+})(HA)^2} \tag{7.30}$$

By taking the log, this equation changes into

$$\log K_{eq} = \log \frac{[M-(2A)](H^+)^2}{(M^{2+})(HA)^2} \tag{7.31}$$

If the activities of HA and M-(2A) chelate are assumed unities at standard state, then

$$\log K_{eq} = \log \frac{(H^+)^2}{(M^{2+})} \tag{7.32}$$

or

$$\log K_{eq} = 2 \log (H^+) - \log (M^{2+}) \tag{7.33}$$

where $\log K_{eq}$ is called the stability constant. It determines the solubility and stability of the metal chelates formed (Tan et al., 1971a,b). The solubility of chelates is low or in other words the stability of chelates is high, if the value of $\log K_{eq}$ is high. These authors have calculated $\log K_{eq}$ values for metal-fulvic acid chelates and some of the data are shown in Table 7.1. These $\log K_{eq}$ values are in the range of those reported by Stevenson (1994) for his 1:1 complexes of Cu and Zn with fulvic and humic acids. The data in the table indicate that comparatively the value of $\log K_{eq}$ increases from Mg-FA → Zn-Fa and to Cu-FA chelates, suggesting that the Cu-FA chelates are more stable

Table 7.1 Stability Constants ($\log K_{eq}$) of Metal-Fulvic Acid Chelates

Stability Constant	Cu-FA	Zn-FA	Mg-FA
$\log K_{eq}$ (pH 3.5)	7.15	5.40	3.42
$\log K_{eq}$ (pH 5.5)	8.26	5.73	4.06

(less soluble) than Zn-FA and/or Mg/FA chelates. The effect of pH is also clearly demonstrated, and supports the opinion stated earlier that chelation will increase at higher pH levels.

The concept of stability constant for humic chelates discussed above is used in relation to the principle in equilibrium reactions that the equilibrium constant represents in general the rate of reaction. As depicted in equation (7.30), the value of K_{eq} increases if the amount of chelates formed increases. Since at high values of K_{eq} reaction (7.29) is not going to the left, the assumption is made that the chelate is not dissociating into its metal and ligand components, hence the chelate is said to be stable. However, other interpretations are presented in the literature, and the stability constant is, for example, understood by other scientists to provide an index of the affinity of the cation for the ligand (Stevenson, 1994). However, affinity is related more to the attractive force of the ligand for the metal with the connotation of a preferential or selective attraction. Although not much information is available in this respect, the present author's unpublished files indicate that, in general, the affinity of humic acid for reaction with cations is in the following decreasing order:

$Al^{3+} > Fe^{3+} > Cu^{2+} > Mn^{2+} \geq Zn^{2+} >> Mg^{2+} \geq Ca^{2+}$

Given the choice, Al^{3+} will be chelated first over Ca^{2+}, though it is then also true that the Al-humic acid chelate exhibits a higher stability constant, K_{eq}, than Ca-humic acid chelate. In other words, Ca-HA chelates are more soluble than Al-HA chelates. Metal ions, such as Al and Fe, that under normal conditions are insoluble, are soluble in chelate form and participate in soil genesis since they can now be translocated and accumulated in different soil horizons in the pedon.

Complex formation and chelation have an important impact on soil formation, soil fertility, and other issues on soils and the environment. Their role in soil formation will be discussed in the next section, whereas their importance for soil fertility will be addressed in more detail in Chapter 9.

7.5.4 Effect on Soil Genesis

Complex reaction and chelation of humic substances may affect a variety of processes related to soil formation. They can accelerate the decomposition of primary minerals and affect formation of clay minerals, soil horizon differentiation, and soil fertility. Their role in the development of soil horizons and in soil fertility is closely related to mobilization-immobilization issues of metal chelates, and will be discussed in more detail in Chapter 9. The following discussion will address the effect of chelation on the dissolution of primary minerals and the formation of clay minerals.

Decomposition of Soil Minerals

The decomposition of soil minerals by humic substances has attracted considerable attention since the early history of soil science. Long before Dokuchaiev formulated his pedological concept, soil organic acids, including humic acids, were expected to play an important role in the decomposition of rocks and minerals (Sprengel, 1826). Since then, conflicting arguments have been presented as to the effectiveness of these acids in rock and mineral weathering. Mainly due to lack of supporting experimental evidence, a large number of scientists have questioned in the past the role of humic matter as a weathering agent (Clarke, 1911; Loughnan, 1969). However, an equally large number of authors can also be found in the literature defending the role of humic substances as a dissolution agent (Graham, 1941; Van der Marel, 1949).

With the increased knowledge of humic acid chemistry, evidence has accumulated now suggesting humic compounds to play a significant role in mineral dissolution. Today's data indicate that the acidity and chelating capacity of the humic substances can bring about the degradation of rocks and minerals (Singer and Navrot, 1976; Schalscha et al., 1967; Baker, 1973). A massive dissolution of Si, Al, Mg, Fe, and K from mica minerals by fulvic acids has been reported by Schnitzer and Kodama (1976). The extremely low pH of 2.5 employed in their analysis is responsible for the rapid dissolution, and was later confirmed by Tan (1980) from his dissolution experiments using microcline, biotite, and muscovite. The latter author notices that fulvic acid is capable of extracting 10 times more silica and 6 times more aluminum from microcline at pH 2.5 than at pH 7.0. This leads him to conclude that both the acidic effect and complex or chelation reactions have been responsible for the dissolution process (Tan, 1986). The acidic effect is especially of importance at pH 2.5, whereas at higher pH chelation and complex reactions seem to be more effective. The dissolution of silica, aluminum and potassium at pH 7.0 is relatively small in the beginning, but increases rapidly with time to reach a maximum at approximately 30 days.

The data above then indicate that by forming metal-humo complexes and/or chelates, humic substances are playing an important role in the weathering process by accelerating the decomposition of rocks and primary minerals

Effect on Formation of Clay Minerals

The dissolution products from the mineral weathering processes by humic substances are believed to promote or inhibit formation of clay minerals. Schwertmann and Taylor (1977) are of the opinion that crystallization of iron oxides can be inhibited by the presence of ferric-fulvic acid complexes. However, the transformation of hematite to goethite is noticed by Schwertmann (1971) to be mediated by iron-fulvic acid complexes. Direct evidence for the inhibitory effect of fulvic acid on the crystallization of iron and aluminum hydroxides has been presented by Schnitzer and Kodama (1977), and Kodama and Schnitzer (1977; 1980). The concentration of fulvic acid and pH are noticed by these authors to have affected the crystallization process of the sesquioxide minerals. Fulvic acid, present at concentrations of 5.0 g/L, is believed to completely inhibit formation of goethite and hematite, whereas a concentration of 0.5 g FA per liter seems to favor the crystallization of the two minerals above. According to these authors gibbsite is formed at pH 6.0 in the absence of fulvic acid, but the addition of fulvic acid seems to delay first and then inhibit its crystallization and formation. The conclusion is presented by Kodama and Schnitzer that it is unlikely for crystallization of sesquioxide minerals to occur in the presence of substantial amounts of Fe-FA and Al-FA complexes in the soil solution.

A similar inhibitory effect of humic substances has also been reported in the formation of allophane. The process, called *antiallophanic effect*, has recently been noticed in the formation of allophane and imogolite in andosols. This effect is presumably attributed to chelation of Al by humic acids, decreasing the free Al concentrations in the soil solution so that it becomes inadequate for the formation of allophane and imogolite (Shoji, 1993; Huang, 1995; Tan, 2000).

7.5.5 Statistical Modeling

As is the case with adsorption and cation exchange reactions, the opinion exists that the processes of complex reactions and/or chelation can also be studied more accurately by using statistical models. It is believed by several scientists that humic matter exhibits a variety of nonidentical binding sites, and a model would be able to describe quantitatively the relative concentrations and the strengths of these binding sites. Three models have been developed lately for this purpose: (1) a *competitive Gaussian Distribution model* by Dobbs et al. (1989), a *Model V* by Tipping and Hurley (1992), and an *NICA* (nonideal competitive adsorption) model by Koopal et al.(1994), also known as the *NICA-Donnan model*. Though Purdue (1998) indicates that the three models are reasonably suitable for the purpose intended, the author notes that the perceived distribution of binding sites is unfortunately related directly to the experimental conditions and to the type of model that is forced on the experimental data.

Purdue (2001) claims that the three models above are assuming humic matter to contain two classes of binding sites, e.g., sites for specific binding, and sites for nonspecific binding of cations. Each sites' class contains multiple sites whose relative amounts are distributed symmetrically around a central log K value, in which K is the equilibrium constant of the complex reaction between the metal ion and humic acid.

In Purdue's (1988a; 2001) opinion, the development of the models above requires a specific knowledge of a complexing capacity (CC) of humic matter. The latter has been discussed and formulated earlier as equation (7.28). For more details on the models reference is made to the specific references stated above.

7.6 BRIDGING MECHANISM

This is another outstanding reaction among those that have made humic substances so important in soils and in the environment. Negatively charged components will normally be repelled by other negatively charged soil constituents, such as clay minerals, making, therefore, clay-humic matter interactions impossible. The interactions between humic acid and clay are, in fact, very important processes and are responsible for most of the accumulation of humic matter in soils. The reaction between the two negatively charged soil materials is made possible especially by the presence of polyvalent cations or water.

The humic molecule and the clay mineral in suspension are ordinarily each surrounded by water molecules, forming so-called hydration shells. One of these water molecules can serve as the connecting linkage between the humic molecule and the clay mineral. The water molecule is dipolar in nature and one of its H atoms can be bonded by the negatively charged humic molecule, whereas the other

Electrochemical Properties of Humic Matter

H atom will then hook up to the negative charge clay surface. The reaction is called water bridging, since the water molecule acts as a bridge between the humic molecule and the clay mineral. An illustration of such a bridging mechanism is provided in Figure 7.5. This type of bridging shows a close relation with hydrogen bonding due to the fact that a hydrogen atom acts as the connecting linkage. Since two hydrogen atoms are involved in the reaction above, it can perhaps be considered a type of *complex hydrogen bonding*. However, water bridging can also take place between a neutral humic molecule and a negatively charged clay surface, as shown in Figure 7.2.

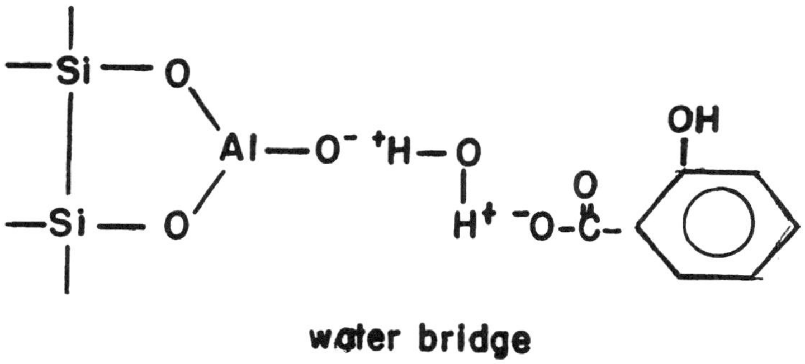

water bridge

Figure 7.5 Water bridge between a negatively charged clay and humic acid.

Polyvalent cations are also capable of acting as bridges between the humic molecule and clay minerals. The reaction is then called cation or metal bridging. In this case, free cations in solution or metals structurally bonded in the octahedrons or tetrahedrons of clays can act as bridges. As noticed from Figure 7.6, the Al- bridge (top) and Ca-bridge (bottom) are formed by Al and Ca ions free in solution. The structural Al-bridge (middle) is produced by octahedral-Al on the surface of the clay mineral allophane.

Bridging is perhaps a very versatile mechanism, since a variety of bridging reactions are possible. As indicated above, bridging mech-

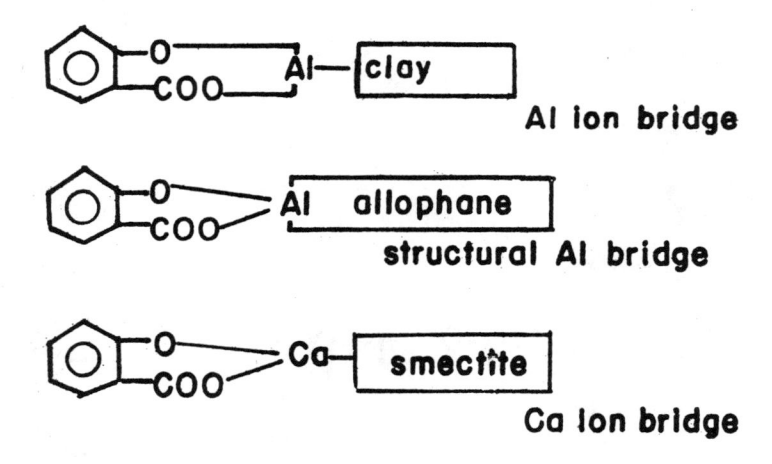

Figure 7.6 Bridging or coadsorption between humic acids and clays through metal ions in solution and structurally bonded atoms.

anisms make the reaction possible between neutral molecules, whereas several combinations between water and cations are also available. Anions can be bonded through bridging reactions by humic molecules, which are significant processes in detoxification reactions of xenobiotics. Several of the harmful pesticides introduced in soils are anionic in nature and can be neutralized by humic matter through binding by

Electrochemical Properties of Humic Matter

the process of metal bridging. Phosphate ions, which are mostly negatively charged, are also bonded either directly or through water and metal bridging by humic matter. These are topics that will be discussed in more detail in the next Chapter.

CHAPTER 8

AGRONOMIC IMPORTANCE OF HUMIC MATTER

8.1 IMPORTANCE IN SOILS

It is well known that soil organic matter has a favorable effect on the physical, chemical, and biological characteristics of soils. With the increased knowledge in humic acid chemistry, this effect is now realized to be caused by the active components of the inorganic and humus fraction.

8.1.1 EFFECT ON SOIL PHYSICAL PROPERTIES

The physical properties of soils are noted to change due to addition of soil organic matter yielding humic substances. These changes are usually interrelated, and the change in one physical property will often be followed by changes in other physical properties. Soils high in organic matter usually exhibit high water-holding capacities, display well-developed structures, have low bulk density values and are often fluffy or friable in consistencies.

Although all the physical properties are no doubt very important for the well-being of the soil ecosystem, it is perhaps soil structure that

Agronomic Importance of Humic Matter

is the most significant in soil formation and soil degradation as well as in plant growth and environmental quality. The formation of soil structure favorable for plant growth is assumed to be caused by the interaction between humic acid and clays and/or by complex reactions between humic acid and Al and other metal ions. Since soil structure also infers a mutual arrangement of the three soil phases, solid, liquid, and gas, a change in soil structure will affect the balance between these three phases. The liquid and gas phases are especially vulnerable, since they are also subject to continuous exchanges with the environment. The effect of humic acid is to create and preserve a stable structure that can provide the proper amounts of pore spaces for the storage of optimum amounts of water and oxygen.

The cementation effect of humic acid has long been considered a major factor in formation of soil structure (Baver, 1963), which is very important in especially sandy soils. These soils are usually very loose and friable and often single grained or structureless. The amounts of clay present are insignificant to cement the sand particles together. The dominant amounts of sand, which are chemically inert, are incapable of reacting with either cations or humic acids. Consequently, aggregation of sand particles can only be enhanced by cementation with humic acid. In contrast, another problem arises in soils rich in silt and clay. Here, crust formation occurs when the soils are low in organic matter contents. For farmers, the occurrence of surface crusts provides a very big problem. Seed germination, soil aeration, and infiltration of water will be inhibited. Enhancing aggregation by adding soil organic matter is often noted to prevent the formation of these surface crusts. The absence of soil organic matter in clayey soils causes in addition the development of structureless conditions, and massive structures are common which inhibit aeration, water penetration, and root growth. By creating granular structures due to the interaction of clay with humic acid the unfavorable physical conditions may be alleviated.

Although its cementation effect plays an important role, the interaction of humic acid with metal ions and clays is believed to be of a more decisive factor in many soils for the creation of stable soil structures. One of the most striking examples in this respect is the unique physical condition in andosols. These soils are known for their black color due to high organic matter contents, low bulk densities and crumb to granular structures. The soils can become very wet due to their extremely high water-holding capacities. However, in such wet conditions, they still display low plasticity and stickiness and are friable in consistence (Tan, 1998; Shoji et al., 1993). The consensus is that the high organic matter content, allophane and Al in andosols have played a major role in the development of the unique physical properties. Although several hypotheses can be presented, it is commonly assumed that the amorphous clay and humic acids are responsible for the exceptional physical properties. Exposed groups of Al and Si on the surfaces of allophane and imogolite are capable of interacting with humic acids forming humo-Al-allophane or humo-Si-imogolite complexes or chelates. The structural Al acts in essence as a connecting bridge, hence the interaction can be called *Al-bridging*. As discussed in Chapter 7, Al ions in the soil solution can react in the same way acting as bridges between humic acids and negatively charged surfaces of the clay mineral (Figure 7.6). Not only will the organic substances be protected from rapid decomposition by such an interaction, but these chelates constitute the nucleus for formation of granular or crumb structures. In turn, crumb structures provide for an abundant amount of macro- and micropore spaces, which together with those present in the amorphous clays account for the high total porosity exhibited by andosols. The accumulation of organic matter (humic acids) known for its high adsorption capacity for water, together with the increased amounts of pore spaces, therefore increases the water-holding capacity of the andosols.

A similar soil physical development can also be noticed in mollisols. However, the difference is that mollisols are neutral soils containing high amounts of Ca and crystalline clays, assisting in the accumulation of organic matter and the creation of excellent physical conditions. The clay fraction of these soils is characterized by smectite with kaolinite as an admixture. Smectite, the dominant clay mineral, does not have surfaces containing exposed Al hydroxyl groups. The planar surfaces are composed of siloxane surfaces and are mostly negatively charged due to isomorphous substitution. Consequently, smectite can only be attracted to humic acid with Ca ions acting as bridges (Figure 7.6). On the other hand, the kaolinite minerals present

Agronomic Importance of Humic Matter

contain on one side surfaces composed of Al octahedrons. However, these surfaces are usually negatively charged due to dissociation of the exposed hydroxyl groups. Hence, the interaction of kaolinite with humic acid is made possible only through the Ca-bridging mechanism. The humo-Ca-clay mineral chelates are considered the reasons for the preservation of soil organic matter (humic acids) and for the development of granular structures and other favorable physical characteristics in mollisols. Soils, containing smectite in their clay fraction, are often experiencing a common problem, manifested in formation of large cracks when dry. This is known to be caused by the high swell and shrink capacity of the expanding 2:1 type of clay minerals, such as smectite. The large cracks formed modify the soil's behavior with respect to aeration and water penetration, and may cause damage to plant roots. The effect of humic acid as discussed above by interacting with the expanding clay is noted in mollisols to reduce the extent of shrinking and swelling.

8.1.2 EFFECT ON SOIL CHEMICAL PROPERTIES

Humic matter can affect the soil chemical properties in various ways, since it can generate a variety of chemical reactions. As indicated before, the chemical behavior of humic matter is in general controlled by two functional groups: the carboxyl and phenolic-OH groups. The carboxyl groups start to dissociate their protons at pH 3.0 (Posner, 1964), and the humic molecule becomes negatively charged (Figure 7.1). At pH < 3.0, the charge is very small, or even zero. At pH 9.0, the phenolic-OH groups also dissociate their protons, and the humic molecule attains a high negative charge. The issue of these pH values in the dissociation of functional groups of humic substances in natural soils has been discussed in Chapter 7.

Since the development of the negative charge is pH dependent, this charge is called *pH-dependent charge* or *variable charge* (Tan, 1998). A number of reactions can take place because of the presence of these charges. At low pH values, the humic molecule is capable of attracting cations, and such electrostatic attraction leads to cation exchange reactions. This kind of reaction will no doubt affect the cation exchange capacity (CEC) in soils. The CEC of humic matter can be estimated from its total acidity values, which are usually very high. Humic acid shows CEC values, in terms of total acidity values, ranging from 500 to 1200 cmol/kg, whereas fulvic acid exhibits a somewhat higher range of 600 to1500 cmol/kg (Tan, 1998; 2000; Schnitzer and Khan, 1972). At high pH values, when the phenolic -OH groups are also dissociated, complex reactions and chelation become of importance (Figure 7.2). Complex reactions are considered to be a weaker bonding mechanism than chelation, due to formation of a coordinate bond with a single donor group. On the other hand, chelation is viewed to be a stronger bonding process because of formation of a chelate ring structure. Both adsorption and complex reactions can also take place by a water or metal bridging process. This was the vehicle for interaction reactions between humic matter and clay as discussed earlier.

It is assumed that the interactions with metals are going to take place first at the sites that form the strongest bonds, e.g., coordinate bonding and chelating sites. As these stronger bonding sites become saturated, attraction to the weaker sites, e.g., electrostatic bonding and water bridging sites, becomes increasingly greater (Stevenson, 1994). However, under certain conditions the assumption above is very difficult to justify. At low pH values, the only sites available are the sites for electrostatic attraction and the sites for water bridging. Neither the complexing nor the chelation sites are ready for reactions.

The complexing and chelation capacity of humic matter is considered today of utmost importance in many environmental quality issues. Depending on several factors, e.g., pH, saturation of sites, and electrolyte concentration, humic matter can form both soluble and insoluble complexes with metals, hence providing for a dual function in the soil ecosystem. In natural conditions most of the chelates may be in insoluble forms due in part to the participation of clays in the reaction process. The fulvic acid fraction is assumed to form the more soluble metal chelates because this humic substance is soluble in water to begin with, is lower in molecular weight and has higher contents of functional groups. The fulvo-metal chelates remaining soluble may serve then as carriers of trace metal elements to be transported to plant roots. On the other hand, humic acid tends to produce more insoluble metal chelates, and the humo-metal chelate is considered to

Agronomic Importance of Humic Matter

serve as a *sink* for toxic metals. Large amounts of free Al in acid soils are made chemically inactive by chelation with humic acid, preventing Al toxicity in crops and plant growth (Tan and Binger, 1986). Hence, humic acid can act as a buffer in alleviating adverse effects of heavy metals and toxic substances such as pesticides and other xenobiotics. However, depending on many factors, such as the type of metallic ion, cationic valences, cation saturation, and degree of dissociation of the humic molecule, humic acid is also capable of forming soluble metal chelates. This is considered of extreme importance by many chemists and environmentalists in the mobilization and concentration of radionuclides in the environment (Gaffney et al., 1996). The chelates, carrying the toxic compounds, can migrate long distances and may pollute the ground water or reappear at other locations to be precipitated. Nonpolar hydrophobic compounds, e.g., DDT and PCB, can be made soluble in this way, hence preventing their accumulation in soils and sediments. However, the creation of insoluble and soluble chelates by humic acids seems to generate controversial problems for the environment. In the form of humo-chelates, these xenobiotics are indeed prevented from adsorption by soil clays, but the interaction with humic acids decreases their rate of decomposition, photolysis, volatilization, and biological uptake. The latter is expected to lengthen their lifetimes or to increase their mean residence time. This is expected to also affect their transport distances in our natural ecosystem.

8.1.3 EFFECT ON THE SOIL REDOX SYSTEM

The reduction and oxidation reactions in soils, called redox reactions, are chemical processes involving electron transfer. They affect formation and accumulation of humic matter. As explained earlier, more humic matter will be formed in reduced than in oxidized environments as exemplified by formation of peats. However, it is also noted that humic matter is capable of inducing reduction and oxidation reactions, hence affecting the redox system in the environment. Humic substances are in fact important components of the soil redox systems, capable of transferring electrons (Flaig, 1988; 1972). They are considered by Ziechmann (1994) *electron donor–acceptor complexes*, with the aromatic structures, containing OH and COOH groups, functioning as electron donors, and quinonoid structures as electron acceptors. However, most of the data presented so far are based on the capacity of humic substances as electron donors with the transition metals at the higher oxidation states serving as the electron acceptors. A reaction of such an electron transfer is illustrated below in Figure 8.1, by which a divalent ion (M^{2+}) is reduced into a monovalent ion by accepting an electron from the humic acid molecule.

In reduction–oxidation chemistry, compounds, capable of donating electrons, are sometimes referred to as *electron-rich* substances embodied by substances in the reduced state, whereas their counterparts, compounds capable of accepting electrons, are called *electron-poor* substances, which are the materials in the oxidized state (Tan, 1998; Sposito, 1989). Electron-rich substances are usually char-

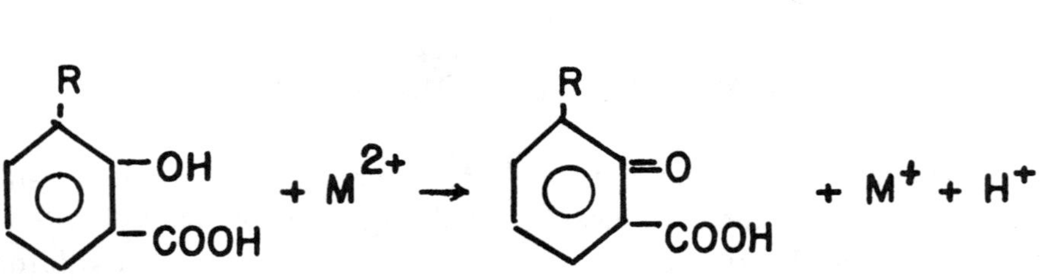

Figure 8.1 A schematic representation of a redox reaction of humic acid showing electron transfer to a metal M^{2+}.

Agronomic Importance of Humic Matter

acterized by *negative pe* values, whereas the electron-poor compounds commonly exhibit *positive pe* values. This parameter pe is often used for measuring the capacity of substances to donate or accept electrons (Tan, 1998). It is derived from a generalized redox equation as follows:

$$\text{Oxidation} + e^- \rightleftarrows \text{Reduction} \tag{8.1}$$

for which the electrochemical potential, measured against a standard hydrogen electrode, is defined as:

$$E_H = E° + \frac{RT}{nF} \ln \frac{\text{Oxidation}}{\text{Reduction}} \tag{8.2}$$

in which $E°$ = standard electrochemical potential, R = gas constant, T = absolute temperature (°Kelvin), n = valence, and F = Faraday constant. Since the reaction is a reduction-oxidation reaction, the electrochemical potential, E_H, is called a *redox potential*. In the above equations, the term oxidation is referring to substances in the oxidized states and reduction to materials in the reduced state. The oxidized materials carry higher valencies than their reduced species, and the electrons have the responsibility for balancing the equation. In the case above, the oxidized substance is one valence higher than the reduced element, which carries one lower valence due to reaction with the electron.

If electron (e^-) activity increases, the reaction shifts to the right, meaning reduction takes place. When, on the other hand, electron activity decreases (no e^- available), the reaction shifts to the left, or in other words oxidation occurs. In analogy to the concept of pH, electron activity can be represented by *pe* as illustrated below (Tan, 1998):

$$pH = -\log (H^+) \tag{8.3}$$

$$pe = -\log(e^-) \tag{8.4}$$

It should be realized that the analogy also includes the fact that neither electrons nor hydrogen ions can exist as free particles in natural conditions or in the soil solution. Both can exist only in association with the solvent or a solute species.

The parameter pe is closely related to the redox potential E_H according to the following equation (Tan, 1998):

$$E_H = 0.059\ pe \quad \text{or} \quad pe = E_H/0.059 \tag{8.5}$$

From the above, it follows that pe can also represent the redox potential, since conversion of E_H into pe (or vice versa) can be accomplished very easily by using equation (8.5). However, it should be realized that pe was not defined as a redox potential, but is by definition the *activity of electrons* (see equation 8.4). The redox potential is the *electrochemical potential* of the reduction-oxidation reaction and has been defined as reflected in equation (8.2).

Electron transfer in redox reactions is sometimes accompanied by a proton transfer as illustrated by the reaction below:

$$MnO_2 + 2e^- + 4H^+ \rightleftarrows Mn^{2+} + 2H_2O \tag{8.6}$$

This leads some scientists to believe that a redox reaction can accordingly be generalized as follows (Bartlett, 1999):

$$\text{Oxidation} + e^- + H^+ \rightleftarrows \text{Reduction} \tag{8.7}$$

The equilibrium constant k is derived by Bartlett (1999) as follows:

Agronomic Importance of Humic Matter

$\log k = \log \text{red} - \log \text{ox} - \log e^- - \log H^+$

or

$\log k = pe + pH$ (8.8)

in which log k = the log of the equilibrium constant k. Considering log k equal to pe+pH by ignoring the most important reaction parameters, log red and log ox, is very confusing and misplaced. Even more mind-boggling is the contention of the author above to consider pe + pH as the redox parameter or redox potential. This is stretching the basics of electrochemical potentials a little bit too far. According to equation (8.8), pe+pH equals the logarithm of the equilibrium constant, but an equilibrium constant is not the electrochemical potential of a redox reaction. The electrochemical potential of a redox reaction is formulated and defined differently, as can be noticed in equation (8.2)

The electrons are added in redox equations for the purpose of balancing the equations by reducing the charges of the oxidized substance and should not be applied for reaction with the H^+ ion. The H^+ ions, added in redox reactions involving oxides, must be accounted for and are often used for convenience to balance the equation by converting them into H_2O, as noted in reaction (8.6). Additional examples are presented below to illustrate more clearly the issue of incompatibility of equation (8.7) for application with ionic components:

$Fe^{3+} + e^- + H^+ \rightleftarrows Fe^{2+}$ is incorrect   (8.9)

$Fe^{3+} + e^- \rightleftarrows Fe^{2+}$ is correct   (8.10)

$Fe^{2+} + 2e^- \rightleftarrows Fe$ is correct   (8.11)

In soil chemistry, standard half-cell reactions involving electron transfer between ionic components are written in accordance with

IUPAC without the hydrogen ions (Weast, 1972). This is the process called electron transfer and the number of electrons transferred is used for the determination of equivalencies or equivalent weights.

The Role of Humic Matter as a Redox Agent

As discussed above, *electron availability* can be used as an indication of the reduction-oxidation status of the soils. It affects the oxidation and reduction states of H, C, N, O, S, Fe, Mn, Cu, Zn and many other elements, and as such controls the solubility and availability of many nutrient elements to plants. Ions in the reduced forms are usually more soluble, hence more available, to plant roots.

Oxidation-reduction reactions of metal ions can occur in soils in several ways. They can be mediated by microorganisms, induced by the presence of organic compounds such as humic acids, and by photochemical reaction. Ultraviolet radiation at the wavelength between 360 and 450 nm is believed to be able to reduce iron. The process, called *photoreduction*, can be illustrated by the following reaction (McKnight et al., 2001; David and David, 1976):

$$Fe(OH)^{2+} \xrightarrow{h\nu} Fe^{2+} + OH^{\cdot} \qquad (8.12)$$

in which hv is the energy of radiation required for the reaction. The reaction is apparently reversible, since the authors claim that the OH-radical can reoxidize Fe^{2+} back into Fe^{3+}. Insoluble Fe(III)oxides can also be reduced into Fe(II)oxides by photochemical reduction. The process is allegedly mediated by the presence of humic substances in the reduced state, which will be photo-oxidized into oxidized humic substances. The reactions are written by Sulzberger et al. (1994) and Sulzberger and Laubscher (1995) as follows:

$$Fe(III)oxide + HA_{reduced} \rightarrow Fe(II)oxide + HA_{oxidized} \qquad (8.13)$$

Agronomic Importance of Humic Matter

$$\text{Fe(II)oxide} \xrightarrow{\text{detachment}} \text{Fe}^{2+} \tag{8.14}$$

$$\text{Fe(II)oxide} + \text{oxidation} \rightarrow \text{Fe(III)oxide} + \text{reduction} \tag{8.15}$$

The release of the Fe^{2+} as an ion (equation 8.14) is called by the authors a *detachment* process. In soil chemistry we call this *dissolution*. The rate of reaction (8.14) versus that of reaction (8.15) is considered by the authors of importance for the overall *quantum yield of photoreduction*. They believe that this quantum yield, in moles of Fe^{2+} produced per photon of incident UV radiation, is higher for dissolved Fe(III) species than for Fe(III) in the oxide form. Increasing degree of crystallinity of Fe oxides is assumed to decrease the quantum yield, since amorphous iron oxides are noted to be substantially more photoreactive than the crystalline goethite or lepidocrocite minerals. The quantum yield also decreases with increasing soil pH, because Fe^{2+} is unstable and cannot exist at high soil pH. It tends to be oxidized and precipitated rapidly into $Fe(OH)_3$ in alkaline soils.

The photochemical dissolution of reduced iron is assumed by McKnight et al. (2001) to be a temporally dynamic process causing its concentration to fluctuate. These authors also claim that sorption of humic substances by Fe oxides enhances the photoreactivity of recently precipitated oxides, but that such sorption decreases the photo-reactivity of aged iron oxides. The latter seems to be in support of the effect of increasing crystallinity on photoreactivity of these iron compounds as discussed above.

This process of photoreduction is, of course, a very interesting topic. However, the use of the terms reduced-HA and oxidized-HA are very mystifying and perhaps misplaced. The presence of humic compounds in a reduced or oxidized form is subject to many arguments. Lately, several chemists seem to advance the idea about the role of humic substances in redox reactions with metals through trans-formation of humic matter from a reduced into an oxidized state. The conversion process is presumably mediated by microorganisms, and the reactions can perhaps be written as follows (Lovley et al., 1996; 1998):

Microorganism + $HA_{oxidized}$ → CO_2 + $HA_{reduced}$ (8.16)

$HA_{reduced}$ + Fe^{3+} $\xrightarrow{abiotic}$ Fe^{2+} + $HA_{oxidized}$ (8.17)

Reaction (8.17) indicates that the humic substance labeled $HA_{reduced}$ is an electron donor, and Fe^{3+} is then the electron acceptor. This reduced humic substance has, therefore, excess electrons and is then negatively charged, because by rules in soil chemistry, compounds with excess electrons are negative in charge. The substance labeled $HA_{oxidized}$ must be a neutral or noncharged humic substance, since only in this way can equation (8.17) be balanced with 2+ charges on both sides.

Though it seems to be a legitimate reaction, equation (8.17) nevertheless creates a lot of questions. Equation (8.16) indicates the production of CO_2, which means that a carbon is lost during the microbially induced electron transfer to the oxidized form of HA. The reaction as written is very misleading, since the CO_2 can be interpreted as being derived from the decomposition of the oxidized-HA compound by the microbes. That this is far from true can be seen from the following explanations gathered by the current author. The iron-reducing microorganisms in fresh water and marine environments are believed to have the ability to use humic matter as electron acceptors (Coates et al., 1998). This opinion is used by Lovley et al. (1996) for advancing their idea that in anoxic conditions humic substances can accept electrons released by microbial oxidation of organic substrates. The latter accounts then for the CO_2 in the equation, which in fact does not participate at all in the reaction as written in equation (8.16), and deleting the CO_2 will not change the balance of reaction as written. By donating the electrons to Fe^{3+}, the humic molecule is considered to be re-oxidized.

The theory above, though interesting, is contrary to the electron donor-acceptor concept presented by Ziechmann (1994) and Stevenson (1994). As indicated earlier, these authors argue that the aromatic structures containing phenolic-OH and carboxyl groups are functioning as electron donors. These functional groups are by nature protonated, hence represent reduced conditions. When these functional groups

Agronomic Importance of Humic Matter

become negatively charged, they are carrying, in other words, excess electrons. On the other hand, the quinoid or quinone units in a humic molecule are usually assumed to be phenol structures in the oxidized state (Schubert, 1965; Flaig et al., 1975; Stevenson, 1994). Humic molecules containing these functional groups represent then humic substances in the oxidized state. However, since they are also considered to be electron acceptors (Ziechmann, 1994), it is rather confusing how reaction (8.17) can take place yielding two types of electron acceptors, e.g.. Fe^{2+} and oxidized humic acid, or $HA_{oxidized}$. The reaction is then unbalanced, unless the term $HA_{oxidized}$ is referring to neutral humic substances as indicated before. However, neutral humic acids are only present when fully protonated, in other words in 'reduced' form.

In the presence of humic substances, a similar reduction process has been reported for MoO_4^{2-} into Mo^{5+} (Lakatos et al., 1977; Goodman and Cheshire, 1972; Skogerboe and Wilson, 1981). This, in addition, supports the idea that humic substances exhibit high negative pe values. Insoluble Mn-oxide (MnO_2) is noted to also be converted into soluble Mn^{2+} by marine humic acid and fulvic acid (Harvey and Boran, 1985). The conversion of insoluble Mn into soluble form is likewise assumed to be a photoreduction process and considered by the authors of high biological importance in the aquatic environment. Manganese is an essential nutrient for marine plankton. Since marine humic acid is the ultimate reducing agent, the formation of humic acid is viewed by Harvey and Boran (1985) as an ecological feedback.

8.1.4 EFFECT ON SOIL BIOLOGICAL PROPERTIES

Humic substances are considered energy rich material and play an important role in plant and microbial growth and the biochemical cycle in soils. They have been formed from phytogenic, animal and microbial substances, and as such are assumed to be relatively stable, at least more stable than carbohydrates or protein. However, since they still contain a lot of energy, the humic compounds are subject to further decomposition and will eventually be broken down by soil organisms into CO_2 and H_2O.

A multitude of interactions between humic matter and soil biota are present in the literature. The most extensive reports have probably been on its effect on plant growth and crop production. This will be the topic of the next section. Less known is the effect of humic matter on microorganisms and biochemical processes, which according to the little information available in the literature can be distinguished into an *indirect* and *direct effect*. Changes brought about by humic matter on biochemical chemical processes may have a pronounced effect on microbial development and activity, hence such an effect can be considered an indirect effect. Two examples of importance are the effect of humic matter on the carbon and nitrogen cycle.

Carbon Cycle

The carbon cycle is the perpetual movement of organic carbon from the air into the soil and back into the air. It is nature's way of cleaning the environment by recycling organic waste (Tan, 2000). In principle, the cycle starts when CO_2 gas in the atmosphere is absorbed by green plants and converted into carbohydrates by a process called photosynthesis. With leaf-fall or when plants die, the vegetative remains are subject to decomposition and mineralization processes, which return the carbon from the soil to the air as CO_2 gas. For more detailed information on decomposition and mineralization, two important biochemical processes assisting the cycle to run, reference is made to Tan (2000) and Stevenson (1986).

Humic matter plays an active role in fixation and releasing organic carbon. By fixation of part of the organic carbon in the form of humic substances, this process is conserving the organic carbon and at the same time reducing the production of CO_2. Humic acids have a carbon content of 50-57%, but most of it is relatively more resistant to microbial attack than that of carbohydrates. This carbon reserve in humic matter worldwide was assumed earlier to amount to 10^{12} metric tons, but it is believed that most of it is unavailable as a direct carbon source for many of the microorganisms in soils (Müller-Wegener, 1988). The resistance of humic matter to biological decomposition has been expressed by Stevenson (1994) in terms of *mean residence time* (MRT),

Agronomic Importance of Humic Matter

which is the age of humic matter from the time of formation to its decomposition in soils. Though MRT values seem to vary considerably from 250 to 1900 years, they, nevertheless, indicate the relative stability of humic substances to microbial attack. It should be realized that MRT does not represent the absolute age of humic substances, but reflects only the average age because of the transient nature of the compounds in soils, where 'old' humic matter is continuously decomposing while new humic material is synthesized. Decomposition of humic matter is also expected to be more rapid in aerobic than in anaerobic environments. Under well-aerated conditions where oxygen is not a limiting factor, oxidative degradation processes are more likely to occur producing CO_2 and H_2O to complete the interrupted carbon cycle. Nitrogen and sulfur compounds are released as byproducts (Stevenson, 1986). The final stage of decomposition is expected to be the breakdown of the more resistant lignoid or phenol part of the humic molecule, in which actinomycetes and fungi are considered to play a major role. In simulated oxidative degradation analyses of humic acids, fatty acids, aliphatic carboxylic acids, phenolic acids, and benzene carboxylic acids are produced (Griffith and Schnitzer, 1989). However, whether similar compounds are produced in the natural process remains a question for argument.

In poorly drained soils and soils of the wetlands, decomposition reactions of humic matter occur at a greatly reduced rate. These soils support a different population of microorganisms which produce different types of end-products, though CO_2 is included for continuation of the carbon cycle. An incomplete decomposition under anaerobic conditions generally yields fermentation products, e.g., methane, mercaptans, nitrosamines and the like, some of which are foul-smelling whereas others are believed to be carcinogenic (Stevenson, 1994). The main decomposition processes in such environments are expected to be hydrolysis and reductive cleavage. A schematic representation of formation of methane through hydrolysis by methanogenic bacteria is given below as an example:

$$(C_6H_5)\text{-COOH} + 6H_2O \rightleftharpoons 3HCOOH + CO_2 + 3CH_4 \qquad (8.12)$$
part of a humic molecule   *formic acid*   *methane*

Whether the processes in a natural environment can be simulated under laboratory conditions are again issues open to questions. Hydrolysis of humic matter in nature is an enzymatic reaction, whereas in the laboratory the reaction is catalyzed by acids or bases, which according to Parsons (1989) involves the cleavage of bonds. However, it seems that only simple and peripheral bonds are broken down (Figure 8.2). More complex bonds, such as aromatic bonds through methylene bridging, have been reported to be very resistant as noted in the analysis by Piper and Posner (1972) and Stevenson(1989) using the reductive Na amalgam method.

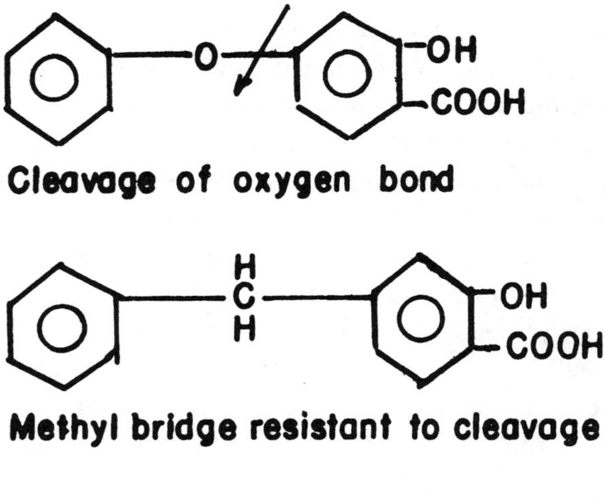

Cleavage of oxygen bond

Methyl bridge resistant to cleavage

Figure 8.2 Decomposition of humic matter by reductive cleavage of aromatic bonds using the Na amalgam procedure (Piper and Posner, 1972; Stevenson, 1989).

Agronomic Importance of Humic Matter

Summarizing, it can be stated that humic matter is an active constituent of the organic cycle in the soil ecosystem. By utilizing organic carbon for formation of humic substances, it is preserving it to the benefit of the physico-chemical condition of the soil ecosystem. It is a form of *soil carbon sequestration*, a process considered of vital importance for the environment (Lal, 2001). Though relatively stable, the stored carbon remains a formidable energy source for many microorganisms. The microbial population is often noted to thrive prolifically in soils rich in humus. By way of enzymatic decomposition and mineralization, the humic substances are eventually broken down into H_2O and CO_2, which completes the cycle.

Nitrogen Cycle

The nitrogen cycle is another indirect effect of humic matter on the biological properties of soils. Very simply defined, it is the movement of nitrogen from the atmosphere through the plants into the soil, before it is returned to the atmosphere in its original gaseous state. The cycle is composed of a sequence of biochemical reactions involving the active participation of the soil microbial population and plant life. For more details on the specific biochemical reactions reference is made to Tan (2000) and Stevenson (1986). As illustrated in Figure 8.3, the nitrogen cycle involves an *outer cycle* representing the overall cycle, and an *inner cycle* which occurs in the soils. In the inner cycle, the NO_3^- is not denitrified, but consumed by plants and soil organisms by a process called *immobilization*. The cycle is completed in the soil and it is here that humic matter plays an active role in affecting the nitrogen cycle. As a result of decomposition of plant residues a variety of nitrogenous compounds are released, e.g., amino acids, amines and peptides, part of which are used in the synthesis of humic matter, leading some authors to regard this as an immobilization process (Müller-Wegener, 1988). However, nitrogen fixation is perhaps a better term, though the mechanism is not similar to NH_4 fixation by expanding clays. It is more the incorporation of nitrogenous compounds in the humic molecule, which perhaps can be likened to N-fixation by nitrogen-fixing bacteria, with the difference that gaseous N is utilized by the microorganisms, whereas solid or ionic compounds

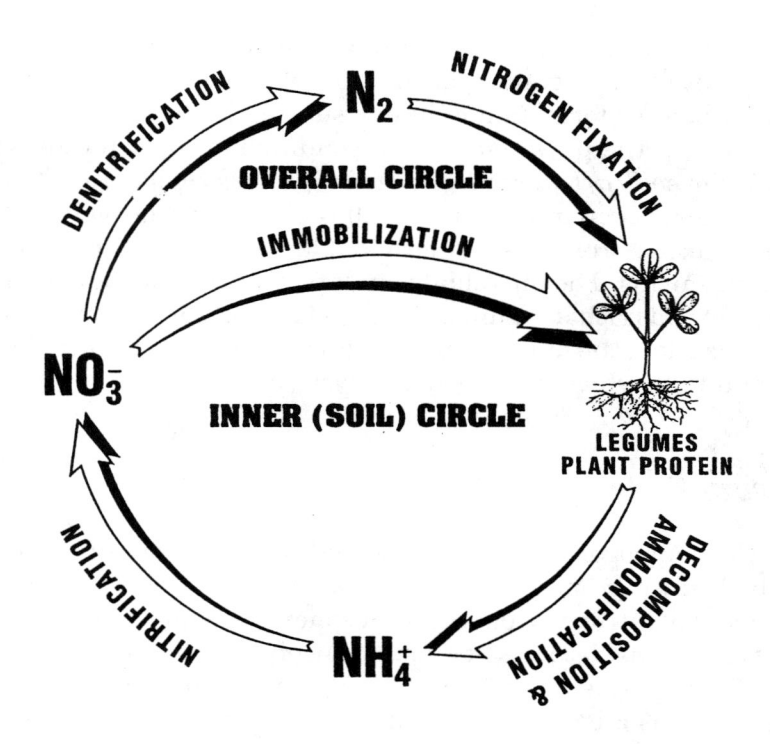

Figure 8.3 Simplified diagram of the nitrogen cycle, showing the overall and the inner cycle. Drawing by W. G. Reeves, art coordinator, University of Georgia.

are involved in the synthesis of humic substances. Because of this, the N content of the humic molecule tends to increase affecting its carbon to nitrogen ratio. Generally the C/N ratio of plant residues varies from 80:1 in wheat straw to 20:1 in legume material (Tan, 2000; Stevenson, 1994). While on the one hand part of the organic carbon is incorporated in the humic molecule, with the remainder being lost as CO_2 in the decomposition process, on the other hand N is being added in the humic structure. These processes cause the C/N ratio of the humic substances to decline to the narrowest ratio at which C and N can exist together in soils. Generally the C/N ratio of humic matter may fall to relatively stable values. A carbon content of 50-57% and a nitrogen content of 4-5%, giving a C/N ratio of 10 to 14, are characteristic for well-developed humic acids. The nitrogen stored in the humic molecule will be released again after decomposition and mineralization of the humic substances. Part of it will be used by the microorganisms, whereas the remainder will be subject to ammonification and nitrification processes in continuation of the nitrogen cycle.

Fixation of Agrochemicals

This is another indirect effect of humic matter on soil organisms. With the agricultural revolution, increasing amounts of inorganic and organic compounds are introduced in the soils as wastes. Most of them will affect the flora and fauna in the soil ecosystem, some of them beneficially, but many others harmfully. Humic matter with its huge cation exchange and chelation capacity can adsorb and detoxify a number of toxic compounds. Reduction of micronutrient toxicity, most resulting from heavy metals such as Fe, Cu, Zn and Mn, has been presented earlier by underscoring alleviation of Al toxicity in plant growth by humic acids (Tan and Binger, 1986; Ahmad and Tan, 1986). From an environmental or ecological standpoint, the chelation of heavy metals by humic substances may also reduce toxic hazards for human beings and animals. This is especially important in view of the huge production and disposal of large amounts of domestic and industrial sewage sludge, notorious for their extremely high contents of heavy metals (Tan, 2000). The adsorption and chelation of these toxic metals by humic acids represent important processes of detoxification.

Of considerable interest in this respect is also the interaction of humic substances with pesticides and their degradation products. In the aforementioned section the possibility has been raised that pesticide residues may form stable complexes with humic matter. Such an interaction may at one hand greatly increase their persistence in soils, but on the other hand may also reduce their activity. Two interactions are suggested to play major roles in the controversial effects: (1) direct linkage of the pesticide to the humic molecule, allowing the preservation of the active sites of the pesticides, and (2) synthesis of humic acid-like compounds using pesticide residues, which brings about the inactivation of the pesticides. Direct linkages occur for example when basic pesticides, such as s-triazine, react with the carbonyl groups of humic acids (Stevenson, 1994), and pesticides containing carbonyl groups react with amino groups of humic acids. Cross-coupling of xenobiotics with humic acids is another possibility. The biodegradation of pesticide residues, on the other hand, yields reactive products that can link with carbonyl, carboxyl, phenolic-OH, and amino groups of the humic molecule to form new humic acid-like substances. Such a process usually results in a deactivation of reactive sites. Because of its incorporation, building up the molecular structure of a new humic substance, the identity and behavior of the pesticide have been completely erased. The importance of such a reaction in environmental quality is without question and many express the opinion that fixation of environmentally relevant xenobiotics by humic acids will have a positive effect on plant and microbial life.

Effect on Enzyme Activity

This effect of humic acids has only recently attracted research attention. Though almost all biochemical processes are enzymatic in nature, their complex and variable structure and most often the extreme difficulties and inability to directly determine enzymes in soils cause many soil researchers to shy away from studying the issue. Enzymes are proteinaceous compounds and can be determined indirectly through their capacity to transform one compound into another. By definition, enzymes are thermolabile catalysts produced by living tissue but capable of action outside the tissue (Gortner, 1949). Hormones, such as auxin, are excluded since they conduct their physiological functions only in the living tissue. A catalyst is then a substance capable of altering the speed of reaction without appearing as part of the final product. The names of enzymes all end with *ase* and are descriptive of the type of compounds broken down. For example, *cellulase* is an enzyme that breaks cellulose into its constituent sugar components, whereas *protease* is important in the splitting of protein into amino acids, and *urease* breaks down urea, found in fertilizers and urine, into ammonia. Some of the enzymes are produced by plant cells as *constitutive enzymes*, whereas others are only produced when a susceptible substrate is present, and such enzymes are called *induced enzymes*. Cellulase is an example of an induced enzyme, while urease is an example of a constitutive enzyme. Free enzymes, enzymes not associated with the microbial biomass, are reported to be accumulated in soils by entrapment (fixation) within intermicellar spaces of expanding lattice clays (Paul and Clark, 1989), though the opinion in clay mineralogy is that most enzymes are too big to penetrate the intermicellar spaces of clay minerals. According to the purpose of this book, in the following discussion the focus will be on enzyme-humic acid interactions only. For more details on clay–enzyme interactions, reference is made to Tan (1998). Complex formation and interaction between enzymes and humic matter are better reasons for the presence of free enzymes in soil humus. In the form of humo-complexes, the enzymes are protected from physical and biochemical attack, and remain stable and active. In this way, the compound ATP (adenosine triphosphate), a very important coenzyme in all biosynthetic and catabolic cell reactions, can be present in soils, though ATP has not been isolated yet from soils. It is these free enzymes that make organic and nutrient cycling in soils possible. Much of the organic C and N entering the soil are compounds, polymeric in nature, and as large organic compounds would not be available for uptake by higher plants and microorganism, unless their molecular masses are reduced by enzymatic depolymerization. It would be an environmental disaster if the enzymes would not have survived in soils. Without the interaction with clay and humic matter, free enzymes tend to be denatured very rapidly by a host of physico-chemical properties, e.g., pH and ionic composition, or tend to serve as substrates for proteolytic microorganisms (Burns, 1986). The possibility has also been offered that the enzymes can be incorporated in the structure of the humic molecule during the synthesis of humic matter. By doing so, humic matter is reported to modify the structure and affect the active sites of the enzymes (Müller-Wegener, 1988).

Although a lot of information is now present on soil enzymes, the effect of humic acids on their activities is still inconclusive, which is attributed to the mixed results produced by the various research conducted on this aspect. In their investigations with a series of proteolytic enzymes, Ladd and Butler (1975) notice that the activities of papain, subtilopeptidase A, termolysin, and ficin were stimulated by humic acids. On the other hand, humic acid has decreased the activities of carboxypeptidase A, trypsin and pronase B. The effect of humic acids on enzyme activities seems also to vary according to plant species. Whereas invertase activity has been reported to decrease in wheat roots, the activity of this enzyme seems to be stimulated in pea (*Pisum sativum*) roots (Malcolm and Vaughan, 1979).

Effect on Organisms

Soil Organisms. – Different opinions are available in the literature on the direct effect of humic matter on the growth and activity of organisms in the environment. Present in the soil as chemically reactive colloids, humic compounds are known to interact with inorganic and organic soil components and metal ions, thus modifying soil conditions for plant growth. Burges and Latter (1960) and Prat (1960) believe that humic acid is a source of food and energy for microorganisms. This is supported by Mathur and Paul (1966), who indicate that *Pseudomonas sinuosa*, *Actinomyces* sp. and other bacteria can use humic acid as a source of C and N. Neelakantan et al. (1970) have reported similar findings with *Aspergillus* sp. and *Streptomyces* sp. Using [^{15}N] labeled humic acid, Andreyuk et al. (1973) find humic acid to be a source of N for *Bacillus megatarium*, *P. fluorescens*, *Actinomyces globisporus* and *Mycobacterium citreum*. By utilizing humic C and N, the microorganisms apparently decompose the humic acid, since Bhardway and Gaur (1972) notice absorption of humic acid by *Rhizobium* sp. and *Azotobacter* sp. cells in experiments with [^{14}C] labeled Na-humate. Decomposition of humic acid is believed to occur more rapidly by mixed than by single cultures of microorganisms (Andreyuk et al., 1973). However, Kononova (1970) seems to dispute the possibility of absorption of humic acid by microorganisms, since cleavage of humic acids in most of the work cited has not been established. McLoughlin and Kuster (1972) also report conflicting evidence by indicating that humic substances have no effect on growth and respiration of the yeast *Candida utilis*. Studies on the effect of

Agronomic Importance of Humic Matter

fulvic acids on the growth of an ectomycorrhizal fungus, *Pisolithus tinctorius*, as conducted by Tan and Nopamornbodi (1979), yield results showing evidence of definite absorption of fulvic acids. Moderate amounts of fulvic acids (640 ppm) seem to have stimulated the growth and dry weight content of the ectomycorrhiza, cultivated at pH 7.0 and 4.0 (Figure 8.4). The mycelia of colonies developed at pH 7.0 are sub-

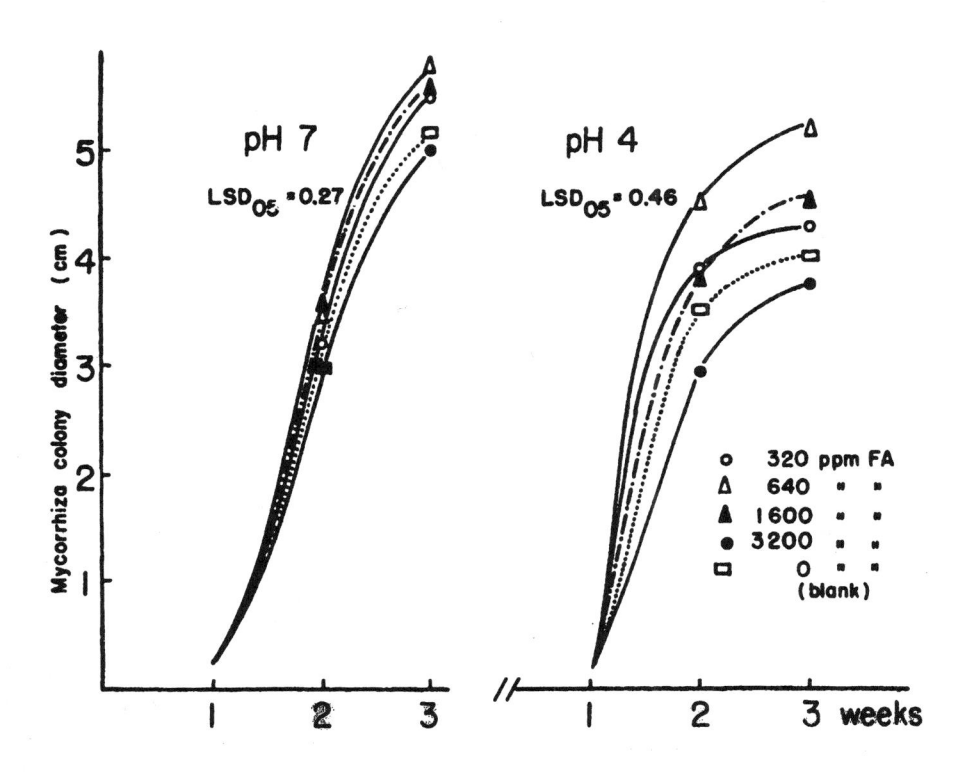

Figure 8.4 Effect of fulvic acid (pH 7.0 and 4.0) on colony growth of the ectomycorrhizal fungus, *Pisolithus tinctorius* (Tan and Nopamornbodi, 1979).

stantially darker in color than the cultures grown at pH 4.0. The possibility of fulvic acid being absorbed by the fungus, as indicated by the distinct discoloration of its mycelia, is supported also by the fact that fulvic acid has been found to be composed of polysaccharides (Tan and Clark, 1968; Clark and Tan, 1969). Fungal cells are known to contain large amounts of polysaccharides, hence it is likely that fulvic acid is considered an important food source by these organisms.

Aquatic Organisms. - In more recent reports, humic substances are also considered important for organisms living in aquatic environments, though the effect is almost the same as that discussed above for soil organisms. Some believe that humic substances cannot serve as a heterotrophic food source, but agree that aquatic microorganisms are capable of using their enzymes to attack the free carboxyl end of humic substances (Harvey and Boran, 1985). Others claim that the growth of these microorganisms in vitro increases in the presence of humic substances at concentrations of 30 mg/L in the growth media (Visser, 1985). This is supported by an earlier report, testifying increased bacterial biomass and activities due to the presence of aquatic humic matter (Stewart and Wetzel, 1982). It is assumed that the humic substances are used as an N source (Claus et al., 1997), or is utilized by a cometabolism mechanism by the aquatic microflora (de Haan, 1977). Cometabolism is a process by which the humic substance is taken up by the organism, but is not used as food or a source of energy. Apparently it is useful for cell growth and in assisting decomposition of substances by the cell. Other modes of action of humic substances are their auxin-like action, decoupling of oxidative phosphorylation and their effect on regulating membrane permeability (Flaig, 1970; Chaminade and Blanchet, 1953). A more recent finding on membrane permeability is given below to illustrate the uncertainties underlining the issue. Though not much is known for sure in this aspect, the little information available, mostly coming from histological research, reveals the likelihood of increased permeability of membranes of aquatic organisms by humic substances. Studies by Tham et al. (1994) on the population of macroinvertebrates in bog streams in Germany indicate that some taxa of the Diptera and Crustacea have been decreased in density with increasing humic matter content in the streams. The possibility is raised that some of the humic matter has penetrated the membranes of the organisms, causing harmful subcellular reactions, an assumption deduced from changes noted in the osmotic reactions of a *Trichoptera sp*. Similar subcellular reactions have also been quoted by the authors on gills of brown trout. Hartung and Allread (1994) seem to corroborate the damaging effect by noting an algicidal effect of humic acid in their research. Results of their pond tests show that peat humic acid can provide algae control at low concentration, though the data are hardly statistically significant. The authors believe that the algicidal reaction is induced by solar radiation. However, no damage by the humic acid is noticed to other forms of aquatic life, e.g., fish, frogs, turtles, snails, daphnids, and worms. The fact that many types of fish, alligators, and other aquatic organisms are thriving well in black water rivers is intriguing and more information has to be collected to definitely confirm the harmful effect of humic acid on some aquatic life.

Another aspect of great environmental significance is the complex reaction between live microorganisms and humic matter and soil clays. Because of their sheer number in soils and aquatic environments, microorganisms are bound to form complexes with clay minerals and humic substances. Numerically, the population of microorganisms is the largest among all soil organisms. Microbial cells are believed to interact with humic matter in much the same way as the interaction between clay and humic acids or humic acid and enzymes as discussed earlier. The difference is only that microorganisms, such as bacteria, are too large to be classified as soil colloids, but nevertheless they behave as colloids. The colloidal behavior is manifested in repulsion between the negatively charged bacterial cells, causing them to disperse in aqueous solutions. Formation of such suspension is a characteristic known to be exhibited only by colloids. Another contrasting difference is that bacteria are living cells, which possess the capability to grow, exhibit metabolism, and move independently. However, the larger (higher) soil flora and fauna will react differently with humic matter or clay. They are generally too large and well insulated to be affected by the charge characteristics of the humic and mineral fractions. In analogy to the preservation of organic matter in soils as a result of complex formation with clays, the interaction of microbial cells with humic matter or clays also ensures the survival and accumulation of certain groups of microorganisms, their enzymes and metabolic products. The adsorption of protein by clays and humic matter and their subsequent protection from decomposition has been known for many years and thoroughly discussed above. Hence, it is reasonable to expect that a microbe-humic complex provides some protection for the organism from physico-chemical attack by other microorganisms. Though the discussion above underscores only the interactions between humic matter and soil organisms, it does not infer that complex reactions between clay and soil organisms are second in importance. It plays an equally significant role in affecting microbial life in the soil ecosystem, and for more details on interaction reactions between clays and microorganisms, their kinetics and mechanisms, reference is made to Tan (1998).

8.2 IMPORTANCE IN PLANT GROWTH

The effect of soil organic matter on plant growth has been known for some time, but only recently has this effect been credited to humic compounds. Humic matter enters into complex reactions with a variety of organic and inorganic components of soils and will influence plant growth and crop production indirectly and directly. Indirectly, it is known to improve soil fertility by modifying physical, chemical and biological conditions in soils. These effects have been discussed in some detail in the aforementioned sections. Therefore, this section will discuss only the direct effect of humic acids. With the increased knowledge of humic acid chemistry, starting a decade ago, increasing amounts of information have been accumulated testifying to the profound influence of humic matter on plants. The reports are formidable not only because of the huge amounts of data but also because of the variety of effects allegedly exerted by humic substances. Almost any aspect of plant life seems to be affected in one way or the other, from plant nutrition and processes in plant physiology to plant growth and crop production.

8.2.1 Effect on Plant Nutrition

The opinion exists that humic acid can serve as plant food and

Agronomic Importance of Humic Matter

much of the older literature attests that small amounts of organic substances can stimulate plant growth.

Humic Matter as a Food Source

This concept finds its origin perhaps from the experiments of John Woodward, who reported late in the 17th century that plants were growing better in muddy water than in rain or river water (Tan, 2000; Brady, 1990). This led his critics to conclude that humus was taken up by the plants, a theory revived in the beginning of the 20th century by results of experiments conducted by people using compost and peat extracts while growing plants. Added in small amounts to nutrient solutions the organic extracts are believed to have a stimulating effect on the growth of duck weed (*Lemna minor* or *Lemna major*) plants (Mockeridge, 1920; Visser, 1986). These results are disputed by Clark (1924), who indicates that the added organic substances have affected the growth of microorganisms capable of producing growth-stimulating substances that cause the better growth of the duck weed. With the increased knowledge of humic acids, starting a few decades ago, more information has accumulated on the direct effect of humic matter on plant growth. Inconclusive reports suggesting at first that small degradation products of humic acid can be taken up by plants (Flaig, 1975) were later defended by claims that penetration of root membranes by humic molecules is made possible by a depolymerization or depolycondensation process (Dell'Agnola and Nardi, 1986; Burns, 1986). Roots are believed to cause a considerable reduction in pH of the soil rhizosphere by releasing a variety of organic acids. These root exudates, e.g., oxalic, fumaric, malic, citric, and succinic acids, act as depolycondensation agents on humic acids. Dell'Agnola and Nardi (1986) are of the opinion that the organic acids cause the transformation of high molecular weight humic acid into humic fractions with low molecular weights (m.w. < 5000). However, it is noted that this depolycondensation process is reversible and that the transformation is strictly dependent on the presence of organic acids and low pH values (pH 2.5). From a more recent investigation it is asserted that root exudates from corn plants (*Zea mays* L.) are capable of converting insoluble humic acids into soluble forms (Nardi et al., 1997). Though Nardi and coworkers didn't elaborate, the present author believes that the increased solubility of humic acid has contributed to its rapid depolymerization. In the acidic environment, the soluble humic matter is an easy target for hydrolysis in breaking down the humic molecule. The contention that small humic fractions can serve as food sources for microorganisms seems to be supported by earlier results. In their experiments growing an ectomycorrhizal fungus (*Pisolithus tinctorius*) with fulvic acid, Tan and Nopamornbodi (1979) observed discoloration of fungal mycelia. Mycelia turning yellow to dark brown demonstrate absorption of the fulvic compound. Known to be the smallest humic fraction, fulvic acid is composed mostly of polysaccharides (Tan and Clark, 1968; Clark and Tan, 1969), a very valuable food source for microorganisms. The ecological significance of this result is enormous. Fulvic acid is expected to be present in soils in amounts larger than sugars. Therefore, it is more readily available than the sugars. Microorganisms that can use fulvic acid as food or as an energy source are less susceptible to fierce competition than those which can only use sugars.

More and more people today seem to accept the idea that these small humic fractions can be taken up by plants. Higher plants are in this respect compared by some as facultative heterotrophic organisms capable of consuming small fragments of humic acid (Popov and Chertov, 1997). The lower molecular weight humic fractions are believed to be active in cell differentiation and in stimulation of enzyme activities.

Humic-N as Plant Nutrient

Humic substances are known to contain appreciable amounts of N. The N content of humic matter is reported to be in the range of 2 –5%, with humic acids generally exhibiting N contents in the higher range and fulvic acids in the lower range (Tan, 1998; Cranwell and Haworth, 1975; Schnitzer and Khan, 1972). Most of this N is assumed to be present as amino acid N, acid insoluble N, and hydrolyzable unknown N (HUN). Lower amounts are available in the form of NH_3-N and amino sugars. Though Stevenson (1994) assumes that one half of the total N in humic matter can be accounted for as amino acid, Rossell et al. (1978) indicate that the bulk of N is in the acid insoluble-N

Agronomic Importance of Humic Matter

fraction. In view of the above, it is expected that most of the N is relatively stable and available only with difficulty for growing plants. However, most scientists agree that after decomposition of the humic molecule, the fixed N can be released and made available to growing plants. Since in laboratory analyses acid hydrolysis is the process by which amino acid N is liberated, hence it can be expected that a similar hydrolysis reaction is playing a major role in natural conditions. Though one theory assumes that the N is released as N_2 gas, the more common concept is that it is released in the NH_3 form with the assistance of heterotrophic microorganisms. Production of N_2 gas would have completed the N-cycle, but at this stage this does not conform with the N-cycle. On the other hand, the production of NH_3 concurs with the flow of the biochemical reactions by allowing this stage to be followed by the nitrification process in continuation of the N cycle (see Figure 2.6). The possibility of such a decomposition of humic-N and its subsequent utilization by microorganisms have recently attracted some investigations by a team of German and US scientists, using aquatic humic acid collected from the St. Mary's river in Georgia, USA. The results of their laboratory experiments seem to indicate that the autochthonous microflora of the river water were capable of utilizing humic-N as a source of N (Claus et al., 1997), as mentioned briefly in an earlier section above on the effect of humic matter on soil organisms.

In addition to being a food source, humic acid has also been reported to increase N-uptake by plants and to enhance plants' capability of utilizing N in nitrogen deficient soils. The increased uptake is believed to have in turn a stimulating effect on the synthesis of N-compounds in plants (Chaminade, 1958; 1966).

Humic-Phosphate Nutrition

Humic matter is not considered to carry P in its molecule in adequate amounts for plant growth, unless one is arguing about biologically fixed phosphates, which will be discussed at the end of this section. However, it is known that humic matter can affect the solubility of insoluble P-compounds in soils. Its chelation capacity is a major force for decomposing rock phosphates and other insoluble forms of phosphates in the soil, e.g., $AlPO_4$ and $FePO_4$. The strong affinity of

Al and Fe ions in aqueous solutions for phosphate ions rapidly converts soluble P fertilizers into insoluble $AlPO_4$ and $FePO_4$ by a process called phosphate fixation. This process is especially important in acid soils where the Al and Fe concentrations are generally very high. The presence of humic matter is believed to increase the solubility of the metal phosphates by reducing the activity of Fe and Al ions by complexation (Del Re et al.,1978). Sinha (1971; 1972) also found that fulvic acid increased the solubility of insoluble phosphates because of complexation of the metal ions by fulvic acid. Such an interaction results in a release of part of the phosphate as free ions, which can be illustrated by the following reactions:

$$Al(OH)_3 + H_2PO_4^- \rightarrow HO-\underset{\underset{OH}{|}}{Al}-H_2PO_4 + OH^- \qquad (8.13)$$

$$HA + Al(OH)_2 H_2PO_4 \rightarrow HA-Al(OH)_2 + H_2PO_4^- \qquad (8.14)$$

This high affinity of humic substances for Al, Fe, and Ca is a major reason why they compete for these elements with phosphate ions. Lobartini et al. (1994) have confirmed the enhanced dissolution of the phosphate mineral due to interaction with humic acid, whereas in a more recent investigation Lobartini et al. (1998) underscore the role of humic acid as a powerful chelator of Al and Fe. The dissolution of $AlPO_4$ and $FePO_4$ by humic acids yields mostly free orthophosphate ions, as detected by their ^{31}P NMR (nuclear magnetic resonance) analyses. Corn plants grown by Lobartini and coworkers in hydroponics, with $AlPO_4$ and $FePO_4$ as the sole source of P, exhibit with the addition of humic acids a similar, if not better, performance than those grown with KH_2PO_4 as the P source. Indications are also present that some of the phosphate is possibly converted into a fulvo-metal-P complex compound. The latter, called coadsorption or metal bridging, has been reported earlier by Weir and Soper (1963), who believe that humo-Fe complexes, in addition to humic matter only, can also attract P yielding humo-Fe-P complexes. The interaction is a metal bridging process and can be illustrated as follows:

Agronomic Importance of Humic Matter

$$HA^- + Fe(OH)_2H_2PO_4 \rightarrow HA-Fe(OH)-H_2PO_4 + OH^- \qquad (8.15)$$

Such an interaction is considered a form of biological fixation of phosphates. Another form of biological fixation of phosphates is the immobilization of phosphates into the cellular material of microorganisms. Although biologically fixed phosphate is biologically stable, it is more soluble than the fixed Al- and Fe-phosphates. Because organic matter is always subject to decomposition and mineralization, the phosphates fixed biologically can be released more easily than their inorganically fixed counterparts. Many of the organically bound phosphates also exist in soluble forms, hence will play a vital role in replenishing inorganic-P in the soil solution. For a more detailed discussion of biological fixation of phosphates, their products and decomposition, reference is made to Tan (1998) and Stevenson (1994).

Humic Macro- and Micronutrient Relations

It should be emphasized perhaps that humic matter cannot be considered as a fertilizer, since it does not contain any of the nutrients as fertilizers do. The plant nutrients are carried only by virtue of its large cation exchange and chelation capacity. In many acid soils, the exchange and chelation sites may be saturated with H^+ ions, and only when these soils are fertilized can these sites be saturated with macro- and micronutrients. Hence the exchange sites serve as storage for large quantities of available nutrients for plant and microbial growth. Plant roots can obtain the adsorbed cations by cation exchange, using H^+ ions, produced as by-products of root respiration, as exchange materials. This exchange phenomenon between humic matter and plant roots is also very important for the salt balance of the soil ecosystem. High salt concentrations, usually toxic to growing plants, can be alleviated by adsorption by humic acid. It is also believed that this exchange facilitated by humic acid allows plants to grow within a wider pH range (Badura, 1965).

Though by rules of soil chemistry, chelated metals are more strongly bonded than those on exchange sites, depending on pH the chelated metals can also be made available to plants by exchange. From their investigations, Lindsay (1974) and Lindsay and Norvell (1969) note that at pH 7.0, Ca^{2+} ions can release Zn and Fe from Zn-EDTA and Fe-EDTA complexes. At low pH, Fe^{3+} ions can displace Zn^{2+} from Zn-EDTA complexes. Since most of the humo-metal chelates remain soluble, they will move with the flow of soil water, hence provide the carrier mechanism for transport to the soil-root interface. The metal chelation properties of humic matter find application today in the fertilizers industry in Europe, which will be discussed more in detail in the section on industrial importance in Chapter 9.

8.2.2 Effect on Plant Physiology

A number of reports indicate that humic matter has in general a stimulating effect on plant respiration and photosynthesis.

Plant Respiration

A significant increase in respiration of barley plants (*Hordeum sativum*) grown in nutrient solutions due to the application of humic substances has been reported by Khristeva (1953). The consumption of O_2 by these plants is noticed to double within a 60 minute period. According to Sladky and Tichy (1959), spraying begonia plants (*Begonia semperflorens*) with humic acid has also affected plant respiration, as displayed by a release of 17 to 30% more CO_2. Other plants, whose respiration processes have been observed to be stimulated one way or the other by humic substances, are tomatoes (*Solanum lycopersicum*), corn (*Zea mays*), wheat (*Triticum vulgare*), and gourd (*Cucurbita maxima*) (Sladky, 1959; Smidová (1962). These increases in endogenous respiration are noticed not to be accompanied by concurrent increases in rate of glycolysis, the breakdown of carbohydrates. While on the one hand respiration processes are apparently increased, and on the other hand, humic substances are reducing glycolysis, the two processes raise concerns for producing an imbalance in metabolite concentrations in plant tissue (Prát, 1970; Visser, 1986).

Photosynthesis

The effect of humic substances on photosynthesis has also attracted some attention in soil science. Sladky (1965) claims that sugar beets (*Beta vulgaris*) exhibit a 22% increase in adsorption of CO_2 as a result of application of humic acid sprays at 300 mg/L. Other people testify that humic matter has increased the chlorophyll content of several plants, e.g., water flaxseed (*Spirodela polyrhiza*), sugar beet, and tomatoes (Visser, 1986). Whether an increase in chlorophyll brings about an increase in photosythesis is still a matter of conjecture. The argument can be presented that the stimulating effect on chlorophyll content of green plants can be used to control chlorosis and hence improve photosynthesis.

Metabolite Content

Humic acid is reported to affect the carbohydrate content in plants. The free sugar content in plant tissue is found to increase in general by humic acid application. This effect of humic acid is believed to be different for the different plant parts. Lower increases have been observed in stems and leaves than in roots, which Činčerová (1962) attributes to the substantial increases in growth of the above-ground vegetative part as a result of humic acid application. The accumulation of sugars may affect other important physiological functions of the plants. Raising the cellular sugar content generally increases the osmotic pressure inside plant cells. Flaig and Saalbach (1959) believe that this high osmotic pressure is the reason for the resistance to wilting of plants grown in the presence of humic matter. Some reports also indicate a beneficial effect of humic acid on alkaloid content of plants. Limited investigations with tobacco (*Nicotiana tabacum*) and several medicinal plants show a stimulating effect of alkaloid contents in the leaves due to humic acid application (Aitken et al., 1964; Tolpa, 1976).

Membrane and Protoplasm

The increased uptake of nutrients by plant roots due to humic acid applications has led many authors to believe that humic matter can increase membrane permeability (Khristeva, 1953; Chaminade, 1956; Lee and Bartlett, 1976). It is believed that by affecting the electrogenic proton pump, the permeability of the plasmalemna is increased (Zientara, 1983). Increased transport of glucose has been observed across the membranes of onion (*Allium cepa*), sugar beets, and sunflower (*Helianthus annuus*) cells due to humic acid applications (Blagowestschenski and Prosorowskaja, 1935). However, used at high concentrations (1500 mg/L), humic acid can affect cell membrane damage (Vaughan and MacDonald, 1976) and increased hydration of protoplasma colloids. Otherwise, humic substances used at low concentrations (10 mg/L) are noticed to increase plasma flow in cell tissues, indicating a reduction of cell protoplasm viscosity.

The results of a more recent investigation provides additional support for the effect of humic acid on increasing cellular plant membrane permeability. Irintoto et al. (1993) notice callus cultures of slash pine (*Pinus elliottii* Engelm) in a Brown and Lawrence medium to increase in weight in the presence of humic acids over the control (0 humic acid). The elemental composition of single callus cells, determined by energy dispersive analysis by x-rays (EDAX), indicates also a substantial increase over the control in P, K, Ca and S contents due to treatments with 160 µg HA/mL (Figure 8.5). The authors above consider this to be attributed to the effect of humic acid in increasing permeability of callus membranes. The increased uptake and consequently greater accumulation of nutrients in callus cells are then the reasons for callus weight increases due to humic acids.

Plant Growth

Many of the investigations with humic acids are limited to studying seed germination, shoot growth, and elongation of very young seedlings or elongation of excised roots in vitro (Tan and Nopamornbodi, 1979; Vaughan, 1974; Poapst and Schnitzer, 1971). The results show invariably an effect similar to a hormonal growth effect. On the other hand, work done on nutrient uptake by Guminski and coworkers in Eastern Europe (Guminski et al. , 1977: Guminski,

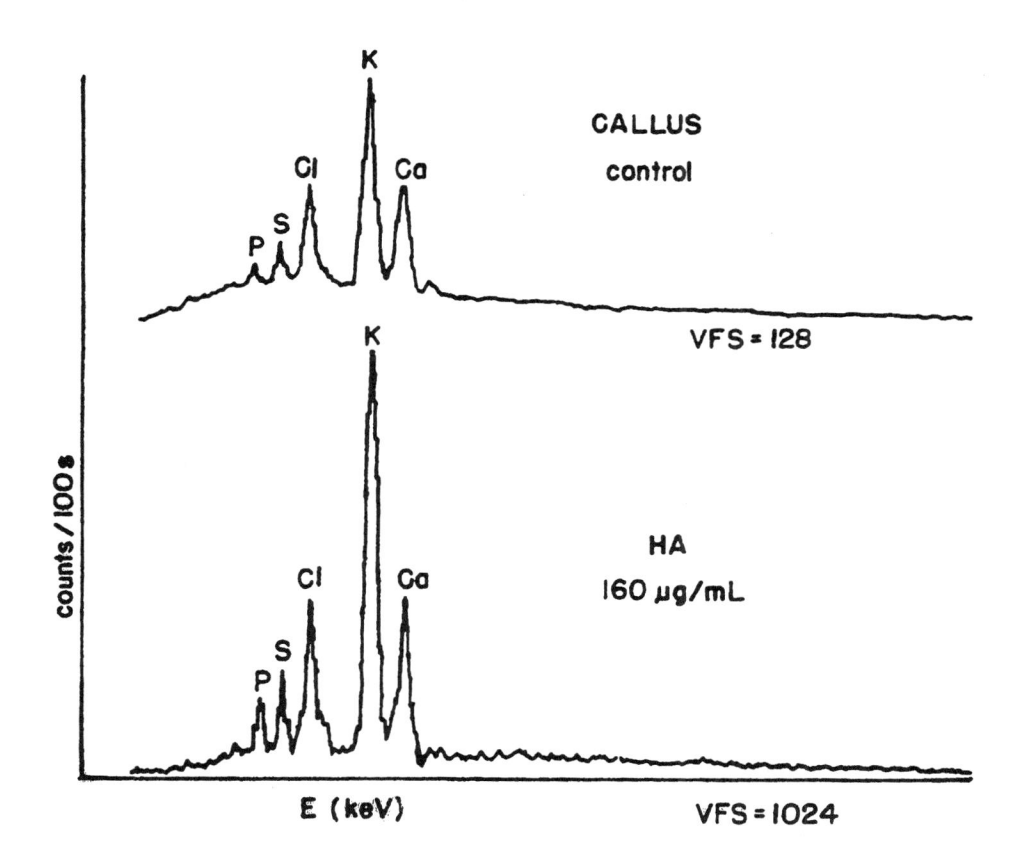

Figure 8.5 Elemental composition of single callus cells of slash pine (*Pinus elliottii* Engelm) as determined by energy dispersive analyis by x-rays (EDAX).

1957) and by Dormaar (1975) in Canada reveals a physiological influence of humic acids on plant growth. From studies on the growth and nutrient uptake of corn plants, Tan and Nopamornbodi (1979) came to the conclusion that humic acid affected plant growth through

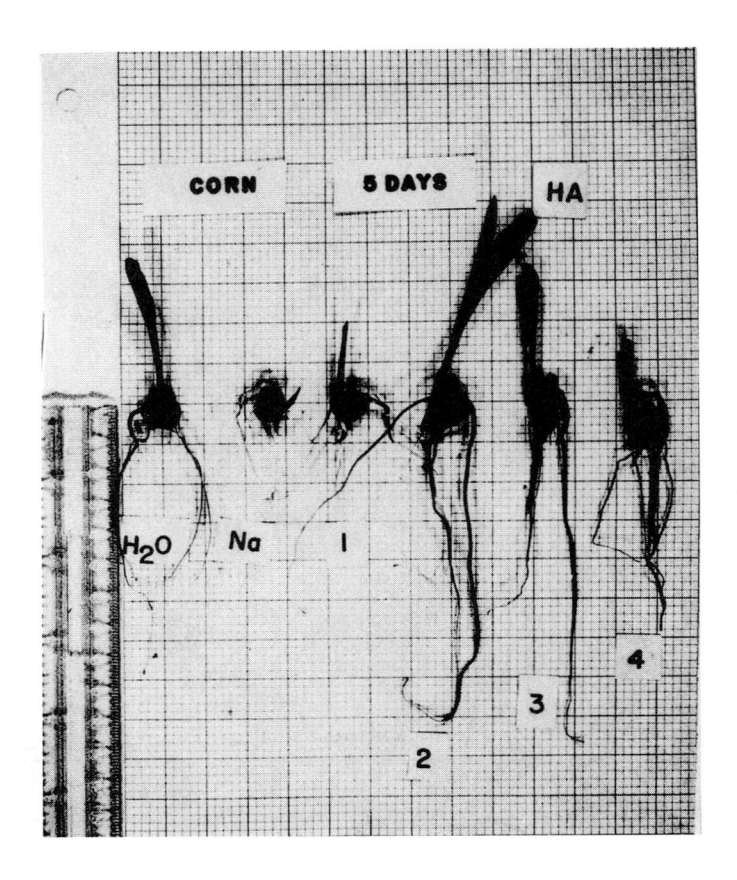

Figure 8.6 Effect of humic acid on germination and growth of 5-days corn (Zea mays L.) seedlings grown in a Hoagland solution to which were added: (H_2O) 0 ppm HA, (Na) blank + 0.66 meq NaOH, (1) 320 ppm HA, (2) 600 ppm HA, (3) 1600 ppm HA, or (4) 3200 ppm HA, respectively.

a combination of the aforementioned processes. Moderate amounts of humic acids were generally beneficial for root and shoot growth of the corn plants (Figure 8.6). At the same time, a significant increase in N content of shoots of the corn seedling was obtained. Dry matter

Agronomic Importance of Humic Matter

production also appeared to be stimulated. Data presented by Visser (1986) seem to support the idea that humic acid can affect beneficially the N contents of plants. In the presence of humic acids, uptake of N by plants is increased significantly, and plants are noted to be able to take up amounts of N at concentrations normally considered toxic to plant growth. Flaig (1953; 1956) and Kononova (1956) indicate that roots are affected more by humic acids than the above-ground vegetative part. Others believe that the effect of humic acids depends on plant species. Based on the effectiveness of humic acids, the following distinction of plants into four groups has been suggested (Khristeva, 1953: Khristeva and Manoilova, 1950):

1. Carbohydrate-rich plants, e.g., potato, sugar beets, carrot, and tomato, show a highly positive response to the application of humic acid.

2. Cereals, e.g., barley, oats, wheat, rice, and corn, show a moderate reaction to application of humic acid.

3. Protein-rich plants, e.g., green beans, peas, and lentils, show a minimal reaction to humic acid applications.

4. Oil-producing plants, e.g., castor-oil plants, cotton, linseed and sunflower, are not affected or will be negatively affected by humic acid.

Whether such a division can be upheld is a matter of conjecture, though the results of a follow-up investigation reveal no effect of humic acid on oil-producing plants, whereas only a small effect is noted in legumes (Chaminade, 1958; 1966).

CHAPTER 9

ENVIRONMENTAL AND INDUSTRIAL IMPORTANCE OF HUMIC MATTER

9.1 IMPORTANCE IN THE ENVIRONMENT

Because of its unique electrochemical and ion exchange behavior, humic matter plays a significant role in the environment. Its function in soils and the biochemical cycles is so vital that many, if not all, biochemical activities will come to a standstill in the absence of humic matter. Degradation of soils starts with the destruction of humic matter, bringing with it all kinds of environmental disasters.

In preceding chapters, humic matter has been discussed as a participant in the carbon and nitrogen cycles, two important cycles for maintaining a healthy environment in the soil's ecosystem. Fixation of organic carbon and proteinaceous compounds in the form of humic matter functions as part of the cycles, enriching soils with much needed organic matter and nitrogen for survival of soil organisms. By later decomposition of humic matter, the fixed carbon and nitrogen are released again, enabling continuation of the respective cycles. Without the synthesis of humic matter, most of the carbon will be lost rapidly

Environmental and Industrial Importance

in the atmosphere in the form of CO_2 gas, leaving the soil void of organic matter. The ensuing disastrous effects of soil degradation can be prevented through management practices aimed at accumulation and maintenance of sufficient amounts of organic matter in soils. For more details reference is made to Tan (2000).

9.1.1 Preservation of Soil Organic Matter

In preceding chapters, it has been pointed out that soil organic matter is perhaps the most important soil component that has attracted a lot of attention since the early days of agriculture. Regardless of opinions to the contrary, it is difficult to dispute the fact that the growth and yield of crops are always better in soils rich in organic matter. This is the reason why soils rich in organic matter, such as the mollisols, can support huge production of corn, wheat, and soybean. Because of this, the regions with mollisols in the United States were given the name *corn belt* in the past, whereas in Russia, the regions occupied by these soils are dubbed as the *wheat belt* of Europe. Large areas of mollisols are also known today to occur in Canada, Argentina, the loessic and steppe regions of China, and Africa. These soils are characterized by thick, black A horizons, containing high amounts of soil organic matter. Under natural conditions, a C_{org} content of 6% has often been reported in the past, which translates into 10 to 12% soil organic matter. Especially tall grass vegetation and semihumid climatic conditions, ensuring an annual turnover of abundant amounts of organic matter and a small degree of leaching, are important factors for the accumulation of organic matter. The humic substances formed are saturated with Ca^{2+} and/or Mg^{2+} ions. These ions are derived from the decomposition of calcareous and loessic parent materials and because of the limited leaching process are accumulating in the soil solution. The Ca-humates can be precipitated as such, or are immobilized after interaction with smectites or montmorillonites through Ca bridging mechanisms (see Chapter 7 and Figure 7.6). In the form of humo-Ca-smectite complexes, preservation of organic matter (humic acids) is ensured and maintained.

Another outstanding example of soils rich in organic matter is the andosols, soils derived from volcanic ash. These soils are also black in color because of high organic matter contents. In contrast to mollisols, andosols are acid soils rich in Al and Fe and amorphous clay minerals, e.g., allophane and imogolite. The general consensus is that these clay minerals together with amorphous silica play a major role in the accumulation of humus, producing the intense black A horizon and the development of physical and chemical properties unique to andosols. Exposed groups of Al and Si on the surfaces of the amorphous minerals are capable of interacting with humic acids, forming humo-Al-allophane or humo-Si-imogolite complexes or chelates (see Figure 7.6). These chelates increase the resistance of humic acids against chemical decomposition and microbial attack. The process is considered one reason for the high organic matter content in andosols.

Soil Nitrogen and Humic Matter Content

The most striking effect of deforestation is the decrease of soil organic matter content, which has an impact on nutrient cycling. Then the process of degradation of soils starts to increase, though it can be slowed down somewhat by the use of proper management practices. Such a system has been known in the tropics for decades under the name *plantation agriculture*, and recently a variation of this method has been introduced in the temperate regions under the name *agroforestry* (Tan, 2000).

A natural rain forest ecosystem has a huge standing biomass and large amounts of nutrients are contained in the plant biomass. These nutrients are usually released upon decomposition of the litter. A dense vegetative cover will add substantially to the organic matter content in soils, which is the main, if not the sole, source of soil nitrogen. It is now an established fact that soil nitrogen content increases with increased organic matter content. A positive correlation exists between organic carbon and total nitrogen contents, which usually takes the form of a linear regression as follows (Tan and Troth, 1981):

$$N = a + bC \tag{8.16}$$

where N = % total N, C = % C_{org}, a = intercept, and b = regression coefficient. The equation above indicates that total nitrogen content increases linearly as organic carbon increases in the soils. As discussed in earlier sections, humic acids contain 4-5% N, hence are expected to be important contributors of soil-N. Therefore, destruction of soil organic matter and especially the humus content may have serious implications for soil nitrogen contents.

9.1.2 Mobilization and Immobilization of Elements

Reduction and oxidation reactions play a significant role in the mobilization and immobilization of soil elements, contributing to soil formation, soil fertility, and decomposition of xenobiotics. In a preceding section, it was discussed that humic matter was capable of donating-accepting electrons, and in doing so, can change the redox status of soil elements, many of which are essential as plant nutrients. The capacity of humic matter in transferring metal ions from their oxidation into the reduced states assists in making them more soluble, hence more available for plant uptake. The mobilization of Al, Fe, and Mn due to changing their redox status is a very important process in soil genesis. Movement of Al, Fe and Mn in the form of humo-chelates from the A to B horizons allows for horizon differentiation in the pedon. The transportation and subsequent deposition as nodules and ores in wetlands and marine strata are additional examples. All of these are vital environmental processes enabling the formation of spodosols, paddy soils, and valuable mineral deposits.

This redox behavior of humic matter is also believed capable of affecting photochemical reactions in the soil environment, which contributes to a photochemical breakdown of a number of xenobiotics. This is a detoxification reaction which will be discussed in more detail in the following section below.

Other important reactions affecting mobilization and immobilization of cations are complex reaction and chelation. The dissolution of many of the metals, e.g., Al and Fe, is usually increased by complex reaction and/or chelation with humic substances. At relatively low 'loads' of cations, leaving the complexing sites still unsaturated, the humo-metal complexes are soluble. These complexes or chelates retain their negative charges, though the latter decrease proportionally by increased complexation of cations, until at saturation all negative charges of the humic molecule become zero. At this point, the humo-metal complexes and/or chelates are precipitated out of the soil solution; in other words, the metals are immobilized.

The processes of mobilization and immobilization of metal ions through formation of chelates and complex compounds have an important bearing on soil genesis and soil fertility. They are considered of utmost importance in the development of specific horizons characterizing the pedon. Current concepts on formation of spodic horizons in spodosols are based on the concept of mobilization and immobilization of Al and Fe. As discussed above, stability and mobility of Al- and/or Fe-chelates depend on the saturation of the reaction sites with the metal ions. If low amounts of Al and/or Fe are available for complex formation, complexes or chelates will be formed in the A horizon with low metal/HA ratios. In this case the amount of Al and/or Fe chelated is insufficient to cause immobilization of the metal-HA compound, which may then move down the pedon with the percolating water (De Coninck, 1980). During the downward migration, the metal chelates may pick up more of the polyvalent cations, resulting in a progressive decrease of the negative charge. The presence of a higher metal ion concentration in the subsoil, or an acidity different from that in the A horizon, may eventually neutralize the remaining charges. The consequent precipitation of the metal-HA chelate gives rise to the development of a spodic horizon.

On the other hand, when sufficient amounts of Al and/or Fe, or other metal ions can be released by mineral weathering so that metal-HA chelates can be formed with high metal/HA ratios, the chelates are then immediately immobilized in the A horizon before they can migrate down the pedon. The formation of mollic epipedons is an example of the immobilization of Ca-HA chelates attributed to saturation of the reaction sites of the humic molecule with Ca ions. Another example is the formation of andic horizons, characterizing andosols, due to immobilization of Al-HA and allophane-Al-HA complexes (Wada and Aomine, 1973).

The role of mobilization and immobilization of metal chelates in soil fertility can be illustrated as follows. Because of chelation and/or complex formation, the metal ions, normally insoluble at the prevailing soil pH, can remain soluble and may then move by diffusion and mass flow to plant roots. Many of the chelated cations are essentially micronutrients, and the chelates are considered to provide the carrier mechanism by which depleted micronutrients in the root zone are replenished (Lindsay, 1974). The chelated nutrients can be made available to plant roots by exchange. This exchange is quite different from the classical or normal exchange as a result of coulombic attraction in a double layer. As discussed in an earlier section, the chelated metal cannot be replaced by for instance K^+ or Na^+, which are not capable of occupying the position of a Cu^{2+} ion or any other transition metal in a chelate. A chelated metal can only be exchanged by another transition metal. Depending on pH, Lindsay (1974) indicates that Fe^{2+} in chelate form can be replaced with relative ease by Zn^{2+} and vice versa.

9.1.3 Biological Detoxification

The exchange and chelation capacity of humic matter provide for a huge buffer capacity in soils and the environment. With the years increasing amounts of a variety of agrochemical and industrial waste find their way into the soils. Only a few of these xenobiotics are nonharmful to the fauna and flora in our environment. Many of the harmful materials are adsorbed or chelated by humic matter and their activities either reduced or canceled. As discussed earlier, the heavy metals, Al, Fe, Cu, Zn, Mn, Co, Hg, and Cd, originating from sewage sludge disposed of in soils, are deactivated due to chelation by humic acids. This role of humic matter is also noted to be very important for detoxification of radioactive elements in relation to nuclear waste disposal. The impact of humic matter is realized in this case as a confinement or as an enhancement of migration of the radioactive elements in the environment (Moulin et al., 1996). However, other scientists are of the opinion that humic substances do indeed have a reducing effect on toxicity and bioaccumulation of heavy metals and organic chemicals. The results of biotesting experiments by Perminova et al. (1997) on the photosynthetic activity of green algae (*Chlorella vulgaris*) show a detoxification effect by humic acids on Cd, Cu, and Pb. A variation in mode of detoxification has also been noticed in that marine humic substances have displayed weaker effects than their soil counterparts. Differences in structural composition are assumed to be the reason for this, since marine humic matter is considered to be more aliphatic in nature which allows for a smaller chelation capacity than the highly aromatic humic substances in soils and peats. Hence, concerns have been raised by the authors that marine environments can be more easily affected by toxic contamination with heavy metals than the soil ecosystem.

Reducing the activity of hazardous materials can apparently also be achieved by humic matter by decreasing bioavailability of the contaminants. Possibilities to this effect have been presented recently by Perminova et al. (1997), who show humic matter restricting bio-availability of polyaromatic carbon compounds to aquatic organisms. Algae, treated with phenanthrine (PHEN) and fluorenthene (FL), have been tested as a food source for the crustacea *Daphnia magna*. In the determination of the feeding activity of the organism, a substantial decrease in algae cells has been observed with the application of humic matter, interpreted by the authors above as an indication for increased consumption of the algae by *Daphnia*. The authors have associated this with the formation of humo-PHEN and humo-FL complexes, decreasing in this way bioavailability of the toxic organics. A decrease in phytotoxicity of the pesticides atrazine and glean to plants has also been obtained and ascribed similarly by the authors above as a decrease in pesticide bioavailability for plant attack.

As discussed earlier, a pesticide residue may also be detoxified by incorporation as an integral part of the humic molecule. However, it should be emphasized here that the use of organic waste in the synthesis of new humic-like substances is not limited to pesticides only, but includes also a variety of other industrial organic wastes. For example, aromatic amines from production of dyes and pharmaceuticals may be released in the environment due to incomplete

Environmental and Industrial Importance

treatments of industrial wastewater. Detoxification of these harmful organics has also been assumed to occur by their incorporation in the humic molecule (Thorn et al., 1996).

New information from more recent studies on biodetoxification of xenobiotics seems to suggest that humic substances of different origin may have different impacts. The effect of humic substances isolated from different soils is noted by Perminova et al. (2001) to vary greatly on detoxification of atrazine and trifluralin. Humic acid from chernozems (mollisols) is reported to be the best in decreasing toxicity of the two herbicides, whereas the least detoxifying effect has been observed for humic acids from podzols (spodosols). As indicated earlier in this book, a number of scientists have suspected the presence of such differences in humic compounds. Kononova (1966) is for example one of the scientists who has hinted at possible differences between humic acids from peat (anaerobic environment) and normal (aerobic) soils. No specific data, other than conjecture, can be presented up till now from research or the literature to support this contention. However, the data from Perminova and coworkers above provide few but concrete leads for confirming the presence of differences in humic acids from different origin. Nevertheless, more information is required to firmly establish that different soils may contain humic acids different in chemistry and behavior.

9.2 Degradation of the Soil Ecosystem

Soil degradation is not just an agricultural issue but an environmental issue as well. It can be distinguished into (1) physical soil degradation, (2) chemical soil degradation, and (3) biological soil degradation. For more details on each of these processes, reference is made to Tan (2000).

The processes of soil degradation are in fact natural processes and all soils will be affected by them. In nature, they are very slow processes, and can perhaps be considered as a process of aging or growing old in terms of animal, plant and human life. However, because of human interference the degradation process in soils is suddenly accelerated, which is often reflected by a process which we call *accelerated erosion*. Accelerated and natural erosion can take different forms, but they have generally the same results, that is leaching and carving the landscape of the earth. The only difference is that it will take natural erosion millions of years to carve the landscape of the earth, whereas accelerated erosion is able to do it, if not within several years, within a person's life span.

The accelerated erosion process starts perhaps as soon as the soil is exposed, because of the rapid destruction of soil organic matter. The exposed soil is subject to attack by rain and wind. In the absence of soil organic matter, the impact of rain drops destroys soil aggregates and disperses clay particles, which clog the soil pore spaces. Runoff, created during heavy and light rains, may carry the individual loose sand, silt, and clay particles and the latter will choke rivers and lakes. In turn, this process reduces the storage capacity of those lakes and rivers. Under dry conditions, the wind will blow dust particles in the air, also reducing air quality. When the surface soil has been stripped off by accelerated erosion, the exposed subsoil, especially in the tropics, which are often rich in iron, may harden and form an iron pan. This pan formation inhibits further vegetative growth.

The notion exists that the processes discussed above can be prevented by encouraging interaction to occur between soil inorganic and organic particles, thereby forming stable soil aggregates that are less susceptible to physical, chemical, biological and weather attack. *Soil aggregation*, sometimes called *granulation*, is often associated with formation of soil structures. Though, the process is strongly influenced by environmental and especially biological factors, good stable soil structures are noticed only to be formed in the presence of large amounts of soil organic matter. For instance, the nice stable crumb structures in mollisols are formed because of the abundance of humic substances, creating the clay-humic acid chelates by Ca-bridging. The granular structures in andosols are additional examples of stable aggregates formed by Al-bridging of clay and humic acid particles. Increased aggregation will in turn increase pore spaces and decrease bulk density, factors beneficial for plant growth. Continuous cropping, however, will decrease the amount of organic matter in soils, and lowering the organic matter content is expected to decrease soil

Environmental and Industrial Importance 301

aggregation. Especially the macropore spaces will be destroyed by the destruction of soil structure, which will harmfully affect water percolation and soil aeration.

9.3 IMPORTANCE IN INDUSTRY

Humic matter can be applied for a variety of purposes in industry, but unfortunately not much has been published. Most of the information in industry is either patented or kept highly classified. This is of course understandable, in view of the fierce competition in the industrial marketplace that can adversely affect the livelihood of the company. However, such an attitude is of disadvantage for the advancement of science and technology. Fortunately, the situation is substantially better in the agricultural and pharmaceutical industries, where the scientists are more at liberty to publish their discoveries and inventions. Nevertheless, several applications of humic substances in industry are mentioned in the older literature. Due to its brown to black color, humic matter has attracted attention for possible use as a staining agent. Patents have been obtained for the production of wood stains and tanning material of leather from humic substances, apparently in the USA for Japanese, Russian and other overseas companies, as quoted from a review by MacCarthy and Rice (1994). The color of humic substances also finds application in the ink industry, and for coloring ceramics, whereas its dispersing property is important for use in drilling fluids and in paints. The usefulness of humic acids in the production of ink and its importance for removing industrial boiler scales have been reported in the older literature (Steelink, 1963; 1967). Humic matter is also believed to be essential in keeping smectite (montmorillonite) slurries in emulsive form. These slurries are used as lubricants in drilling-heads, hence it is necessary to prevent them from becoming aggregated and hard.

Today complex formation between silicic acid and organic compounds finds application in industry in the production of *silicones*. The general formula of silicone is R_2SiO_2, in which R is an organic radical. Usually, the radical is a simple organic compound, a methyl group, CH_3, or an ethyl group, C_2H_5. The suggestion is presented here that humic matter can also be used as a radical. These silicones can be linear or cyclic in structure (Figure 9.1) and silicones composed of cross-linked silica polymers are also possible. Depending on the molecular complexity, silicones have the appearance of oily, greasy, or rubber-like substances. Consequently, they are used as lubricants, hydraulic fluids, and electrical insulators. Antifreeze for automobiles contains silicone for lubricating the water pump. In medical science, silicone finds application as filling material and implants in human bodies. The controversial use of silicone for breast implants has received a lot of attention in the media. Whether the use of humic matter in silicones will affect their behavior is a matter for future investigation, and humic acid is so far known to be nontoxic.

9.3.1 Production of Agrochemicals

The metal chelation properties of humic matter also finds application today in the fertilizers industry. Considerable efforts have been spent in Europe to produce and market fertilizers treated with humic acids. It is believed that mineral fertilizer grains coated with humic substances reduce the entropic action of the soil, enhancing in this way the fertilizer efficiency (Savoini, 1986). Humus treated with ammonium hydroxide solution, called nitrogen-amended humus by MacCarthy and Rice (1994), has been used as a fertilizer and was found to have a stimulating effect on plant growth. However, some caution should be exercised, because according to the authors above large applications of NH_4OH cause the mixture to become harmful. The chemical reaction between ammonium and humus is called NH_3 *fixation* by Stevenson (1994). It is not related to fixation of ammonium by expanding clays. This capacity to adsorb NH_3 has been known for some time, and is exhibited in fact by an array of organic compounds, such as lignin, peat, sawdust and other organic residues, e.g., corn cobs. Over the years, a number of patents have been given for the production of biofertilizers from treatment of these residues with NH_3.

Since humic matter is also a powerful chelator of phosphates, the possibility arises that humic acids can find application in the man-

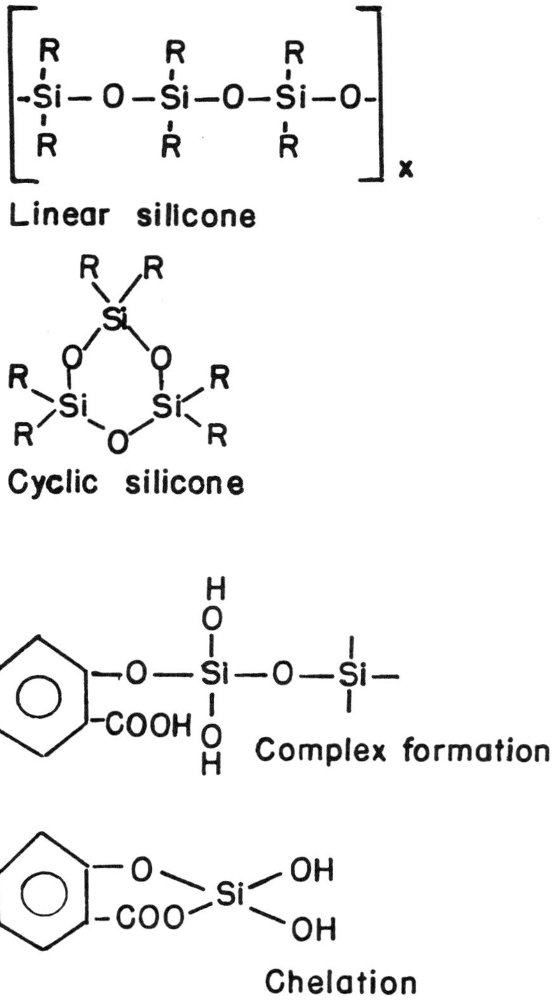

Figure 9.1 The structure of silicone. Top two: linear and cyclic silicone. Bottom two: Proposed silicone produced from humic acid and silicic acid.

ufacture of *bio-superphosphates*. In the production of superphosphate fertilizers, the sulfuric acid needed to solubilize P from the rock-phosphate, can perhaps be replaced by humic acids. Though not particularly strong, the acidic properties of humic acids can rival that of sulfuric acid in making part of the rock phosphate soluble. In view of the excellent record of humo-phosphate chelates in stimulating plant growth, the resulting bio-superphosphate or humo-superphosphate fertilizer can prove to be superior to its inorganic counterpart.

In Europe, iron, zinc, manganese, and copper chelates have also been produced in the form of capsules or pellets for use in alleviating micronutrient deficiencies in plants. The humo-metal complexes provide for easy transport and distribution of the micronutrients. Savoini (1986) indicates that these chelates exhibit a slow and gradual solubility in water in addition to a strong resistance to degradation by light and biophysical attack. They can also be mixed with fertilizers since the slow solubility of Fe chelate, for example, makes the Fe-humate compatible for mixing with a wide range of fertilizers. Humo-micronutrient chelates in liquid form applied as foliar sprays have proven to be more efficient in controlling micronutrient deficiencies than the plain micronutrient elements alone.

The application of humic matter is not limited to production of fertilizers, but similarly as with the bio- N-fertilizers, the manufacture of bio-pesticides is open for consideration and is already appearing at the near horizon. Adsorption of pesticide by humic matter is considered by Stevenson (1994) a key factor in determining its bioactivity and other behavioral properties, e.g., biodegradability. This is a very controversial issue, since in one aspect the bio- or humo-pesticide complex can become more efficient in attacking its target, but on the other hand, such a complex can stimulate its persistence in soils and increases the danger to nontarget materials.

9.3.2 Production of Commercial Humates

Today humic acids are applied in the production of a variety of compounds for use in soils and agriculture. Some of these applications

Environmental and Industrial Importance 305

have been discussed in the previous section. In the United States and several other countries the kinds of humic products in question are called by many of the producers *commercial humates* or plain *humates*. Scientifically, it is a proper name, since they are in essence the salts of humic or fulvic acids. As technical grade chemicals, these humates are seldom in purified forms, as are also the fertilizers. Purification to make them analytical grade reagents is unnecessary for their use in field conditions, and may even increase the cost considerably, making them unaffordable for farmers and growers alike.

Most scientists in the past considered the humates to be 'dirt,' hardly worth research attention, and many today still dislike them. However, like it or not, these commercial humates are here to stay, either due to really observable benefits to soil and plants, or to market persistence. They appear now to be a multi-million dollar business venture, and the tendency is that their importance will increase in the near future. Companies producing and marketing humates are noticed to mushroom worldwide. They come and go, because some seem to be out of business after a duration of a few short years due to business competition and/or other reasons.

Origin and Types of Commercial Humates

Companies, such as Dinosoil, Alpine, TX; Earth Greens, Dallas, TX; GreenSense Humate, Garland, TX; CropChem, Decatur, IL; American Colloid Company, Chicago, IL; Humate International, Jacksonville, FL; Home Hydroponics, Ontario, Canada; Biotechnica, Reading, UK; Australian Humates, Melbourne, Victoria, Australia; and Chemapex, Praha, Czech Republic are now offering a wide range of humates. They are marketed in solid or liquid form under a variety of generic names, e.g., dinosoil, humate, Menefee humate, enersol, humate Ls/Ag, liquid gold humate, biohumate, organic humate, K-humate, humigel and the like. The listing of companies above is not an endorsement of the particular company or its product, but is only a statement of their current presence in the world market. Many other companies, not listed above, are also present offering similar humates.

Most of the produced humates are from lignite-based humic material. Some are in the form of powdered or granulated lignite only (Lobartini et al., 1992; Burdick, 1965), whereas several are lignite humic acid in liquid form dissolved in NaOH, KOH or NH_4OH. A few of them constitute mixtures of lignite humic acid and seaweed or other marine life residues, whereas many more may be fortified by the addition of essential macro- and micronutrients for plant growth. Less known are the humates, produced by inoculation of lignite or lignite-humic acid with microorganisms or humates mixed with bacteria, fungi, and actinomyces species.

Depending on the extraction reagents used, the humates are either Na-humate, K-humate, or NH_4-humate. Because of the absence of a proper purification process, the commercial humates are high in pH, and pH values of 10-12 are not uncommon. The K-humate and NH_4-humate are expected to be relatively high in K and NH_4, respectively, and by origin humic acids already contain 4-5% N, small amounts of S and perhaps traces of P. The extra amount of macro- and micronutrients added to fortify the humates are sometimes quite large. For instance, Humate Fertilizer PK and Humate Fertilizer N, offered by Chemapex, Praha, listed contents of 14% P_2O_5, 12% K_2O, and 12% N, whereas the Organic Humate from Australian Humate listed 24 mg L^{-1} Bo, 200 mg L^{-1} Fe and 40 mg L^{-1} Mn. The very small amounts of micronutrients required by plants is perhaps the reason why the fortified humates may in some instances yield beneficial results, and in other instances no results. To erase the issue of mixed results, a thorough research investigation is perhaps required for testing and determination of the proper balanced content of micronutrients in the humates.

The Issue of Mixing with Microorganisms

Of interest to note are also the humates, produced by mixing or inocculating them with a variety of microorganisms and organic compounds, referred to as biosystems, bactolife or biostimulants. Several of the 'bio-system' microbial products specify the inclusion of

Environmental and Industrial Importance

a broad spectrum of bacteria, fungi, and actinomycetes with the humate. It may perhaps work in sterile environments, but pending more information from scientific research some caution should be exercised in introducing an assortment of foreign life organisms into the soil. In agricultural operations, it is common to add an organism for a known purpose, such as adding *Rhizobium* strains for inoculation of legume plants to enhance biological N_2-fixation, or by applying mycorrhizal fungus, e.g., *Pisolithus tinctorius*, to pine seedlings to improve their growth (Tan, 2000). Seldom is an indiscriminate application of a large variety of microorganisms warranted for a general purpose, since the soil may already contain most of the organisms. The organisms foreign to the soil may either be killed by the indigenous microbial population or they may cause damage to the latter.

Mixing humate with saponin as in the case of Biohumate of the Australian Humate company is perhaps more warranted. It is believed to function as a biostimulant, but saponins in general are known to reduce surface tension, like soap, hence will add the surfactant property to the humate, which is of utmost importance for humate applications as foliar-sprays. In biochemistry, saponins are grouped with the glycosides, which play a role in blocking the activity of toxic substances in the plant tissue so that plant cells are not injured. However, some of the saponins are toxic to some animals and have been used in Africa, Asia, and South America to catch fish in streams and lakes. The plants containing saponins are beaten to pulp in the river and the poisoned fish, edible for humans, rise to the surface.

The Issue of Humates as Fertilizers

The opinion of many companies that humates can be considered as fertilizers raises a lot of questions, since most soil scientists prefer to classify them as soil amendments By origin, humic acids do not carry nutrient elements in the same amounts as fertilizers do. Their N content, as stated above, is also too small to qualify them as fertilizers. The use of KOH and NH_4OH in the production of humates will indeed increase their K and N contents, respectively, but not to the extent that they can easily be called fertilizers. Nevertheless, plants have been noticed to respond quickly to the application of these K- or NH_4-humates. The NH_4-humates may perhaps be of more advantage to plant growth than the K-humates, due to the more visible effect of N in causing a healthy appearance of plants by producing greener leaves. The effect of K-humates is expected to be less conspicuous since K affects carbohydrate production, leaf-tip growth and some root formation. Not much research has been conducted in this respect.

In spite of all the above, the definition of a fertilizer states that a fertilizer is any material applied to soils to improve growth, yield, quality and nutritive value of crops (Tisdale at al., 1993; Jones, 1962; Tan, 1981). According to such a concept, commercial humates should fall in the category of fertilizers due to the many claims that they enhance germination, root development, nutrient uptake, growth and yield of plants. Cow manure, stable manure, compost and the like are called organic fertilizers, hence why shouldn't the humates not be considered as such? The only difference is that the commercial humates do not contain N, P, K, Ca , Mg, and other plant nutrients in the same amounts as do cow and stable manure, unless they are fortified with these nutrients during the manufacturing process.

Other claims that humates improve microbial activity, soil structure development and water-holding capacity are also scientifically well-founded, as is the action of humates as chelators. Some humates are shown to contain 'appreciable' amounts of cytokinins, gibberelins and auxins, empowering them consequently with the capabilities of growth hormones. This is especially of importance with humates produced from or mixed with seaweeds. Whether the latter will be found to be true or not is a matter of investigation, since as discussed in Chapter 4, some truth may be present in all these allegations. Hormones can be released by decomposition from plant materials and incorporated in the humic molecular structure during the humification process. Other growth promoting substances, such as vitamins, are believed to also be included in the synthesis of humic matter. These then are the reasons for the growth-stimulating effects often reported for humic substances.

Environmental and Industrial Importance

Rate of Application

A lot of research is apparently still needed to establish the rate of humate application, since the latter shows considerable variations from one to another brand of humate. Amounts of 100 to 300 kg/ha of the solid form of humates have been suggested for application in field conditions by US companies, but 600 kg and 5 tons/ha have been mentioned by the Australian Humate Company.

The liquid form of humates provides fewer problems in application and according to several scientific studies a rate of 400 to 600 mg L^{-1} humic acid is optimum. Higher rates of ≥ 1000 mg L^{-1} can be beneficial for one, but harmful, though not toxic, for other crops (see Chapter 8).

9.4 IMPORTANCE AS PHARMACEUTICALS

The idea of using humic matter in the pharmaceutical industry is derived from a number of reasons. The oldest reason comes perhaps from peat, which has been known for a long time in Europe for its therapeutic properties. Peat baths were taken in the old days for therapy of gynecological and rheumatic diseases, and even today mud baths are still offered in many European health clinics and spas. In our modern cosmetic industry, face muds are offered at expensive prices for curing or removing old skin. These healing properties have been currently traced to humic acid, the dominant component of peat.

Today, it is the adsorption, dispersion or emulsification property and other attributes of humic matter that attract the interest of medical science and the pharmaceutical industry. The potential for use of humic acids as antiviral, anti-inflammatory, estrogenic, profibrinolytic and anti-coagulatory agents has been under serious investigations for some time at the Medical School in Erfurt, Germany. A successful treatment of thrombophlebitis is credited by Klöcking (1994) to the anticoagulatory property of humic acid. In a more recent investigation, the author above reinforces his contention by demonstrating humic acid at dilute concentrations (0.5 to 6.2 µmol L^{-1})

to function as a very strong inhibitor of thrombin (Klöcking, 1997). Its antiviral potency is revealed by Thiel et al. (1977), who claim to have successfully treated infection of herpes simplex viruses with dilute concentrations (0.5 mg L^{-1}) of ammonium humate. Clinical tests in Hungary show, in addition, possibilities for humic acids to be used in cancer therapy and in healing crushed, cuts, and especially burn-wounds (Jurscik, 1994).

In Russia, a medicine prepared from peat extracts, called *torfot*, is used at the Ukrainian Institute of Eye Diseases at Odessa as a topical treatment for myopia, opacification of the cornea, and early retinal degeneration (Fuchsman, 1980). Perhaps, it can also prove to be useful one day for retarding or slowing the growth of cataracts.

The cation exchange and chelation capacity of humic acids have been applied in veterinary medicine. Intoxication of rats due to consumption of the heavy metals, Cd and Pb, is reported to be controlled by the application of low concentrations (0.1%) of humic acids in their food supply (Ridwan, 1977). This opens possibilities for use of humic acids in treatments of other stomach ailments in the line of the over-the-counter medicine called Kaopectate or Pepto-Bismol. The adsorption capacity of humic matter is far greater than that of kaolinite, and so far humic substances are not known to be toxic to soil organisms. Fish and other aquatic organisms are noticed to thrive prolifically in black water environments, as indicated earlier.

The application of humic acids in medicine and in the pharmaceutical industry in the United States is currently lagging far behind that in European countries. Mention should perhaps be made of the presence of Morganics Inc., Scottsdale, AZ, a company that has just started to recognize fulvic acid as a powerful antioxidant (personal communications from Craig Ricketts, CEO). However, its product with an effect closest to humate is perhaps *Earth Force*, which Morganics claims to act as as a biological activator for flowering and potted plants and also enhance seed germination. Once again it should be emphasized that no endorsement whatsoever is presented here.

A final note in the medicinal application of humic acids is the product Humate-P, a biological compound approved by the US Food and Drug Administration in 1999 (FDA Talk Paper, T99-15, April 1, 1999) for therapy of a clotting disorder. Humate-P has been cleared for

Environmental and Industrial Importance

treatment of patients with hemophelia A and/or von Willebrand disease, a bleeding disorder. The reasoning for choosing such a name is everybody's guess in soil science, since humate-P is neither a humic acid derivative nor has it any relation to soil humic matter. The P does not stand for phosphorus, but stands for 'pasteurization' needed for sterilization of the product. It is a purified product derived from pooled human blood plasma, hence should better carry the name Plasma-P or Plasmate-P. Using the name Humate-P brings only confusion to humic acid chemistry and science unless the plasma can be proven to contain humic acids.

APPENDIX A

GREEK ALPHABET

Greek letter	Greek name	Greek letter	Greek name
A α	Alpha	N ν	Nu
B β	Beta	Ξ ξ	Xi
Γ γ	Gamma	O o	Omicron
Δ δ	Delta	Π π	Pi
E ϵ	Epsilon	P ρ	Rho
Z ζ	Zeta	Σ σ	Sigma
H η	Eta	T τ	Tau
Θ θ	Theta	Υ υ	Upsilon
I ι	Iota	Φ φ	Phi
K κ	Kappa	X χ	Chi
Λ λ	Lambda	Ψ ψ	Psi
M μ	Mu	Ω ω	Omega

APPENDIX B

Atomic Weights of the Major Elements in Soils

Element	Symbol	Atomic number	Atomic weight
Aluminum	Al	13	027.0
Antimony	Sb	51	121.8
Argon	Ar	18	039.9
Arsenic	As	33	074.9
Barium	Ba	56	137.3
Beryllium	Be	04	009.0
Bismuth	Bi	83	209.0
Boron	B	05	010.8
Bromine	Br	35	079.9
Calcium	Ca	20	040.1
Carbon	C	06	012.0
Cesium	Cs	55	132.9
Chlorine	Cl	17	035.5
Chromium	Cr	24	052.0
Cobalt	Co	27	058.9
Copper	Cu	29	063.5
Fluorine	F	09	019.0
Gallium	Ga	31	069.7
Germanium	Ge	32	072.6
Gold	Au	79	197.0

Appendix B

Element	Symbol	Atomic number	Atomic weight
Helium	He	02	004.0
Hydrogen	H	01	001.0
Iodine	I	53	126.9
Iridium	Ir	77	192.2
Iron	Fe	26	055.9
Krypton	Kr	36	083.8
Lanthanum	La	57	138.9
Lead	Pb	82	207.2
Lithium	Li	03	006.9
Magnesium	Mg	12	024.2
Manganese	Mn	25	054.9
Mercury	Hg	80	200.6
Molybdenum	Mo	02	095.9
Nickel	Ni	28	058.7
Nitrogen	N	07	014.0
Oxygen	O	08	016.0
Phosphorus	P	15	031.0
Platinum	Pt	78	195.1
Potassium	K	19	039.1
Radon	Rn	86	222.0
Radium	Ra	88	226.1
Rhodium	Rh	45	102.9
Rubidium	Rb	37	085.5
Selenium	Se	34	079.0
Silicon	Si	14	028.1
Silver	Ag	47	107.9
Sodium	Na	11	023.0
Strontium	Sr	38	087.6
Sulfur	S	16	032.1
Tantalum	Ta	03	180.9
Tellurium	Te	52	127.6

Element	Symbol	Atomic number	Atomic weight
Thallium	Tl	81	204.4
Thorium	Th	90	232.1
Tin	Sn	50	118.7
Titanium	Ti	22	047.9
Tungsten	W	74	183.9
Uranium	U	92	238.0
Vanadium	V	23	050.9
Xenon	Xe	54	131.3
Yttrium	Y	39	088.9
Zinc	Zn	30	065.4
Zirconium	Zr	40	091.2

REFERENCES AND ADDITIONAL READINGS

Achard, F. K. 1786. Chemische untersuchung des torfs. Crell's Chem. Annal. 2:391-403.
Abrams, I. M., 1975. Macroporous condensate resins as absorbents. Industrial Eng. Chem. Prod. Res. Dev. 14:108-112.
Ahmad, F., and K. H. Tan. 1986. Effect of lime and organic matter on soybean seedlings grown in aluminum toxic soil. Soil Sci. Soc. Am. J. 50:656-661.
Aiken, G. R. 1985. Isolation and concentration techniques for aquatic humic substances. Pp.363-385. In: *Humic Substances in Soil, Sediment, and Water. Geochemistry, Isolation, and Characterization*, G. R. Aiken, D. M. McKnight, R. L. Wershaw, and P. MacCarthy (eds.). Wiley-Interscience, New York, NY.
Aiken, G. R. 1988. A critical evaluation of the use of macroporous resins for the isolation of aquatic humic substances. Pp.15-27. In: *Humic Substances and their Role in the Environment*, F. H. Frimmel and R. F. Christman (eds.). Report Dahlem Workshop on Humic Substances and their Role in the Environment, Berlin 1987, March 29 - April 3. Wiley-Interscience, New York, NY.
Aiken, G. R., D. M. McKnight, R. L. Wershaw, and P. MacCarthy. 1985. An introduction to humic substances in soil, sediment, and water. Pp.1-9. In: *Humic Substances in Soil, Sediment, and Water. Geochemistry, Isolation, and Characterization*, G. R. Aiken, D. M. McKnight, R. L. Wershaw, and P. MacCarthy (eds.). Wiley-Interscience, New York, NY.
Aitken, J. B., B. Acock, and T. L. Senn. 1964. The characteristics and

effects on humic acids derived from leonardite. South Carolina Agri. Expt. Sta. Tech. Bull. 1015. Clemson University, Clemson, SC.

Alberts, J. J., and Z. Filip. 1994. Effect of organic solvent pre-extraction of source substrates on elemental composition, Fourier transform infrared spectra and copper binding in estuarine humic and fulvic acid. Pp.781-790. In: *Humic Substances in the Global Environment and Implications on Human Health*, N. Senesi and T. M. Miano (eds.). Proc. 6th Intern. Meeting of the Intern. Humic Substances Soc., Monopoli, Bari, Italy, Sept. 20-25, 1992. Elsevier, Amsterdam.

Almendros, G., and F. J. Gonzalez-Vile. 1987. Degradative studies on a soil humic fraction-sequential degradation of inherited humin. Soil Biol. Biochem. 19:513-520.

Andreyuk, E. I., S. A. Gordienko, I. N. Konoto, and V. A. Martynenko. 1973. Assimilation of humic acid nitrogen by microorganisms. Mikrobiologichnyi Zhurnal 35:139-142.

Arnolds, O., C. T. Hallmark, and L. P. Wilding. 1995. Andisols from four different regions of Iceland. Soil Sci. Soc. Am. J. 59:161-169.

Badura, L. 1965. On the mechanism of the stimulating influence of Na-humate upon the process of alcoholic fermentation and multiplication of yeast. Acta Soc. Bot. Pol. 34:287-328.

Bailey, G. W., and J. L. White. 1970. Factors influencing the adsorption and movement of pesticides in soils. Pp.29-92. In: *Residue Reviews*, Vol.32, F. A. Gunther and J. D. Gunther (eds.). Springer-Verlag, New York, NY.

Baker, W. E. 1973. The role of humic acids from Tasmanian podzolic soils in mineral degradation and metal mobilization. Geochim. Cosmochim. Acta 37:269-281.

Bardway, K. K., and A. C. Gaur. 1972. Growth stimulating effect of humic acid on bacteria. Zentralblatt fur Bakteriologie, Parasitenkunde, Infektionskrankheiten und Hygienee. Abteilung I, Referate 126:649-694.

Barker, S. A., M. H. B. Hayes, R. G. Simmonds, and M. Stacey. 1967. Studies on soil polysaccharides. I. Carbohydr. Res. 5:13-24.

Bartlett, R. J. 1999. Characterizing soil redox behavior. Pp.371-397. In: *Soil Physical Chemistry*, Second edition, D. L. Sparks (ed.),

CRC Press, Boca Raton, FL.

Baver, L. D. 1963. The effect of soil organic matter on soil structure. Pontif. Acad. Sci. Scr. Varia 32:383-413.

Beck, K. C., J. H. Reuter, and E. M. Purdue. 1974. Organic and inorganic geochemistry of some coastal plain rivers of the southeastern United States. Geochim. Cosmochim. Acta 38:341-364.

Bedrock, C. N., M. V. Cheshire, J. A. Chudek, B. A. Goodman, and C. A. Shand. 1994. ^{31}P NMR studies of humic acid from a blanket peat. Pp.227-232. In: *Humic Substances in the Global Environment and Implications on Human Health*, N. Senesi and T. M. Miano (eds.). Proc. 6th Intern. Meeting of the Intern. Humic Substances Soc., Monopoli, Bari, Italy, Sept. 20-25, 1992. Elsevier, Amsterdam.

Bellar, T. A., J. J. Lichtenberg, and R. C. Kroner. 1974. The occurrence of organohalides in chlorinated drinking water. J. Am. Waterworks Assoc. 66:703-706.

Berzelius, J. J. 1839. *Lehrbuch der Chemie*. Third edition (translated by Wohler). Arnoldische Buchhandlung, Dresden, Leipzig.

Black, A. P., and R. F. Christman. 1963. Characteristics of colored surface waters. J. Am. Water Works Assoc. 55:753-770.

Bortiatynski, J. M., P. G. Hatcher, and H. Knicker. 1996. NMR techniques (C, N, and H) in studies of humic substances. Pp.57-77. In: *Humic and Fulvic Acids. Isolation, Structure, and Environmental Role,* J. S. Gaffney, N. A. Marley, and S. B. Clark (eds.). ACS Symposium Series 651. American Chemical Society, Washington, DC.

Bottomley, W. 1920. The effect of organic matter on the growth of various plants in culture solutions. Annal. Bot. (London) 34:353-365.

Boysen-Jensen, P., 1936. *Growth Hormones in Plants*. Translated into English by G. S. Avery and P. R. Burkholder. McGraw-Hill Book Co., New York, NY.

Brady, N. C. 1990. *The Nature and Properties of Soils*. Tenth edition. Macmillan Publ. Co., New York, NY.

Breault, R. F., J. A. Colman, G. R. Aiken, and D. M. McKnight. 1996. Copper speciation and binding by organic matter in copper-contaminated stream water. Environ. Sci. Technol. 30:3477-3486.

Breger, I. A. 1963. Origin and classification of naturally occurring carbonaceous substances. Pp.50-86. In: *Organic Geochemistry*, I. A. Breger (ed.). Pergamon Press, Elmsford, NY.

Breger, I. A. 1976. Geochemical interrelationships among fossil fuels, marsh, and landfill gas. Pp.913-918. In: *The Future Supply of Nature-made Petroleum and Gas Technical Report*, R. F. Meyer (ed.). Pergamon Press, Elmsford, NY.

Bremner, J. M. 1950. Some observations on the oxidation of soil organic matter in the presence of alkali. J. Soil Sci. 1:198-204.

Bremner, J. M. 1954. A review of recent work on soil organic matter. II. J. Soil Sci. 5:214-232.

Burauel, P., and F. Führ. 2001. Long-term fate of organic chemicals in lysimeter experiments and the role of the bound residue fraction. Pp.289-302. In: *Humic Substances and Chemical Contaminants*, C. E. Clapp, M. H. B. Hayes, N. Senesi, P. R. Bloom, and P. M. Jardine (eds.). Proc. Workshop and Symposium Int. Humic Substances Soc., Soil Sci. Soc. Am., and Am. Soc. Agronomy, Anaheim, CA, 26-27 Oct., 1997. Soil Sci. Soc. Am., Inc., Madison, WI.

Burdick, E. M. 1965. Commercial humates for agriculture and the fertilizer industry. Econ. Bot. 19:152-156.

Burges, A., and P. Latter. 1960. Decomposition of humic acid by fungi. Nature, London 186:404-405.

Burns, R. G. 1986. Interactions of enzymes with soil minerals. Pp.429-451. In: *Interactions of Soil Minerals with Natural Organics and Microbes*, P. M. Huang and M. Schnitzer (eds.). SSSA Special Publ. No. 7. Soil Sci. Soc. Am., Madison, WI.

Calvin, M., and J. A. Bassham. 1962. *The Photosynthesis of Carbon Compounds*. Benjamin, Menlo Park, CA.

Cameron, R. S., B. K. Thornton, R. S. Swift, and A. M. Posner. 1972. Molecular weight and shape of humic acid from sedimentation and diffusion measurements on fractionated extracts. J. Soil Sci. 23:394-408.

Campbell, C. R. 1973. *Establishment and Growth of Corn in Chicken Litter Fertilized Tall Fescue*. MS Thesis, University of Georgia,

References

Athens, GA. pp.83.

Chaminade, R. 1946. Sur une methode de dosage de la fraction humifique de la matière organique des sols. C. R. Acad. Agric. 32:131-134.

Chaminade, R. 1956. Action de l'acide humique sur le développement et la nutrition minérale des végétaux. Trans. 6th Int.Congr. Soil Sci. 4:443-448.

Chaminade, R. 1958. Influence de la matière organique humifée sur l'efficacité de l'azote. Annal. Agron. (Paris) 9:167-192.

Chaminade, R. 1966. Effet physiologique des constituants de la matière organique des sols, sur la métabolisme des plantes, la croissance et le rendement. Pp.35-47. In: *The Use of Isotopes in Soil Organic Matter studies*. Report FAO/IAEA Tech. Meeting, Brunswick Volkenrode, 1963. Pergamon Press, Oxford.

Chaminade, R., and R. Blanchet. 1953. Mécanisme de l'action stimulante de l'humus sur la nutrition minérale des végétaux. C. R. Acad. Sci. Paris 237:1768-1770.

Chefetz, B., J. Tarchitzky, A. P. Desmukh, P. G. Hatcher, and Y. Chen. 2002. Structural characterization of soil organic matter and humic acids in particle-size fractions of an agricultural soil. Soil Sci. Soc. Am. J. 66:129-141.

Chen, Y., and M. Schnitzer. 1976. Scanning electron microscopy of a humic acid and its metal and clay complexes. Soil Sci. Soc. Am. J. 40:682-686.

Chen, Y., N. Senesi, and M. Schnitzer. 1977. Information provided on humic substances by E_4/E_6 ratios. Soil Sci. Soc. Am. J. 41:352-358.

Cheng, K. L. 1977. Separation of humic acid with XAD resins. Microchim. Acta II:389-396.

Cheshire, M. V., M. L. Berrow, B. A. Goodman, and C. M. Mundie. 1977. Metal distribution and nature of some Cu, Mn, and V complexes in humic and fulvic acid fractions of soil organic matter. Geochim. Cosmochim. Acta 41:1131-1138.

Choudri, M. B., and F. J. Stevenson. 1957. Chemical and physico-chemical properties of soil colloids. III. Extraction of organic matter from soils. Soil Sci. Soc. Am. Proc. 21:508-513.

Choudry, G. G. 1983. *Humic Substances: Structural, Photophysical,*

Photochemical, and Free Radical Aspects, and Interactions with Environmental Chemicals. Garden Beach Science Publ., New York, NY.

Christman, R. F., and E. T. Gjessing (eds.). 1983. *Aquatic and Terrestrial Humic Materials*. Ann Arbor Sci., The Butterworth Group, Ann Arbor, MI.

Christman, R. F., D. L. Norwood, Y. Seo, and F. H. Frimmel. 1989. Oxidative degradation of humic substances from freshwater environments. Pp.34-67. In: *Humic Substances II. In Search of Structure*, M. H. B. Hayes, P. MacCarthy, R. L. Malcolm, and R. S. Swift (eds.). Wiley-Interscience, New York, NY.

Cieślewicz, J., E. Niedzwiecki, M. Protasowicki, and S. S. Gonet. 1997. Humus properties of bottom sediments from the Szczecin lagoon (North-West Poland). Pp.553-558. In: *The Role of Humic Substances in the Ecosystems and in Environmental Protection*, J. Drozd, S. S. Gonet, N. Senesi, and J. Weber (eds.). Proc. 8th Meeting Intern. Humic Subst. Soc., Wroclaw, Poland, Sept. 9-14, 1996. IHSS-Polish Soc. Humic Substances, Wroclaw, Poland.

Činčerová, A. 1962. Uber den Einfluss der Huminsäure auf die Veränderungen der freien Zucker in Weizenpflanzen. Pp.47-62. In: *Studies about Humus*. Symposium Czechoslovak Academy of Sciences, Prague.

Clapp, C. E., M. H. B. Hayes. 1997. Isolation of humic substances from an agricultural soil using a sequential and exhaustive extraction process. Pp.3-11. In: *Humic Substances and Organic Matter in Soil and Water Environments: Characteristics, Transformation, and Interactions*, C. E. Clapp, M. H. B. Hayes, N. Senesi, and S. M. Griffith (eds.). Proc. 7th Conf. Intern. Humic Substances Soc., Univ. West Indies, St. Augustine, Trinidad, and Tobago, 3-8 July 1994. Published by International Humic Substances Soc., Inc., Dept. Soil, Water, and Climate, Univ. Minnesota, St. Paul, MN.

Clapp, C. E., M. H. B. Hayes, N. Senesi, P. R. Bloom, and P. M. Jardine (eds). 2001. *Humic Substances and Chemical Contaminants*. Proc. Workshop and Symposium Int. Humic Substances Soc., Soil Sci. Soc. Am., and Am. Soc. Agronomy, Anaheim, CA, 26-27 Oct., 1997. Soil Sci. Soc. Am., Inc.,

References

Madison, WI.

Clark, F. E., and K. H. Tan. 1969. Identification of a polysaccharide ester linkage in humic acid. Soil Biol. Biochem. 1:75-81.

Clark, N. A. 1924. The soil organic matter and growth promoting accessory substances. Industr. Engin. Chem. 16:249-250.

Clarke, F. W. 1911. The data of geochemistry. 2nd edition. U. S. Geol. Survey Bull. 491. U. S. Government Printing Office, Washington, DC.

Claus, H., Z. K. Filip, and J. J. Alberts. 1997. Microbiological utilization and transformation of riverine humic substances. Pp.561-566. In: *The Role of Humic Substances in the Ecosystems and in Environmental Protection*, J. Drozd, S. S. Gonet, N. Senesi, and J. Weber (eds.). Proc. 8th Meeting Intern. Humic Subst. Soc., Wroclaw, Poland, Sept. 9-14, 1996. IHSS-Polish Soc. Humic Substances, Wroclaw, Poland.

Coates, J. D., D. J. Ellis, E. L. Blunt-Harris, C. V. Gaw, E. E. Roden, and D. R. Lovley. 1998. Recovery of humic substance-reducing bacteria from a diversity of environments. Appl. Environ, Microbiol. 64:1504-1509.

Conn, E. E., and P. K. Stumpf. 1967. *Outlines of Biochemistry*. Second edition. John Wiley & Sons, New York, NY.

Cranwell, P. A., and R. D. Hayworth. 1975. The chemical nature of humic acids. Pp.13-18. In: *Humic Substances. Their Structure and Function in the Biosphere*, D. Povoledo and H. L. Golterman (eds.). Proc. Intern. Meeting, Nieuwersluis, The Netherlands, May 29-31, 1972. Centre for Agric. Publishing and Documentation, Wageningen, The Netherlands.

Csicsor, J., J. Gerse, and A. Tikos. 1994. The biostimulant effect of different humic substance fractions on seed germination. Pp.557-562. In: *Humic Substances in the Global Environment and Implications on Human Health*, N. Senesi and T. M. Miano (eds.). Proc. 6th Intern. Meeting of the Intern. Humic Substances Soc., Monopoli, Bari, Italy, Sept. 20-25, 1992. Elsevier, Amsterdam.

David, F., and P. David. 1976. Photoredox chemistry of Fe(III) chloride and iron(III) perchlorate in aqueous media, a comparative study. J. Phys. Chem. 80:579-583.

Dawson, H. J., B. F. Hrutfiord, R. J. Zakoski, and F. C. Ugolini. 1981. The molecular weight and origin of yellow organic acids. Soil Sci. 132:191-199.

Dębska, B. 1997. The effect of green manure on the properties of soil humic acids. Pp.260-282. In: *The Role of Humic Substances in the Ecosystems and in Environmental Protection*, J. Drozd, S. S. Gonet, N. Senesi, and J. Weber (eds.). Proc. 8th Meeting Intern. Humic Subst. Soc., Wroclaw, Poland, Sept. 9-14, 1996. IHSS-Polish Soc. Humic Substances, Wroclaw, Poland.

De Coninck, F. 1980. Major mechanisms in fomation of spodic horizons. Geoderma 24:101-128.

De Haan, H. 1977. Effect of benzoate on microbial decomposition of fulvic acids in Tjeukemeer (The Netherlands). Limnol. Oceanogr. 22:38-44.

Dell'Amico, C., G. Masciandaro, A. Ganni, B. Ceccanti, C. Garcia, T. Hernandez, and F. Costa. 1994. Effects of specific humic fractions on plant growth. Pp.563-566. In: *Humic Substances in the Global Environment and Implications on Human Health*, N. Senesi and T. M. Miano (eds.). Proc. 6th Intern. Meeting of the Intern. Humic Substances Soc., Monopoli, Bari, Italy, Sept. 20-25, 1992. Elsevier, Amsterdam.

Del Re, A. A., C. Bartoletti, and P. Fontana. 1978. Solubilita del fosfato di alluminio amorfo ed effetto degli acidi umici. Agrochimica 23:426-433.

De Saussure, T. 1804. Recherches chemiques sur la végétation. Paris Annal. 12:162.

Dobbs, J. C., W. Susetyo, L. A. Carreira, and L. V. Azarraga. 1989. Competitive binding of protons and metal ions in humic substances by lanthanide ion probe spectroscopy. Anal. Chem. 61:1519-1524.

Döbereiner, J. W. 1822. Zuer Pneumatischen Chemie.III. Phytochemie, pp.64-74.

Dormaar, J. F. 1975. Effects of humic substances from Chernozemic A_h horizons on nutrient uptake by *Phaseolus vulgaris* and *Festuca sabrella*. Can. J. Soil Sci. 55:111-118.

Dragunov, S. S. 1958. Structure of humic acids and preparation of humic fertilizers (In Russian). Tr. Moskovsk. Torfyan. In-ta

References

No.8.

Dubach, P., N. C. Mehta, and H. Deuel. 1963. Schonende Extraktion von Huminstoffen und Isolierung der Fulvosäure-Fraktion aus verschiedenen Bodentypen. Z. Pflanzenernähr. Dung. Bodenk. 102:1-7.

Dudas, M. J., and S. Pawluk. 1970. Chernozem soils of the Alberta Parklands. Geoderma 3:19-36.

Felbeck, Jr., G. T. 1965. Structural chemistry of soil humic substances. Adv. Agron. 17:327-368.

Filip, Z., K. Haider, H. Beutelspacher, and J. P. Martin. 1974. Comparison of IR spectra from melanins of microscopic soil fungi, humic acids and model phenol polymers. Geoderma 11:37-52.

Filip, Z., J. Semotan, and M. Kutilek. 1976. Thermal and spectrophotometric analysis of some fungal melanins and soil compounds. Geoderma 15:131-142.

Flaig, W. 1953. Über den Einfluss von Spuren von Humusstoffen bzw. der Modellsubstanzen auf das Pflanzenwachstum. Landwirtsch. Forsch. 8:133-139.

Flaig, W. 1956. Zur Chemie der Huminsäuren und deren Modellsubstanzen. Trans. 6th Int. Congr. Soil Sci. II:471-478.

Flaig, W., 1967. Chemical composition and physical properties of humic substances. Pp.81-112. In: *Studies about Humus*, B. Novak and V. Rypacek (eds.). Trans. Intern. Symposium "Humus et Planta IV." Prague.

Flaig, W. 1968. Uptake of organic substances from soil organic matter by plant and their influence on metabolism. Pontif. Acad. Sci. Scri. Varia 32:723-770.

Flaig, W. 1970. Effect of humic substances on plant metabolism. Proc. 2nd Int. Peat Congress, Leningrad, pp.579-606.

Flaig, W. 1975. An introductory review on humic ssubstances: aspects of research on their genesis, their physical and chemical properties, and their effect on organisms. Pp.19-42. In: *Humic Substances. Their Structure and Function in the Biosphere*, D. Povoledo and H. L. Golterman (eds.). Proc. Intern. Meeting, Nieuwersluis, The Netherlands, May 29-31, 1972. Centre for Agric. Publishing and Documentation, Wageningen, The Netherlands.

Flaig, W. 1988. Generation of model chemical precursors. Pp. 75-111. In: *Humic Substances and their Role in the Environment*, F. H. Frimmel and R. F. Christman (eds.). Report Dahlem Workshop on Humic Substances and their Role in the Environment, Berlin 1987, March 29 - April 3. Wiley Interscience, New York, NY.

Flaig, W., and H. Beutelspacher.1951. Electron microscope investigations on natural and synthetic humic acids. Z. Pflanzenernähr. Düng. Bodenk. 52:1-21.

Flaig, W., and H. Beutelspacher. 1968. Investigations of humic acids with the analytical ultracentrifuge. Pp.23-30. In: *Use of Isotopes and Radiation in Soil Organic Matter Studies*. Int. Atomic Energy Agency, Vienna.

Flaig, W., and E. Saalbach. 1959. Über den Einfluss von im Boden gerottetem Stroh auf das Wachstum und die Nährstoffaufnahme von Roggen und Keimpflanzen. Z. Pflanzenernähr. Düng. Bodenk. 87:229-235.

Flaig, W., H. Beutelspacher, and E. Rietz. 1975. Chemical composition and physical properties of humic substances. Pp.1-211. In: *Soil Components*, Vol. 1. *Inorganic Components*, J. E. Gieseking (ed.). Springer-Verlag, New York, NY.

Flaig, W., U. Schobinger, and H. Deuel. 1959. Umwandlung von Lignin in Huminsäuren bei der verrottung von Weizenstroh. Chem. Ber. 92:1973-1982.

Forsyth, W. G. C. 1974. 1974. Studies on the more soluble complexes of soil organic matter. I. A method of fractionation. Biochem. J. 41:176-181.

Frissel, M. J. 1961. The adsorption of some organic compounds, especially herbicides, on clay minerals. Verslag Landbouwk. Onderzoek 76:3-14.

Fuchs, W. 1930a. Huminsäuren. Kolloid Zeitschr. 52:248-252.

Fuchs, W. 1930b. Huminsäuren. Kolloid Zeitschr. 53:124-126.

Fuchs, W. 1931. *Die Chemie der Khole*. Springer, Berlin

Fuchsman, C. H. 1980. *Peat: Industrial Chemistry and Technology*. Academic Press, New York, NY.

Gaffney, J. S., N. A. Marley, and S. B. Clark. 1996. Humic and fulvic acids and organic colloidal materials in the environment. Pp.2-16. In: *Humic and Fulvic Acids. Isolation, Structure, and*

References

Environmental Role, J. S. Gaffney, N. A. Marley, and S. B. Clark (eds.). ACS Symposium Series 651. American Chemical Society, Washington, DC.

Gaffney, J. S., N. A. Marley, and K. A. Orlandini. 1996a. The use of hollow-fiber ultrafilters for the isolation of natural humic and fulvic acids. Pp.26-40. In: *Humic and Fulvic Acids. Isolation, Structure, and Environmental Role,* J. S. Gaffney, N. A. Marley, and S. B. Clark (eds.). ACS Symposium Series 651. American Chemical Society, Washington, DC.

Giles, C. H., T. H. McEwan, S. N. Nakhwa, and D. Smith. 1960. Studies in adsorption. Part XI. A system of classification of solution adsorption isotherms, and its use in diagnosis of adsorption mechanisms and in measurement of specific surface areas of solids. J. Chem. Soc. p.3973.

Glonek, T., T. O. Henderson, R. L. Hildebrand, and T. C. Meyers. 1970. Biological phosphonates: Determination of phosphorus-31 nuclear magnetic resonance. Science 169:192-194.

Goodman, B. A., and M. V. Cheshire. 1972. A Mossbauer spectroscopic study of the effect of pH on the reaction between iron and humic acid in aqueous media. J. Soil Sci. 30:85-91.

Gortner, R. A. 1949. *Outlines of Biochemistry.* Third edition, edited by R. A. Gortner, Jr., and W. A. Gortner. John Wiley & Sons, Inc., New York, NY

Gosh, K., and M. Schnitzer. 1980a. Macromolecular structures of humic substances. Soil Sci. 129:266-276.

Gosh, K., and M. Schnitzer. 1980b. Fluorescence excitation spectra of humic substances. Canad. J. Soil Sci. 60:375-379.

Gosh, K., and M. Schnitzer. 1980c. Effect of pH and neutral electrolyte concentration on free radicals in humic substances. Soil Sci. Soc. Am. J. 44:975-978.

Graham, E. R. 1941. Colloidal organic acids as factors in the weathering of anorthite. Soil Sci. 52:291-295.

Greenland, D. J., G. R. Lindstrom, and J. P. Quirk. 1961. Role of polysaccharides in stabilization of natural soil aggregates. Nature 191:1283-1284.

Greenland, D. J., G. R. Lindstrom, and J. P. Quirk. 1962. Organic materials which stabilize natural soil aggregates. Soil Sci. Soc.

Am. Proc. 26:366-371.

Griffith, S. M., and M. Schnitzer. 1989. Oxidative degradation of soil humic substances. Pp.69-98. In: *Humic Substances II. In Search of Structure*, M. H. B. Hayes, P. MacCarthy, R. L. Malcolm, and R. S. Swift (eds.). Wiley-Interscience, New York, NY.

Guminski, S. 1957. The mechanism and conditions of the physiological actions of humic substances on the plant. Pochvovedenie 12:36.

Guminski, S., and Z. Guminska. 1953. Studies on the activity of humus on plants. Acta Soc. Bot. Pol. 22:45-54.

Guminski, S., D. Augustin, and J. Sulej. 1977. Comparison of some chemical and physico-chemical properties of natural and model sodium humates and biological activity of both substances in tomato water cultures. Acta Soc. Bot. Pol. XIVI:437-448.

Hadzi, D., C. Klofutar, and S. Oblak. 1968. Hydrogen bonding in some adducts of oxygen bases with acids. Part IV. Basicity in hydrogen bonding and ionization. J. Chem. Soc. (A):905-918.

Haider, K. 1994. Advances in the basic research of the biochemistry of humic substances. Pp.91-107. In: *Humic Substances in the Global Environment and Implications on Human Health*, N. Senesi and T. M. Miano (eds.). Proc. 6th Intern. Meeting of the Intern. Humic Substances Soc., Monopoli, Bari, Italy, Sept. 20-25, 1992. Elsevier, Amsterdam.

Haider, K., J. P. Martin, Z. Filip, and E. Fustec-Mathon. 1975. Contribution of soil microbes to the formation of humic compounds. Pp.71-85. In: *Humic Substances. Their Structure and Function in the Biosphere*, D. Povoledo and H. L. Goleman (eds.). Proc. Intern. Meeting, Nieuwersluis, The Netherlands, May 29-31, 1972. Center for Agric. Publishing and Documentation, Wageningen, The Netherlands.

Hargitai, L. 1955. Comparative organic matter investigations in various soil types with optical methods. Agrartud. Egy. Agronomy Kar. Kiadv. (Budapest) 2:1-27.

Hargitai, L. 1997. Applications of humification indexes, reactive nitrogen forms, and chemical topology for considerations of sorption of xenobiotic compounds by soil organic matter. an agricultural soil using a sequential and exhaustive extraction process. Pp.3-11. In: *Humic Substances and Organic Matter in*

References

Soil and Water Environments: Characteristics, Transformation, and Interactions, C. E. Clapp, M. H. B. Hayes, N. Senesi, and S. M. Griffith (eds.). Proc. 7th Conf. Intern. Humic Substances Soc., Univ. West Indies, St. Augustine, Trinidad, and Tobago, 3-8 July 1994. Published by International Humic Substances Soc., Inc., Dept. Soil, Water, and Climate, Univ. Minnesota, St. Paul, MN.

Hartung, H. A., and P. G. Allread. 1994. The algicidal action of peat humic substance and its copper chelate in ponds. Pp.1317-1324. In: *Humic Substances in the Global Environment and Implications on Human Health*, N. Senesi and T. M. Miano (eds.). Proc. 6th Intern. Meeting of the Intern. Humic Substances Soc., Monopoli, Bari, Italy, Sept. 20-25, 1992. Elsevier, Amsterdam.

Harvey, G. R., and D. A. Boran. 1985. Geochemistry of humic substances in seawater. Pp.233-247. In: *Humic Substances in Soil, Sediment, and Water. Geochemistry, Isolation, and Characterization*, G. R. Aiken, D. M. McKnight, R. L. Wershaw, and P. MacCarthy (eds.). Wiley-Interscience, New York, NY.

Hatcher, P. G., R. Rowan, and M. A. Mattingly. 1980. ^{1}H and ^{13}C NMR of marine humic acids. Org. Chem. 2:77-85.

Hatcher, P. G., G. E. Maciel, and L. W. Dennis. 1981. Aliphatic structure of humic acids: A clue to their origin. Org. Geochem. 2:43-48.

Hatcher, P. G., I. A. Breger, G. E. Maciel, and N. M. Szeverenyi.1985. Geochemistry of humin. Pp.275-302. In: *Humic Substances in Soil, Sediment, and Water. Geochemistry, Isolation, and Characterization*, G. R. Aiken, D. M. McKnight, R. L. Wershaw, and P. MacCarthy (eds.). Wiley-Interscience, New York, NY.

Hatcher, P. G., J-L Faulon, D. A. Clifford, and J. P. Matthews. 1994. A three-dimensional structural model for humic acids from oxidized soil. Pp.133-138. In: *Humic Substances in the Global Environment and Implications on Human Health*, N. Senesi and T. M. Miano (eds.). Proc. 6th Intern. Meeting of the Intern. Humic Substances Soc., Monopoli, Bari, Italy, Sept. 20-25, 1992. Elsevier, Amsterdam.

Hatcher, P. G., M. Schnitzer, L. W. Dennis, and G. E. Maciel. 1981. Aromaticity of humic substances in soils. Soil Sci. Soc. Am. J.

45:1089-1094.

Hayes, M. H. B. 1985. Extraction of humic substances from soil. Pp.329-362 . In: *Humic Substances in Soil, Sediment, and Water. Geochemistry, Isolation, and Characterization*, G. R. Aiken, D. M. McKnight, R. L. Wershaw, and P. MacCarthy (eds.). Wiley-Interscience, New York, NY.

Hayes, M. H. B., and R. L. Malcolm. 2001. Considerations of compositions and aspects of the structures of humic substances. Pp.3-39. In: *Humic Substances and Chemical Contaminants*, C. E. Clapp, M. H. B. Hayes, N. Senesi, P. R. Bloom, and P. M. Jardine (eds.). Proc. Workshop and Symposium Int. Humic Substances Soc., Soil Sci. Soc. Am., and Am. Soc. Agronomy, Anaheim, CA, 26-27 Oct., 1997. Soil Sci. Soc. Am., Inc., Madison, WI.

Hayes, M. H. B., P. MacCarthy, R. L. Malcolm, and R. S. Swift. 1989. The search for structure: Setting the scene. Pp.34-67. In: *Humic Substances II. In Search of Structure*, M. H. B. Hayes, P. MacCarthy, R. L. Malcolm, and R. S. Swift (eds.). Wiley-Interscience, New York, NY.

Hayes, T. M., M. H. B. Hayes, J. O. Skjemstad, R. S. Swift, and R. L. Malcolm. 1997. Isolation of humic substances from soil using aqueous extractants of different pH and XAD resins, and their characterization by ^{13}C-NMR. Pp.13-24. In: *Humic Substances and Organic Matter in Soil and Water Environments: Characteristics, Transformation, and Interactions*, C. E. Clapp, M. H. B. Hayes, N. Senesi, and S. M. Griffith (eds.). Proc. 7th Conf. Intern. Humic Substances Soc., Univ. West Indies, St. Augustine, Trinidad, and Tobago, 3-8 July 1994. Published by International Humic Substances Soc., Inc., Dept. Soil, Water, and Climate, Univ. Minnesota, St. Paul, MN.

Hedges, J. I. 1988. Polymerization of humic substances in natural environments. Pp.45-57. In: *Humic Substances and their Role in the Environment*, F. H. Frimmel and R. F. Christman (eds.). Report Dahlem Workshop on Humic Substances and their Role in the Environment, Berlin 1987, March 29 - April 3. Wiley Interscience, New York, NY.

Hermann, R. 1845. Bemerkungen zu Mulder's Untersuchungen uber Modersubstanzen und Ackererde. J. Prakt. Chemie 34:156-163.

Hobson, R. P., and H. J. Page. 1932a. Studies on the carbon and nitrogen cycles of the soil. J. Agric. Sci. 22:297-299.

Hobson, R. P., and H. J. Page. 1932b. Studies on the carbon and nitrogen cycles of the soil. J. Agric. Sci. 22:497-515.

Hobson, R. P., and H. J. Page. 1932c Studies on the carbon and nitrogen cycles of the soil. J. Agric. Sci. 22:516-526.

Hoffman, I., and M. Schnitzer. 1968. A thermogravimetric investigation of the effect of soil oxidizing agents on the ignition of carbonaceous materials. Fuel 47, No.3.

Holmgren, G. G. S., and C. S. Holzhey. 1984. A simple colorimetric measurement for humic acid in spodic horizons. Soil Sci. Soc. Am. J. 48:1374-1378.

Hoppe-Syeler, F. 1889. Uber Humin Substanzen, ihre Emstenung und ihre Eigenschaften. 2. Physiol. Chemie 13:66-121.

Huang, P. M. 1995. The role of short-range order mineral colloids in abiotic transformation of organic components in the environments. Pp.135-167. In: *Environmental Impact of Soil Component Interactions. Natural and Anthropogenic Organics.* Vol.1, P. M. Huang, J. Berthelin, J.-M. Bollag, W. B. McGill, and A. L. Page (eds.). CRC, Lewis Publ., Boca Raton, FL.

Ishiwatari, R. 1985. Geochemistry of humic substances in lake sediments. Pp.147-180. In: *Humic Substances in Soil, Sediment, and Water. Geochemistry, Isolation, and Characterization*, G. R. Aiken, D. M. McKnight, R. L. Wershaw, and P. MacCarthy (eds.). Wiley-Interscience, New York, NY.

Ivins, M. 2001. Home, home on the latrine. Time Magazine, August 6, 158:26.

Jackson, T. A. 1975. Humic matter in natural waters and sediments. Soil Sci. 119:56-64.

Jones, U. S. 1982. *Fertilizers and Soil Fertility.* Second ed. Reston Publ. Co., Reston, VA.

Jurcsik, I. 1994. Possibilities of applying humic acids in medicine (wound healing and cancer therapy). Pp.1331-1340. In: *Humic Substances in the Global Environment and Implications on Human Health*, N. Senesi and T. M. Miano (eds.). Proc. 6th Intern. Meeting of the Intern. Humic Substances Soc., Monopoli, Bari, Italy, Sept. 20-25, 1992. Elsevier, Amsterdam.

Kalle, K. 1938. Zum Problem der Meereswasserfarbe. Ann. Hydrogr. Mar. Meteorol. 66:1-13.

Kalle, K. 1966. The problem of the gelbstoff in the sea. Oceanography and Marine Annal. Review, H. Barns (ed.) 4:91-104.

Kappler, H., and W. Ziechmann. 1969. Ein mathematisches Modell zur Beschreibung von Humifizierungs vergängen. Brennstoff-Chemie 50:348-351.

Kemp, A. L. W., and A. Mudrochova. 1975. Nitrogen in sedimented organic matter from lake Ontario. Pp.137-157. In: *Humic Substances. Their Structure and Function in the Biosphere*, D. Povoledo and H. L. Golterman (eds.). Proc. Intern. Meeting, Nieuwersluis, The Netherlands, May 29-31, 1972. Centre for Agric. Publishing and Documentation, Wageningen, The Netherlands.

Khan, S. U., and M. Schnitzer. 1971. Sephadex gel filtration of fulvic acid.: The identification of major components in two low molecular weight fractions. Soil Sci. 112:231-238.

Khristeva, L. A. 1953. The participation of humic acids and other organic substances in the nutrition of higher plants. Pochvivedenie 10:46-59.

Khristeva, L. A., and A. Manoilova. 1950. The nature of the direct effect of humic acids on the growth and development of plants. Dokl. vsesoyuz. Akad. s.-kh. Nauk Lenina 11:10-16.

Klavins, M., and E. Apsite. 1997. Aquatic humic substances from surface waters in Latvia. Pp.269-282. In: *The Role of Humic Substances in the Ecosystems and in Environmental Protection*, J. Drozd, S. S. Gonet, N. Senesi, and J. Weber (eds.). Proc. 8th Meeting Intern. Humic Subst. Soc., Wroclaw, Poland, Sept. 9-14, 1996. IHSS-Polish Soc. Humic Substances, Wroclaw, Poland.

Klöcking, H. P. 1994. Influence of humic acids and humic acid-like polymers in fibrinolytic and coagulation system. Pp.1337-1340. In: *Humic Substances in the Global Environment and Implications on Human Health*, N. Senesi and T. M. Miano (eds.). Proc. 6th Intern. Meeting of the Intern. Humic Substances Soc., Monopoli, Bari, Italy, Sept. 20-25, 1992. Elsevier, Amsterdam.

Klöcking, H. P. 1997. Antigoagulatory efficacy of poly(hydroxy)-

carboxylates. Pp.951-953. In: *The Role of Humic Substances in the Ecosystems and in Environmental Protection*, J. Drozd, S. S. Gonet, N. Senesi, and J. Weber (eds.). Proc. 8th Meeting Intern. Humic Subst. Soc., Wroclaw, Poland, Sept. 9-14, 1996. IHSS-Polish Soc. Humic Substances, Wroclaw, Poland.

Klöcking, R. 1994. Humic substances as potential therapeutics. Pp.1245-1257. In: *Humic Substances in the Global Environment and Implications on Human Health*, N. Senesi and T. M. Miano (eds.). Proc. 6th Intern. Meeting of the Intern. Humic Substances Soc., Monopoli, Bari, Italy, Sept. 20-25, 1992. Elsevier, Amsterdam.

Klöcking, R., B. Helbig, G. Schötz, and P. Wutzler. 1997. A comparative study of the antiviral activity of low-molecular phenolic compounds and their polymeric humic acid-like oxidation products. Pp.955-960. In: *The Role of Humic Substances in the Ecosystems and in Environmental Protection*, J. Drozd, S. S. Gonet, N. Senesi, and J. Weber (eds.). Proc. 8th Meeting Intern. Humic Subst. Soc., Wroclaw, Poland, Sept. 9-14, 1996. IHSS-Polish Soc. Humic Substances, Wroclaw, Poland.

Kodama, H., and M. Schnitzer. 1977. Effect of fulvic acid on the crystallization of Fe(III) oxides. Geoderma 19:279-291.

Kodama, H., and M. Schnitzer. 1980. Effect of fulvic acid on the crystallization of aluminum hydroxides. Geoderma 24:195-205.

Kononova, M. M. 1956. Humus der Boden der U.D.S.S.R.; seine Natur und Rolle in Bodenbildungsprozessen. Trans. 6th Int. Congr. Soil Sci. II:557-565.

Kononova, M. M. 1961. *Soil Organic Matter*. T. Z. Nowakowski and G. A. Greenwood (transl.). Pergamon Press, Oxford, England.

Kononova, M. M. 1966. *Soil Organic Matter*. Pergamon Press, Elmsford, NY.

Kononova, M. M. 1970. *Microorganisms and Organic Matter of Soils*. Israel Program for Scientific Translations. Kester Press, Jerusalem.

Koopal, L. K., W. H. Van Riemsdijk, J. C. M. de Wit, and M. F. Benedetti. 1994. Analytical isotherm equations for multicomponent adsorption to heterogeneous surfaces. J. Colloid Interface Sci. 166:51-60.

Kumada, K. 1955. Absorption spectra of humic acids. Soil Plant Food 1:29-30.

Kumada, K. 1965. Studies on the colour of humic acids. Part 1. On the concepts of humic substances and humification. Soil Sci. Plant Nutr. 11:151-156.

Kumada, K. 1987. *Chemistry of Soil Organic Matter.* Japan Sci. Soc. Press, Tokyo.

Kumada, K., H. M. Hurst. 1967. Green humic acid and its possible origin as a fungal metabolite. Nature 214:5-88.

Kumada, K., and E. Miyara. 1973. Sephadex gel filtration of humic acids. Soil Sci. Plant Nutr. 19:255-263.

Kumada, K., and O. Sato. 1962. Chromatographic separation of green humic acid from podzol humus. Soil Sci. Plant Nutr. 31:611-623.

Ladd, J. N., and J. H. A. Butler. 1975. Humus-enzyme systems and synthetic organic polymer-enzyme analogs. Pp.143-194. In: *Soil Biochemistry,* E. A. Paul and A. D. McLaren (eds.). Vol.4. Marcel Dekker, Inc. New York, NY.

Lakatos, B., J. Meisel, and G. Mady. 1977. Biopolymer-metal complex systems. I. Experiments for the preparation of high purity peat humic substances and their metal complexes. Acta Agron. Sci. Hun. 26:259-271

Lal, R. (ed.). 2001. *Soil Carbon Sequestration and the Greenhouse Effect.* Special Publ. No.57. Soil Sci. Soc. Am., Madison, WI.

Lamar, W. L. 1968. Evaluation of organic color and iron in natural surface waters. U.S. Geol. Survey Prof. Paper 600D, D24-D29.

Lee, Y. S., R. J. Bartlett. 1976. Stimulation of plant growth by humic substances. Soil Sci. Soc. Am. J. 40:876-879.

Leenheer, J. A. 1985. Fractionation techniques for aquatic humic substances. Pp.409-429. In: *Humic Substances in Soil, Sediment, and Water. Geochemistry, Isolation, and Characterization,* G. R. Aiken, D. M. McKnight, R. L. Wershaw, and P. MacCarthy (eds.). Wiley-Interscience, New York, NY.

Leenheer, J. A., G. K. Brown, P. MacCarthy, and S. E. Cabiniss. 1998. Models of metal binding structures in fulvic acid from the Suwannee river, Georgia. Environ. Sci. Technol. 32:2410-2416.

Lindsay, W. L. 1974. Role of chelation in micronutrient availability. Pp.507-524. In: *The Plant Root and Its Environment,* E. W.

Carson (ed.). University Press of Virginia, Charlottesville, VA.
Lindsay, W. L., and W. A. Norvell. 1969. Equilibrium relationship of Zn^{2+}, Fe^{3+}, Ca^{2+}, and H^+ with EDTA and DTPA in soils. Soil Sci. Soc. Am. Proc. 33:62-68.
Lobartini, J. C., G. A. Orioli, and K. H. Tan. 1997. Characteristics of soil humic acid fractions separated by ultrafiltration. Comm. Soil Sci. Plant Anal. 28:787-796.
Lobartini, J. C., K. H. Tan, and C. Pape. 1989. Biochemical release of P from apatite by humic acid. Ninth Int. Symp. Environm. Biochem., Moscow, Russia, p.341.
Lobartini, J. C., K. H. Tan, and C. Pape. 1994. The nature of humic acid-apatite interaction products and their availability to plant growth. Comm. Soil Sci. Plant Anal. 25:2355-2369.
Lobartini, J. C., K. H. Tan, and C. Pape. 1998. Dissolution of aluminum and iron phosphate by humic acids. Comm. Soil Sci. Plant Anal. 29:535-544.
Lobartini, J. C., K. H. Tan, L. E. Asmussen, R. A. Leonard, D. Himmelsbach, and A. R. Gingle. 1991. Chemical and spectral differences in humic matter from swamps, streams and soils in the southeastern United States. Geoderma 49:241-254.
Lobartini, J. C., K. H. Tan, J. A. Rema, A. R. Gingle, C. Pape, and D. S. Himmelsbach. 1992. The geochemical nature and agricultural importance of commercial humic matter. Science Total Environment 113:1-15.
Loughnan, F. C. 1969. *Chemical Weathering of Silicate Minerals.* Elsevier Scientific Publ. Co., New York, NY.
Lovley, D. R., J. D. Coates, E. L. Blunt-Harris, E. J. P. Phillips, and J. C. Woodward. 1996. Humic substances as electron acceptors for microbial respiration. Nature 382:445-448.
Lovley, D. R., J. L. Fraga, E. L. Blunt-Harris, L. A. Hayes, E. J. P. Phillips, and J. D. Coates. 1998. Humic substances as a mediator for microbially catalyzed metal reduction. Acta Hydrochim, Hydrobiol. 26:152-157.
MacCarthy, P., and J. A. Rice. 1985. Spectroscopic methods (other than NMR) for determining functionality in humic substances. Pp.527-559. In: *Humic Substances in Soil, Sediment, and Water. Geochemistry, Isolation, and Characterization*, G. R. Aiken, D.

M. McKnight, R. L. Wershaw, and P. MacCarthy (eds.). Wiley-Interscience, New York, NY.

MacCarthy, P., M. J. Peterson, R. L. Malcolm, and E. M. Thurman. 1979. Separation of humic substances by pH gradient desorption from a hydrophobic resin. Anal. Chem. 51:2041-2043.

Maillard, L. C. 1916. Synthese de matières humiques par action des acides amines sur les sucres reducteures. Annal. Chem. Phys. 5:528-317.

Malcolm, R. L. 1985. Geochemistry of stream fulvic and humic substances. Pp.181-209. In: *Humic Substances in Soil, Sediment, and Water. Geochemistry, Isolation, and Characterization*, G. R. Aiken, D. M. McKnight, R. L. Wershaw, and P. MacCarthy (eds.). Wiley-Interscience, New York, NY.

Malcolm, R. E., and D. Vaughan. 1979. Effect of humic acid fractions on invertase activities in plant tissues. Soil Biol. Biochem. 11:65-72.

Martin, J. P., and K. Haider. 1971. Microbial activity in relation to soil humus formation. Soil Sci. 111:54-63.

Martin, J. P., K. Haider, and E. Bondietti. 1975. Properties of model humic acids synthesized by phenoloxidase and autoxidation of phenols and other compounds formed by soil fungi. Pp.171-186. In: *Humic Substances. Their Structure and Function in the Biosphere*, D. Povoledo and H. L. Golterman (eds.). Proc. Intern. Meeting, Nieuwersluis, The Netherlands, May 29-31, 1972. Centre for Agric. Publishing and Documentation, Wageningen, The Netherlands.

Mathur, S. P., and S. R. Farnham. 1985. Geochemistry of humic substances in natural and cultivated peatlands. Pp.53-85. In: *Humic Substances in Soil, Sediment, and Water. Geochemistry, Isolation, and Characterization*, G. R. Aiken, D. M. McKnight, R. L. Wershaw, and P. MacCarthy (eds.). Wiley-Interscience, New York, NY.

Mathur, S. P., and E. A. Paul. 1966. A microbiological approach to the problem of soil humic acid structure. Nature, London 212:646-647.

Mayer, L. M. 1985. Geochemistry of humic substances in estuarine environments. Pp. 211-232. In: *Humic Substances in Soil,*

Sediment, and Water. Geochemistry, Isolation, and Characterization, G. R. Aiken, D. M. McKnight, R. L. Wershaw, and P. MacCarthy (eds.). Wiley-Interscience, New York, NY.

McKnight, D. M., D. T. Scott, D. C. Hrncir, and D. R. Lovley. 2001. Pp.351-369. In: *Humic Substances and Chemical Contaminants*, C. E. Clapp, M. H. B. Hayes, N. Senesi, P. R. Bloom, and P. M. Jardine (eds.). Proc. Workshop and Symposium Int. Humic Substances Soc., Soil Sci. Soc. Am., and Am. Soc. Agronomy, Anaheim, CA, 26-27 Oct., 1997. Soil Sci. Soc. Am., Inc., Madison, WI.

McLoughin, A. J., and E. Kuster. 1972. The effect of humic substances on the respiration and growth of microorganisms. Plant Soil 37:17-25.

Mehlich, A. 1960. Charge properties in relation to sorption and desorption of selected cations and anions. Trans. Int. Soil Sci. Conf., Madison, WI, Vol.II/III: 292-302.

Mellor, D. P. 1964. Historical background and fundamental concepts. Pp.1-50. In: *Chelating Agents and Metal Chelates,* F. P. Dwyer and D. P. Mellor (eds.). Academic Press, New York, NY.

Minson, D. J., and J. R. Wilson. 1980. Comparative digestibility of tropical and temperate forage – A contrast between grasses and legumes. J. Austr. Inst. Agri. Sci. 46:247-249.

Miller, R. W., and D. T. Gardiner. 1998. *Soils in Our Environment*. Eight edition. Prentice Hall, Upper Saddle River, NJ.

Mills, M. S., E. M. Thurman, J. Ertel, and K. A. Thorn. 1996. Organic geochemistry and sources of natural aquatic foams. Pp.151-192. In: *Humic and Fulvic Acids. Isolation, Structure, and Environmental Role,* J. S. Gaffney, N. A. Marley, and S. B. Clark (eds.). ACS Symposium Series 651. American Chemical Society, Washington, DC.

Mockeridge, F. A. 1920. The occurrence and nature of the plant growth-producing substances in various organic manural composts. Biochem. J. 14:432-450.

Mortenson, J. L., D. M. Anderson, and J. L. White. 1965. Infrared spectroscopy. Pp.743-770. In: *Methods of Soil Analysis. Part I,* C.A. Black (ed-in-chief). Agronomy series nine. Am. Soc. Agronomy, Inc., Publ., Madison, WI.

Mortland, M. M., J. J. Fripiat, J. Claussidon, and J. Uijterhoeven. 1963. Interaction between ammonia and expanding lattices of mont-morillonite and vermiculite. J. Phys. Chem. 67:248-253.
Moulin, V. M., C. M. Moulin, and J-C Dran. 1996. Role of humic substances and colloids in the behavior of radiotoxic elements in relation to nuclear waste disposal. Pp.259-271. In: *Humic and Fulvic Acids. Isolation, Structure, and Environmental Role*, J. S. Gaffney, N. A. Marley, and S. B. Clark (eds.). ACS Symposium Series 651. American Chemical Society, Washington, DC.
Mudgal, V. G. 1972. *Characterization of Poultry Litter Extracts and their Adsorption by Soils*. Ph.D. dissertation, University of Georgia, Athens, GA. pp.63.
Mulder, G. J. 1840. Untersuchungen uber Humusartigen Materien. Annal. Chem. Pharm. 36:243-295.
Mulder, G. J. 1862. Die Chemie der Ackerkrume. Miller, Leipzig, Berlin
Muller, H. D. 1975. Poultry Waste. Georgia's 30 Million Dollar Forgotten Crop. Coop. Extension Service, Leaflet 206. College of Agri., University of Georgia, Athens, GA.
Müller, P. E. 1878. Studien über die näturlichen Humusformen. Tidsskr. für Skovbrug 3:1-14.
Müller-Wegener, U. 1988. Interaction of humic substances with biota. Pp.179-192. In: *Humic Substances and their Role in the Environment*, F. H. Frimmel and R. F. Christman (eds.). Report Dahlem Workshop on Humic Substances and their Role in the Environment, Berlin 1987, March 29 - April 3. Wiley Interscience, New York, NY.
Murmann, R. K. 1964. *Inorganic Complex Compounds*. Holt-Reinhold, New York, NY.
Nardi, S., G. Concheri, F. Reniero, and G. Dell'Agnola. 1997. Humic substance mobilization by root exudates of two different maize hybrids. Pp.981-986.In: *The Role of Humic Substances in the Ecosystems and in Environmental Protection*, J. Drozd, S. S. Gonet, N. Senesi, and J. Weber (eds.). Proc. 8th Meeting Intern. Humic Subst. Soc., Wroclaw, Poland, Sept. 9-14, 1996. IHSS-Polish Soc. Humic Substances, Wroclaw, Poland.
Neelakantan, S., M. M. Misira, H. K. Tewari, and S. R. Vyas. 1970.

Characterization and microbial utilization of humic acid in Hissar soil. Agrochimica 14:341-344.
Newman, R. H., and K. R. Tate. 1980. Soil phosphorus characterization by ^{31}P nuclear magnetic resonance. Comm. Soil Sci. Plant Anal.11:835-842.
Nissenbaum, A., and J. R. Kaplan. 1972. Chemical and isotopic evidence of the *in situ* origin of marine humic substances. Limnol. Oceanogr. 17:570-582.
Oden, S. 1914. Zuer Kolloidchemie der Humusstoff. Kolloid Zeitschr. 14:123-130.
Oden, S. 1919. Die Huminsaueren. Kolloidchemie Beih. 11:75-260.
Ogner, G. 1983. ^{31}P-NMR spectra of humic acids: A comparison of four different raw humus types in Norway. Geoderma 29:215-219.
Orioli, G. A., and N. R. Curvetto. 1980. Evaluation of extractants for soil humic substances. I. Isotachophoretic studies. Plant Soil 55:353-361.
Orlov, D. S. 1985. *Humus Acids of Soils*. Moscow University Press. Translated from Russian (K.H. Tan, ed.) Amerind Publ. New Delhi, India
Page, H. J. 1930. Studies on the carbon and nitrogen cycles in the soil. J. Agric. Sci. 20:455-459.
Parsons, J. W. 1989. Hydrolytic degradation of humic substances. Pp.99-120. In: *Humic Substances II. In Search of Structure*, M. H. B. Hayes, P. MacCarthy, R. L. Malcolm, and R. S. Swift (eds.). Wiley-Interscience, New York, NY.
Passer, M. 1957. Analytical procedure for the organic constituents of peat. Report No. 14. Chemical Products from Peat Project. University of Minnesota, Minneapolis, MN.
Paul, E. A., and F. E. Clark. 1989. *Soil Microbiology and Biochemistry*. Academic Press, Inc., San Diego, CA.
Perminova, I. V., N. Yu. Gretschishcheva, V. S. Petrosyan, M. A. Anisimova, N. A. Kulikova, G. F. Lebedeva, D. N. Matorin, and P. S. Venediktov. 2001. Impact of humic substances on toxicity of polycyclic aromatic hydrocarbons and herbicides. Pp.275-287. In: *Humic Substances and Chemical Contaminants*, C. E.Clapp, M. H. B. Hayes, N. Senesi, P. R. Bloom, and P. M. Jardine (eds.). Proc. Workshop and Symposium Int. Humic Substances Soc.,

Soil Sci. Soc. Am., and Am. Soc. Agronomy, Anaheim, CA, 26-27 Oct., 1997. Soil Sci. Soc. Am., Inc., Madison, WI.

Perminova, L. V., D. V. Kovalevsky, N. Yu. Yaschenko, N. N. Danchenko, A. V. Kudryavtsev, D, M., Zilin, V. S. Petrosyan, N.A. Kulikova, O. I. Philippova, and G. F. Lebedeva. 1997. Humic substances as natural detoxicants. Pp.399-406. In: *Humic Substances and Organic Matter in Soil and Water Environments: Characteristics, Transformation, and Interactions*, C. E. Clapp, M. H. B. Hayes, N. Senesi, and S. M. Griffith (eds.). Proc. 7th Conf. Intern. Humic Substances Soc., Univ. West Indies, St. Augustine, Trinidad, and Tobago, 3-8 July 1994. Published by International Humic Substances Soc., Inc., Dept. Soil, Water, and Climate, Univ. Minnesota, St. Paul, MN.

Pfeffer, P. E., and W. V. Gerasimowicz. 1989. *Nuclear Magnetic Resonance in Agriculture*. CRC Press, Boca Raton, FL.

Piccolo, A., and F. J. Stevenson. 1981. Infrared spectra of Cu^{2+}, Pb^{2+}, and Ca^{2+} complexes of soil humic substances. Geoderma 27:195-208.

Pierce, R. H., Jr., and G. T. Felbeck. 1975. A comparison of three methods of extracting organic matter from soils and marine sediments. Pp.217-232. In: *Humic Substances. Their Structure and Function in the Biosphere*, D. Povoledo and H. L. Golterman (eds.). Proc. Intern. Meeting, Nieuwersluis, The Netherlands, May 29-31, 1972. Centre for Agric. Publishing and Documentation, Wageningen, The Netherlands.

Poapst, P. A., and M. Schnitzer. 1971. Fulvic acid and adventitious root formation. Soil Biol. Biochem. 3:215-219.

Poapst, P. A., C. Genier, and M. Schnitzer. 1970. Effect of a soil fulvic acid on stem elongation in peas. Plant Soil 32:367-372.

Poapst, P. A., C. Genier, and M. Schnitzer. 1971. Fulvic acid and adventitious root formation. Soil Biol. Biochem. 3:367-372.

Popov, A. I., and O. G. Chertov. 1997. On humic substances as a direct nutritive components of plant-soil trophic system. Pp.993-998. In: *The Role of Humic Substances in the Ecosystems and in Environmental Protection*, J. Drozd, S. S. Gonet, N. Senesi, and J. Weber (eds.). Proc. 8th Meeting Intern. Humic Subst. Soc.,

References

Wroclaw, Poland, Sept. 9-14, 1996. IHSS-Polish Soc. Humic Substances, Wroclaw, Poland.

Posner, A. M. 1964. Titration curves of humic acid. Trans. 8th Int. Congr. Soil Sci., Bucharest, Romania, 1964. Academy of the Socialist Republic of Romania, Bucharest 11:161-164.

Prát, S. 1960. Distribution of the humus substance function in plants. Biologia Plantarum (Praha) 2:308-312.

Prát, S. 1970. Permeability and the effect of humic substances on plant cells. Pp.607-610. In: Proceeding Second Inter. Peat Congress, Leningrad, R. A. Robertson (ed.). HMSO, Edinburgh.

Preston, C. M., S. P. Mathur, and B. S. Rauthan. 1981. The distribution of copper, amino compounds, and humic fractions in organic soils of differing copper contents. Soil Sci. 131:344-352.

Purdue, E. M. 1988. Measurements of binding site concentrations in humic substances. Pp.135-154. In: *Metal Speciation – Theory, Analysis, and Application*, J. R. Kramer, and H. E. Allen (eds.). Lewis Publ., Chelsea, MI.

Purdue, E. M. 1998a. Metal binding by humic substances in surface waters – Experimental and modeling constraints. Pp.169-190. In: *Metals in Surface Waters*, H. E. Allen et al. (eds.). Ann Arbor Publishers, Chelsea, MI.

Purdue, E. M. 2001. Modeling concepts in metal-humic complexation. Pp.305-316. In: *Humic Substances and Chemical Contaminants*, C. E. Clapp, M. H. B. Hayes, N. Senesi, P. R. Bloom, and P. M. Jardine (eds.). Proc. Workshop and Symposium Int. Humic Substances Soc., Soil Sci. Soc. Am., and Am. Soc. Agronomy, Anaheim, CA, 26-27 Oct., 1997. Soil Sci. Soc. Am., Inc., Madison, WI.

Ranville, J. F., and R. L. Schmiermund. 1998. An overview of environmental colloids. Pp.25-45. In: *Perspectives in Environmental Chemistry*, D. L. Macalady (ed.). Oxford University Press, New York, NY.

Rice, J. A., and P. MacCarthy. 1989. Isolation of humin by liquid-liquid partitioning. Sci. Total Environment 81/82:61-69.

Richie, G. S. P., and A. M. Posner. 1982. The effect of pH and metal binding on the transport properties of humic acids. J. Soil Sci. 33:233-247.

Ridwan, F. N. J. 1977. *Untersuchungen zum Einfluss von Huminsäuren auf die Blei- und Cadmium-Absorption bei Ratten.* Dissertation, Universität Göttingen, Germany.

Riffaldi, R., and M. Schnitzer. 1972. Electron spin resonance spectrometry of humic substances. Soil Sci. Soc. Am. Proc. 36:301-307.

Rossell, R. A., J. C. Salfeld, and H. Sochtig. 1978. Organic components in Argentine soils. I. Nitrogen distribution and their humic acids. Agrochimica 22:98-105.

Roulet, N., N. Mehta, P. Dubach, and H. Deuel. 1963. Abtrennung von Kohlenhydraten und Stickstoff-verbindungen aus Huminstoffen durch Gelfiltration und Ionenaustausch-chromatographie. Z. Pflanzenernähr. Dung. Bodenk. 103:1-9.

Russell, J. E., and E. W. Russell. 1950. *Soil Conditions & Plant Growth.* Eight Edition. Longmans, Green and Co., London.

Saiz-Jiminex, C., K. Haider, and J. P. Martin. 1975. Anthraquinones and phenols as intermediates in the formation of dark colored humic acid-like pigments by *Eurotium echinulatum.* Soil Sci. Soc. Am. Proc. 39:649-662.

Savoini, G. 1986. General conclusions on the application of humic substances in agriculture. Pp.40-53. In: *Humic Substances. Effect on Soil and Plants.* Proc. Congress on Humic, Acids EniChem Agricoltura, Milano, 1986. Publ. by Ramo Editoriale degli Agricoltura, Milano.

Schalscha, E. B., H. Appelt, and A. Schatz. 1967. Chelation as a weathering mechanism: I. Effect of complexing agents on the solubilization of iron from minerals and granodiorite. Geochim. Cosmochim. Acta 31:587-596.

Scharpenseel, W. H. 1966. Aufbau und Bindungsform der Ton-Huminsauere-Organo-mineralische Komplexen Untersuchungen am Roentgengerat, I.R. Spectrometer und Elektronenmikroskop. Zeitschr. Pflanzenern. Dung. bodenk. 114:3-8.

Scheffer, F., und B. Ulrich. 1960. *Humus und Humusdüngung.* Ferdinand Enke, Stuttgart.

Schnitzer, M. 1965. The application of infrared spectroscopy to investigations on soil humic compounds. Canad. Spectrosc. 10:121-127.

Schnitzer, M. 1965a. Contribution of organic matter to the cation exchange capacity of soil. Nature 207:667-669.

References

Schnitzer, M. 1967. Humic-fulvic acid relationships in organic soils and humification of the organic matter in these soils. Can. J. Soil Sci. 47:245-250.

Schnitzer, M. 1975. Chemical, spectroscopic, and thermal methods for the classification and characterization of humic substances. Pp.293-310. In: *Humic Substances. their Structure and Function in the Biosphere*, D. Povoledo and H. L. Golterman (eds.). Proc. Intern. Meeting Humic Substances, Nieuwersluis, The Netherlands, May 29-31, 1972. Centre for Agric. Publishing and Documentation, Wageningen, The Netherlands.

Schnitzer, M. 1976. The chemistry of humic substances. Pp.89-107. In: *Environmental Biogeochemistry*. Vol.1. *Carbon, Nitrogen, Phosphorus, Sulfur, and Selenium Cycles*, J. O. Nriagu (ed.). Proc. 2nd Int. Symp. Environ. Biogeochem., Hamilton, Ontario, Canada, April 8-12, 1975. Ann Arbor Science, Ann Arbor, MI.

Schnitzer, M. 1977. Recent findings on the characterization of humic substances extracted from soils from widely differing climatic zones. Proc. Symposium Soil Organic Matter Studies, Braunschweig. Int. Atomic Energy Agency, Vienna. pp. 117-131.

Schnitzer, M. 1978. Humic substances: Chemistry and Reactions. Pp.1-64. In: *Soil Organic Matter*, M. Schnitzer and S. U. Khan (eds.). Elsevier, Amsterdam and New York, NY.

Schnitzer, M. 1982a. Que vadis organic matter research? Trans. 12th Intern. Congress Soil Sci., New Delhi 4:67-78.

Schnitzer, M. 1982b. Organic matter characterization. Pp.581-594. In: *Methods of Soil Analysis, Part 2*, A. L. Page, R. H. Miller, and D. R. Keeney (eds.). Agronomy series. No.9. Am. Soc. Agronomy and Soil Sci. Soc. Am., Inc., Publ., Madison, WI.

Schnitzer, M. 1986. Binding of humic substances by soil mineral colloids. Pp.77-101. In: *Interactions of Soil Minerals with Natural Organics and Microbes*, P. M. Huang and M. Schnitzer (eds.). SSSA Special Publ. No. 17. Soil Sci. Soc. Am., Inc. Madison, WI.

Schnitzer, M. 1994. A chemical structure for humic acid. Chemical, ^{13}C NMR, colloid chemical, and electron microscopic evidence. Pp.57-69. In: *Humic Substances in the Global Environment and Implications on Human Health*, N. Senesi and T. M. Miano (eds.). Proc. 6th Intern. Meeting of the Intern. Humic Substances Soc., Monopoli, Bari, Italy, Sept. 20-25, 1992. Elsevier, Amsterdam.

Schnitzer, M. 2000. A lifetime perspective on the chemistry of soil organic matter. Adv. Agron. 68:3-58.

Schnitzer, M., and K. Ghosh. 1982. Characterization of water-soluble fulvic acid-copper and fulvic acid-iron complexes. Soil Sci.124:354-363.

Schnitzer, M., and S. U. Khan. 1972. *Humic Substances in the Environment*. Marcel Dekker, Inc., New York, NY.

Schnitzer, M., and H. Kodama. 1975. An electron microscopic examination of fulvic acid. Geoderma 13:279-287.

Schnitzer, M., and H. Kodama. 1976. The dissolution of micas by fulvic acid. Geoderma 15:381-391.

Schnitzer, M., and H. Kodama. 1977. Reactions of minerals with soil humic substances. Pp.741-770. In: *Minerals in Soil Environments*, J. B. Dixon, S. B. Weed, J. A. Kittrick, M. H. Milford, and J. L. White (eds.). Soil Sci. Soc. Am., Madison, WI.

Schnitzer, M., and C. M. Preston. 1986. Analysis of humic acids by solution and solid-state carbon-13 nuclear magnetic resonance. Soil Sci. Soc. Am. J. 50:326-331.

Schnitzer, M., and S. I. M. Skinner. 1968. Gel filtration of fulvic acid, a soil humic compound. Pp.41-48. In: *Isotopes and Radiation in Soil Organic Matter Studies*. Int. Atomic Energy Agency, Vienna.

Schnitzer, M., H. Kodama, and J. A. Ripmeester. 1991. Determination of aromaticity of humic substances by x-ray diffraction analysis. Soil Sci. Soc. Am. J. 55:745-750.

Schnitzer, M., D. A. Shearer, and J. R. Wright. 1959. A study in the infrared of high molecular-weight organic matter extracted by various reagents from a podzolic B horizon. Soil Sci. 87:252-257.

Schubert, W. J. 1965. *Lignin Biochemistry*. Academic Press, New York, NY.

Schulten, H.-R. 1994. A chemical structure for humic acid. pyrolysis-gas chromatography/mass spectrometry, and pyrolysis-soft ionization mass spectrometry evidence. Pp.43-56. In: *Humic Substances in the Global Environment and Implications on*

Human Health, N. Senesi and T. M. Miano (eds.). Proc. 6th Intern. Meeting of the Intern. Humic Substances Soc., Monopoli, Bari, Italy, Sept. 20-25, 1992. Elsevier, Amsterdam.

Schulten, H.-R. 1996. A new approach to the structural analysis of humic substances in water and soils. Pp.42-56. In: *Humic and Fulvic Acids. Isolation, Structure, and Environmental Role*, J. S. Gaffney, N. A. Marley, and S. B. Clark (eds.). ACS Symposium Series 651. American Chemical Society, Washington, DC.

Schulten, H.-R. 2001. Models of humic structures: Association of humic acids and organic matter in soils and water. Pp.73-87. In: *Humic Substances and Chemical Contaminants*, C. E. Clapp, M. H. B. Hayes, N. Senesi, P. R. Bloom, and P. M. Jardine (eds.). Proc. Workshop and Symposium Int. Humic Substances Soc., Soil Sci. Soc. Am., and Am. Soc. Agronomy, Anaheim, CA, 26-27 Oct., 1997. Soil Sci. Soc. Am., Inc., Madison, WI.

Schulten, H.-R. 2002. New approaches to the molecular structure and properties of soil organic matter: Humic-, xenobiotic-, biological-, and mineral-bonds. Pp.351-381. In: *Soil Mineral-Organic Matter-Microorganisms Interactions and Ecosystem Health. Dynamics, Mobility and Transformation of Pollutants and Nutrients*, A. Violante, P. M. Huang, J.-M. Bolag, and L. Gianfreda (eds.). Development in Soil Science 28A. Elsevier, Amsterdam.

Schulten, H.-R, and M. Schnitzer. 1993. A state-of-the-art structural concept for humic substances. Naturwissenschaften 80: 29-30.

Schulten, H.-R, and M. Schnitzer. 1995. Three dimensional models for humic acids and soil organic matter. Naturwissenschaften 82: 487-498.

Schwertmann, U. 1971. Transformation of hematite to goethite in soils. Nature (London) 232:624-625.

Schwertmann, U., and R. M. Taylor. 1977. Iron Oxides. Pp.145-180. In: *Minerals in Soil Environments*, J. B. Dixon and S. B. Weed (eds.). Soil Sci. Soc. Am., Madison, WI.

Senesi, N., and M. Schnitzer. 1977. Effects of pH, reaction time, chemical reduction, and irradiation on ESR spectra of fulvic acid. Soil Sci. 123:224-234.

Senesi, N., and G. Sposito. 1984. Residual copper (II) complexes in purified soil and sewage sludge fulvic acids: Electron spin resonance study. Soil Sci. Soc. Am. J. 48:1247-1253.

Senesi, N., S. M. Griffith, M. Schnitzer, and M. G. Townsend. 1977. Binding of Fe^{3+} by humic materials. Geochim. Cosmochim. Acta 41:969-976.

Shapiro, J. 1957. Chemical and biological studies on the yellow organic acids of lake water. Limnol. Oceanogr. 2:161-179.

Shoji, S., M. Nanzyo, and R. A. Dahlgren. 1993. *Volcanic Ash Soils. Genesis, Properties, and Utilization.* Elsevier, Amsterdam.

Simon, K., and H. Speichermann. 1938. Beiträge zur Humusuntersuchungs methodik. Bodenkunde Pflanzenernährung 8:129-152.

Singer, A., and J. Navrot. 1976. Extraction of metals from basalt by humic acids. Nature (London) 262:479-481.

Sinha, M. K. 1971. Organo-metallic phosphates. I. Interaction of phosphorus compounds with humic substances. Plant Soil 35:471-434.

Sinha, M. K. 1972. Organo-metallic phosphates. IV. The solvent action of fulvic acids on insoluble phosphates. Plant Soil 37:457-467.

Skogerboe, R. K., and S. A. Wilson. 1981. Reduction of ionic species by fulvic acid. Anal. Chem. 53:228-232.

Sladky, Z. 1959. The effect of extracted humus substances on growth of tomato plants. Biol. Plant. (Prague) 1:142-150.

Sladky, Z. 1965. Die durch Blattdüngung mit Humusstoffen hervorgerufenen anatomischen une physiologischen Veränderungen der Zuckerrübe. Biol. Plant. (Prague) 7:251-260.

Sladky, Z., and V. Tichy. 1959. Application of humus substances to overground organs of plants. Biol. Plant. (Prague) 1:9-15.

Šmídová, M. 1960. The influence of humic acid on the respiration of plant roots. Biol. Plant. (Prague) 2:152-164.

Soil Survey Staff. 1990. *Keys to Soil Taxonomy.* SMSS Techn. Monograph No.19. Fourth edition. Virginia Polytechnic Institute and State University, Blacksburg, VA.

Somani, L. L., and S. N. Saxena. 1982. Distribution of humus fractions in some soil groups of Rajasthan. Agrochimica 26:93-103.

Sowden, F. J. 1957. Note on the occurrence of amino groups in soil organic matter. Canad. J. Soil Sci. 37:143-144.

Sowden, F. J., and D. I. Parker. 1953. Amino nitrogen of soils and certain fractions isolated from them. Soil Sci. 76:201-208.
Sowden, F. J., S. M. Griffith, and M. Schnitzer. 1976. The distribution of nitrogen in some highly organic tropical volcanic soils. Soil Biol. Biochem. 8:55-60.
Spigarelli, S. A. 1994. Stimulation of onion root growth by peat humic substances: Effects of extraction temperature and pH. Pp.513-520. In: *Humic Substances in the Global Environment and Implications on Human Health*, N. Senesi and T. M. Miano (eds.). Proc. 6th Intern. Meeting of the Intern. Humic Substances Soc., Monopoli, Bari, Italy, Sept. 20-25, 1992. Elsevier, Amsterdam.
Sposito, G. 1989. *The Chemistry of Soils*. Oxford University Press, New York, NY.
Sposito, G., K. M. Holtzclaw, and C. S. Levesque-Madore. 1978. Calcium ion complexation by fulvic acid extracted from sewage sludge-soil water mixtures. Soil Sci. Soc. Am. J. 41:600-606.
Sposito, G., K. M. Holtzclaw, and C. S. Levesque-Madore. 1981. Trace metal complexation by fulvic acid extracted from sewage sludge: I. Determination of stability constants and linear correlation analysis. Soil Sci. Soc. Am. J. 45:465-468.
Sprengel, C. 1826. Über Pflanzenhumus, Humussaüre und Humussaüre Salze. Kastner's Arch. Ges. Natürlehre (Nürnberg) 8:145-220.
Springer, U. 1934. Farbtiefe und Farbcharakter von Humusextrakten in ihrer Abhandigkeit von der Alkalikonzentration, zugleich ein Beitrag zur Kenntniss der Humustypen. Z. Pflanzenernähr. Düng. Bodenk. 34:1-18.
Springer, U. 1938. Der Heutige Stand der Humusuntersuchungsmethodik mit besonderer Berücksichtung der Trennung, Bestimmung und Characterisierung. Bodenk. D. Pflanzenernähr. 6:312-373.
Stach, E. 1975. *Coal Petrology*. Gebruder Borntraeger, Stuttgart, Germany.
Steelink, C. 1963. What is humic acid. J. Chem. Educ. 40:379-384.
Steelink, C. 1964. Free radical studies of lignin, lignin degradation products and soil humic acids. Geochim. Cosmochim. Acta 28:1615-1622.

Steelink, C. 1967. Humic acid. Encyclopedia of Polymer Sci. and Technology 7:530-539.

Steelink, C. 1985. Implications of elemental characteristics of humic substances. Pp.457-476. In: *Humic Substances in Soil, Sediment, and Water. Geochemistry, Isolation, and Characterization*, G. R. Aiken, D. M. McKnight, R. L. Wershaw, and P. MacCarthy (eds.). Wiley-Interscience, New York, NY.

Steelink, C., and G. Tolling. 1967. Free radicals in soils. Pp.147-173. In: *Soil Biochemistry*, A. D. McLaren and G. H. Peterson (eds.). Marcel Dekker, Inc., New York, NY.

Steinberg, C., and U. Muenster. 1985. Geochemistry and ecological role of humic substances. Pp.105-145. In: *Humic Substances in Soil, Sediment, and Water. Geochemistry, Isolation, and Characterization*, G. R. Aiken, D. M. McKnight, R. L. Wershaw, and P. MacCarthy (eds.). Wiley-Interscience, New York, NY.

Stevenson, F. J. 1965. Gross chemical fractionation of organic matter. Pp.1409-1421. In: *Methods of Soil Analysis, Part I*, C. A. Black, D. D. Evans, J. L. White, L. E. Ensminger, and F. Clark (eds.). Agron. Series No.9. Am. Soc. Agronomy, Madison, WI.

Stevenson, F. J. 1976a. Stability constants of Cu^{2+}, Pb^{2+}, and Cd^{2+} complexes with humic acids. Soil Sci. Soc. Am. J. 40:665-672.

Stevenson, F. J. 1976b. Binding metal ions by humic acids. Pp.519-540. In: *Environmental Biogeochemistry*. Vol.1. *Carbon, Nitrogen, Phosphorus, Sulfur, and Selenium Cycles*, J. O. Nriagu (ed.). Proc. 2nd Int. Symp. Environ. Biogeochem., Hamilton, Ontario, Canada, April 8-12, 1975. Ann Arbor Science, Ann Arbor, MI.

Stevenson, F. J. 1982. *Humus Chemistry, Genesis, Composition, Reactions*. John Wiley & Sons, Inc., New York, NY.

Stevenson, F. J. 1986. *Cycles of Soil Carbon, Nitrogen, Phosphorus, Sulfur, Micronutrients*. John Wiley & Sons, New York, NY.

Stevenson, F. J. 1989. Reductive cleavage of humic substances. Pp.121-142. In: *Humic Substances II. In Search of Structure*, M. H. B. Hayes, P. MacCarthy, R. L. Malcolm, and R. S. Swift (eds.). Wiley-Interscience, New York, NY.

Stevenson, F. J. 1994. *Humus Chemistry. Genesis, Composition, Reactions*. Second Edition. John Wiley & Sons, New York, NY.

References

Stevenson, I. L., and M. Schnitzer. 1982. Transmission electron microscopy of extracted fulvic and humic acids. Soil Sci.133:179-185.

Stewart, A. J., and R. G. Wetzel. 1982. Influence of dissolved humic material on carbon assimilation and alkaline phosphatase activity in natural agalbacterial assemblages. Freshwater Biol. 12:369-380.

Sulzberger, B., and H. Laubscher. 1995. Reactivity of various types of iron(III) oxides toward light induced dissolution. Mar. Chem. 50:103-115.

Sulzberger, B., H. Laubscher, and J. Karametaxes. 1994. Photoredox reaction at the surface of iron(III) hydroxides for natural water systems. Pp.53-73. In: *Aquatic and Surface Photochemistry*, G. Heltz (ed.). CRC Press, Boca Raton, FL.

Swain, F. M. 1963. Geochemistry of humus. Pp.81-147. In: *Organic Chemistry*, J. A. Breger (ed.). Pergamon Press, New York, NY.

Swain, F. M. 1975. Biogeochemistry of humic compounds. Pp.337-360. In: *Humic Substances. Their Structure and Function in the Biosphere*, D. Povoledo and H. L. Golterman (eds.). Proc. Intern. Meeting Humic Substances, Nieuwersluis, The Netherlands, May 29-31, 1972. Centre for Agric. Publishing and Documentation, Wageningen, The Netherlands.

Swift, R. S. 1985. Fractionation of soil humic substances. Pp.387-408. In: *Humic Substances in Soil, Sediment, and Water. Geochemistry, Isolation, and Characterization*, G. R. Aiken, D. M. McKnight, R. L. Wershaw, and P. MacCarthy (eds.). Wiley-Interscience, New York, NY.

Tan, K. H. 1964. The andosols in Indonesia. Meeting on the Classification and Correlation of Soils from Volcanic Ash, Tokyo, Japan, June 11-27, 1964. FAO/UN. World Soil Resources Report No.14, pp.30-35.

Tan, K. H. 1965. The andosols in Indonesia. Soil Sci. 99:375-378.

Tan, K. H. 1975. Infrared absorption similarities between hymatomelanic acid and methylated humic acid. Soil Sci. Soc. Am. Proc. 39:70-73.

Tan, K. H., 1976. Contamination of humic acid by silica gel and sodium bicarbonate. Plant Soil 44:691-695.

Tan, K. H. 1977. High and low molecular weight fractions of humic and fulvic acids. Plant Soil 48:89-101.
Tan, K. H. 1978. Variations in soil humic compounds as related to regional and analytical differences. Soil Sci. 125:351-358.
Tan, K. H. 1980. The release of silicon, aluminum, and potassium during decomposition of soil minerals by humic acid. Soil Sci. 129:5-11.
Tan, K. H. 1981. *Basic Soils Laboratory*. Burgess Publ. Co., Minneapolis, MN.
Tan, K. H. (ed.). 1984. *Andosol*. A Hutchinson Ross Benchmark book. Van Nostrand Reinhold Co., New York, NY.
Tan, K. H. 1985. Scanning electron microscopy of humic matter as influenced by methods of preparation. Soil Sci. Soc. Am. J. 49:1185-1191.
Tan, K. H. 1986. Degradation of soil minerals by organic acids. Pp.1-27. In: *Interactions of Soil Minerals with Natural Organics and Microbes*, P. M. Huang and M. Schnitzer (eds.). SSSA Special Publ. No.17. Soil Sci. Soc. Am., Madison, WI.
Tan, K. H. 1996. *Soil Sampling, Preparation, and Analysis*. Marcel Dekker, Inc., New York, NY.
Tan, K. H. 1998. *Andosols* (In Indonesian). Syllabus of lectures. Grad. School, University of North Sumatra, Medan, Indonesia.
Tan, K. H. 2000. *Environmental Soil Science*. Second Edition. Marcel Dekker, New York, NY.
Tan, K. H., and A. Binger. 1986. Effect of humic acid on aluminum toxicity in corn plants. Soil Sci. 141:20-25.
Tan, K. H., and F. E. Clark. 1968. Polysaccharide constituents in fulvic and humic acids extracted from soil. Geoderma 2:245-255.
Tan, K. H., and J. E. Giddens. 1972. Molecular weights and spectral characteristics of humic and fulvic acids. Geoderma 8:221-229.
Tan, K. H., and R. M. McCreery. 1975. Humic acid complex formation and intermicellar adsorption by bentonite. Pp.629-641. Proc. Int. Clay Conf. Mexico City, Mexico, July 16-23, 1975, S. W. Bailey (ed-in-chief). Applied Publ. Ltd., Wilmette, IL.
Tan, K. H., and V. Nopamornbodi. 1979. Effect of different levels of humic acids on nutrient content and growth of corn (*Zea mays* L.). Plant and Soil 51:283-387.

References

Tan, K. H., and V. Nopamornbodi. 1979a. Fulvic acid and the growth of the ectomycorrhizal fungus, *Pisolithus tinctorius*. Soil Biol. Biochem. 11:651-653.

Tan, K. H., and J. Rema. 1992. *The Biochemical Characteristics of Lignite derived Humate Mixtures and their Humate Fractions*. Report of Earth Tones, Inc., Atlanta, GA. Dept. Crops and Soil Sci., University of Georgia, Athens, GA.

Tan, K. H., and J. Rema. 1993. *Chemical Analysis of Lignite Samples from Utah*. Report of the Meridian Company, Marietta, GA. Dept. Crops and Soil Sci., University of Georgia, Athens, GA.

Tan, K. H., and J. Van Schuylenborgh. 1959. On the classification and genesis of soils derived from andesitic volcanic material under a monsoon climate. Neth. J. Agric. Sci. 7:1-22.

Tan, K. H., and J. Van Schuylenborgh. 1961. On the organic matter in tropical soils. Neth. J. Agric. Sci. 9:174-180.

Tan, K. H., D. S. Himmelsbach, and J. C. Lobartini. 1992. The significance of solid-state ^{13}C NMR spectroscopy of whole soil in the characterization of humic matter. Comm. Soil Sci. Plant Anal. 23:1513-1532.

Tan, K. H., L. D. King, and H. D. Morris. 1971. Complex reactions of zinc with organic matter extracted from sewage sludge. Soil Sci. Soc. Am. Proc. 35:748-751.

Tan, K. H., V. G. Mudgal, and R. A. Leonard. 1975. Adsorption of poultry litter extracts by soil and clay. Environ. Sci. Technol. 9:132-135.

Tan, K. H., J. A. Rema, and J. C. Lobartini. 1988. *The Biochemical and Spectral Characteristics of Unicam and Its Humic Fraction*. Report to the Unicamp Company, Chemical Group, Humate Products, Jacksonville, FL. Dept. Agronomy, University of Georgia, Athens, GA.

Tan, K. H., P. Sihanonth, and R. L. Todd. 1978. Formation of humic acid like compounds by the ectomycorrhizal fungus, *Pisolithus tinctorius*. Soil Sci. Soc. Am. J. 42:906-908.

Tan, K. H., R. A. Leonard, A. R. Bertrand, and S. R. Wilkinson. 1971a. The metal complexing capacity and the nature of the chelating ligands of water extracts of poultry litter. Soil Sci. Soc. Am. Proc. 35:265-269.

Tan, K. H., J. C. Lobartini, D. S. Himmelsbach, and J. E. Asmussen. 1991. Composition of humic acids extracted under air and nitrogen atmosphere. Comm. Soil Sci. Plant Anal. 22:861-877.
Tan, K. H., D. S. Himmelsbach, J. C. Lobartini, and G. R. Gamble. 1994. The issue of artifacts in NaOH extraction of humic matter. Pp.109-114. In: *Humic Substances in the Global Environment and Implications on Human Health*, N. Senesi and T. M. Miano (eds.). Proc. 6th Intern. Meeting of the Intern. Humic Substances Soc., Monopoli, Bari, Italy, Sept. 20-25, 1992. Elsevier, Amsterdam.
Tan, K. H., R. A. Leonard, L. E. Asmussen, J. C. Lobartini, and A. R. Gingle. 1990. The geochemistry of black water in selected coastal streams of the Southeastern United States. Comm. Soil Sci. Plant Nutr. 21:1999-2016.
Tan, K. H., J. C. Lobartini, D. S. Himmelsbach, and L. E. Asmussen. 1991. Composition of humic acids extracted under air and nitrogen atmosphere. Comm. Soil Sci. Plant Anal. 22:867-877.
Tham, J., W. Jansen, and H. Rahman. 1997. Effects of humic material on aquatic invertebrates in streams of a raised bog complex. Pp.929-935. In: *The Role of Humic Substances in the Ecosystems and in Environmental Protection*, J. Drozd, S. S. Gonet, N. Senesi, and J. Weber (eds.). Proc. 8th Meeting Intern. Humic Subst. Soc., Wroclaw, Poland, Sept. 9-14, 1996. IHSS-Polish Soc. Humic Substances, Wroclaw, Poland.
Theng, B. K. G. (ed.). 1980. *Soils with Variable Charge*. Soil Bureau, Dept. Scientific and Industrial Research, New Zealand Soil Science Soc., Lower Hutt, New Zealand.
Thiel, K. D., R. Klöcking, H. Schweizer, and M. Sprössig. 1977. Zentral-blatt für Bakterilogie, Parasitenkunde, Infektionskrankheiten und Hygiene. Abteilung I: Originale Reih A: Mediziche Mikrobiologie und Parasitologie 234:304-321.
Thorn, K. A. 1989. Nuclear-magnetic resonance spectrometry investigations of fulvic and humic acids from the Suwannee River. In: Humic Substances in the Suwannee River, Georgia: Interactions, Properties, and Proposed Structures, R. C. Averett, J. A. Leenheer, D. M. McKnight, and K. A. Thorn (eds.). U.S. Geol. Survey Open-File Report 87-557.

Thorn, K. A., D. W. Folan, and P. MacCarthy. 1989. Characterization of the International Humic Substances Society standard and reference fulvic and humic acids by solution state carbon-13 (^{13}C) and hydrogen-1 (^{1}H) nuclear magnetic resonance spectrometry. Water-Resources Investigations Report 89-4196. U.S. Geol. Survey, Denver, CO.

Thorn, K. A., W. S. Goldenberg, S. J. Younger, and E. J. Weber. 1996. Covalent binding of aniline to humic substances. Pp.299-326. In: *Humic and Fulvic Acids. Isolation, Structure, and Environmental Role,* J. S. Gaffney, N. A. Marley, and S. B. Clark (eds.). ACS Symposium Series 651. American Chemical Society, Washington, DC.

Thurman, E. M., and R. L. Malcolm. 1981. Preparative isolation of aquatic humic substances. Environ. Sci. Technol. 15:463-466.

Thurman, E. M., and R. L. Malcolm. 1983. Structural study of humic substances: New approaches and methods. Pp.1-23. In: Aquatic and Terrestrial Humic Materials, R. F. Christman and E. T. Gjessing (eds.). Ann Arbor Science, Ann Arbor, MI.

Thurman, E. M., R. L. Malcolm, and G. R. Aiken. 1978. Prediction of capacity factors for aqueous organic solutes on a porous acrylic resin. Anal. Chem. 50:775-779.

Tinsley, J., and C. H. Walker. 1964. The composition of organic matter extracted from soil with formic acid. Trans. 8th Intern. Congr. Soil Sci., Bucharest, 2:149-152.

Tipping, E., and M. A. Hurley. 1992. A unified model of cation binding by humic substances. Geochim. Cosmochim. Acta 56:3627-3641.

Tisdale, S. L., W. L. Nelson, J. D. Beaton, and J. L. Havlin. 1993. *Soil Fertility and Fertilizers.* Macmillan Publ. Co., New York, NY.

Tolpa, S. 1976. Progress in studies on active compounds isolated from peat. In: *Peat and Peatlands in the Natural Environment Protection.* Proc. 5th Intern. Peat Congress, Poznan, Poland, 1976. Vol.1: 413-416.

Tsusuki, K., and S. Kuwatsuka. 1978. Composition of oxygen-containing functional groups of humic acids. II. Soil Sci. Plant Nutr. 24:547-560.

U.S. Geol. Survey Staff. 1989. Humic Substances in the Suwannee River, Georgia. Interactions, Properties, and Proposed

Structures. Open File Report 87-557. U.S. Geol. Survey, Denver, CO.

Vandenbroucke, M., R. Pelet, and Y. Debyser. Geochemistry of humic substances in marine sediments. Pp.249-273. In: *Humic Substances in Soil, Sediment, and Water. Geochemistry, Isolation, and Characterization*, G. R. Aiken, D. M. McKnight, R. L. Wershaw, and P. MacCarthy (eds.). Wiley-Interscience, New York, NY.

Van der Marel, H. W. 1949. Mineralogical composition of a heath podzol profile. Soil Sci. 67:193-207.

Van Krevelen, D. W. 1963. Geochemistry of coal. Pp.183-247. In: *Organic Geochemistry*, I. A. Breger (ed.). Pergamon Press, Oxford, England.

Vaughan, D. 1974. A possible mechanism for humic acid action on cell elongation in root segments of *Pisum sativum* under aseptic conditions. Soil Biol. Biochem. 6:241-247.

Visser, S. A. 1963. Electron microscopic and electron-diffraction patterns of humic acids. Soil Sci. 96:353-356.

Visser, S. A. 1983. Application of Van Krevelen's graphical-statistical method for the study of aquatic humic matter. Environ. Sci. Technol. 17:412-417.

Visser, S. A. 1985. Physiological action of humic substances on microbial cells. Soil Biol. Biochem. 17:457-462.

Wada, K., and S. Aomine. 1973. Soil development on volcanic materials during the quaternary. Soil Sci. 116:170-177.

Waksman, S. A. 1932. *Humus*. Williams and Wilkins, Baltimore, MD.

Waksman, S. A. 1936. *Humus*. Williams and Wilkins, Baltimore, MD.

Waksman, S. A. 1938. *Humus–Origin, Chemical, Composition, and Importance in Nature*. Second edition. Williams and Wilkins, Baltimore, MD.

Weast, R. C. 1971. *Handbook of Chemistry and Physics. 52nd edition*. The Chemical Rubber Co. (CRC), Cleveland, OH.

Weber, J. B. 1970. Mechanism of adsorption of s-triazines by clay colloids and factors affecting plant availability. Pp.93-130. In: *Residue Reviews*, Vol.32, F. A. Gunther and J. D. Gunther (eds.). Springer-Verlag, New York, NY.

Wilson, M. A. 1981. Application of nuclear magnetic resonance spectroscopy in the study of the structure of soil organic matter. J. Soil Sci. 32:167-186.

Wright, J. R., and M. Schnitzer. 1960. Oxygen-containing functional groups in the organic matter of the A_o and B_h horizon of a Podzol. Proc. 7th Inter. Congr. Soil Sci., Madison, WI, 2:120-127.

Yoshida, M., K. Sagami, R. Hamada, and T. Kurobe. 1978. Studies on the properties of organic matter in buried humic horizons derived from volcanic ash. I. Humus composition of a buried humic horizon. Soil Sci. Plant Nutr. 24:277-287.

Zachara, J. M., and J. C. Westall. 1998. Chemical models of ion adsorption in soils. Pp.47-95. In: *Soil Physical Chemistry*, D. L. Sparks (ed.). CRC Press, Boca Raton, FL.

Ziechmann, W. 1994. *Humic Substances*. George August Universität Göttingen, Bibliographischer Institut, Wissenschaftsverlag, Mannheim, Germany.

Zientara, M. 1983. Effect of sodium humate on membrane potential in internodal cells of *Nitellopsis obtusa*. Acta Soc. Bot. Pol.(Poland) 52:271-277.

INDEX

Absorbance
  in color ratio, 122
Acetate-malonate pathway
  in formation
    of humic substances, 87
    of phenols and
      quinones, 118
Acetyl-acetone
  in extraction of humic
    substances, 41
Acid
  Arrhenius, 242
  Brønsted-Lowry, 242
  humus, 2
  Lewis, 242
Actinomycetes
  in phenol formation, 85
Adrenaline, 94
Adsorption
  beneficial effect of, 226
  characteristics, 228
  definition of, 226
  differences for use in
    fractionation, 50
  forces of, see forces and
    mechanisms of adsorption

  of metals, 240
  nonspecific, 226, 228
  pseudo, 228
  specific, 226, 228
  temperature effect on, 226,
    228, 229
Adsorption models
  empirical approach of, 230
  scientific approach of, 230
  semiempirical approach of,
    231
  S, L, C, and H, 231
  SC-EDL, surface complexa-
    tion–electric double
    layer, 231
  SCNEM, surface complex-
    ation nonelectrostatic,
    231
Aeration
  soil, 255
Aerobic
  conditions/environments,
    16, 22
  decomposition of saccha-
    rides, 96
  humin, 72-73
Affinity, 241

356

Index

of Al and Fe for phosphate, 283-284
of COOH for Fe, 241
of humic acid for metals, 246, 284-285

Agar
in gel chromatography of humic substances, 43

Aggregation, 255
effect on bulk density, 300
effect on pore space, 300
of soil, 300

Agricultural waste
humic matter in, 27

Agrochemicals
importance of humic acids in, 302
types of, 302, 304

Agroforestry, 294

Alanine, 80

Alcohol
ethyl, 97
solid, 102

Alcoholic hydroxyl, 218

Aldose, 96

Al-fulvate, 16
in andosols, 21
in oxisols, 21
in ultisols, 21

Algae, 25
effect of humic substances on, 279, 287

Al-humate, 16
in andosols, 21

Alkaloid
in tobacco, 288

Allochthonous, 17, 25
humic matter, 31

Allophane, 294
humo-Al, 256, 294

Alpha, α-, humic acid, 16, 61, 69

Aluminum oxide
in gel chromatography of humic substances, 43

Amadori rearrangement, 120

Amicon cell
for ultrafiltration, 56

Amine, 94, 120
detoxification of aromatic, 298
glucosyl, 120
types of, 94
physiological effect of, 94

Amino acid, 4, 7, 10, 13, 90, 113
decarboxylase, 94
decomposition of, 92-93
definition of, 91
determination of, 218
protonation of, 218
source of positive charge, 217-218
types of, 91
used to differentiate humic acid from fulvic acid, 91

Amino sugars
definition of, 98
forms/types of, 98
functions of, 99
polymers of, 98
structure of, 99

Ammonia fixation, 84
by humus, 271, 302
by lignin, peat and sawdust, 302

Ammonification, 272, 273
Amphiphatic property of humic matter, 61
Amphoteric compound, 45, 210
Anaerobic
　decomposition of saccharides, 96
　humin, 72-73
　systems/environments in formation of humic substances, 16, 22, 25, 76
Andic horizon, 296
Andosols, 294
　effect of humic acids on properties of, 256, 296
　fulvic acid/humic acid ratio in, 20
Angiosperm
　dicotyledonous, 30
　monocotelydenous, 30
Anion exchange resins
　criteria for proper, 45
　fractioning of humic substances by, 54
　in gel chromatography of humic substances, 43
　for sorption and isolation of humic substances, 44
　types of, 44
Anoxic condition
　in electron transfer, 266
Anthracite, 26
Anthropogenic humic matter, 29, 33
　in polluted waterways, 33
　sources of, 33
Anthropogenic origin, 20

Anti-allophanic effect, 249
Antioxidant
　in extraction of humic matter, 46
　role of vitamin C as, 107
Apocrenic acid, 6, 14, 15, 152
　formula weight of, 11
Aquatic humic matter, 17, 23, 29, 31
　extraction by freeze drying, 60
　extraction by XAD chromatography, 58-59
　freeze concentration and freezedrying of, 42, 43
　gel chromatography of, 43
　IHHS method of extraction of, 59
　methods of collection and concentrating of, 42-43
　methods of extraction of, 58-59
　role of carbohydrate in formation of, 95
　role in geochemical cycle, 23
　role in organic C cycle, 23
　superspeed centrifugation of, 43
　types of, 31
　ultra-filtration of, 43
　vacuum distillation of, 42
Aquifer
　Ogallala, 29
　Paluxy, 29
　Trinity, 29
　contamination with fulvic acid of, 29

Index

Arabinose, 96
Aridisols,
  organic matter content in, 21
Aromaticity
  definition of, 149
  determination of, 150-151
  importance of, 150
  of fulvic acid, 150, 196, 198
  of humic acid, 150, 196, 198
Aromatization
  definition, 78
  dehydration process in, 78, 79, 80, 80
Ascomycetes
  in decomposition of lignin, 85
Ascorbic acid, 107
Aspergillus
  in phenol formation, 85
  niger, 105
Atomic percentages, 134-135
Atomic ratios, 13, 138-139
  identification of nonhumic substances with, 139
  importance of, 138-139
  of major elements in humic and fulvic acids, 140-141
ATP, adenosine triphosphate, 275
Autochthonous, 25
  humic matter, 31, 120
  microflora, 283
Autolysis, 97
Auxin, 105
  -a, and b, 105
  hetero, 105
  -like effect of humic acid, 274, 309
  plant, 105
  structure of, 106
  synthetic, 105

Bacteria
  methanogenic, 269
Bactolife, see humate
Basidiomycetes
  in decomposition of lignin, 82, 85
Benzene
  definition of, 84
  relation with phenol of, 84
  structure and properties of, 84
Benzoic acid
  amino, 107
  hydroxy, 86
  oxidation of, 90
Beta, β-,
  carotene, 107
  humic acid, 16, 57, 70
  indoleacetic acid, 105
  indolebutyric acid, 105
  naphthalene acetic acid, 105
B_h horizon, 21
Bidentate, 240
Bioaccumulation
  of metals by humic acids, 298
Bioavailability
  of contaminants, 298, 299
  effect of humic acids on,

298-299
of pesticides as affected by humic acids, 298
of phosphates as affected by humic acids, 284
Biochemical cycle
effect of humic matter on, 267, 268
Biodetoxification
effect of humic acids on, 299
of pesticides, 298, 299
of xenobiotics, 299
Biofertilizers, 304
of N, 304
Biogel, 43
Biological uptake
of pesticides by humic acids, 259
of xenobiotics by humic acids, 259
Biomass, 1. 22
Biopolymer degradation theory, 75, 112, 113, 115, 118, 141
Biostimulant, see humate
Biosystem, see humate
Biotin, 107
Bitumen, 74
Black water, 24, 60
distribution of, 24
DOC content of, 24, 59
fulvic and humic acid content of, 24
fulvic acid/humic acid ratio in, 25
Issue of health hazard, 24
Okefenokee swamp, 49
Bogs, 22, 26, 32

Boltzmann constant, 222
Bridging
aluminum, 256, 301
calcium, 256-257, 301
cation, 251, 252
importance of, 250, 252
mechanism, 250
metal, 225, 252, 284
methylene, 270
water, 217, 251
Brown coal, 16
Brown humic acid fraction, 16, 52, 70
Braunhuminsäuren, 16
Brown-rot fungi, 83
B-type humic acid, 18
Bouger-Lambert-Beer law, 179
Buffer capacity of humic matter, 259

CAFOS, confined-animal
feeding operations, 29
deposits, 29
issue of contamination of groundwater by, 29
source for fulvic acid, 29
Ca-fulvate
in mollisols, 21
Ca-humate
in mollisols, 21, 256, 293
Callus
cells, 289
elemental composition of, 289
Carbohydrate, 4, 10, 13

definition, 95
importance as food source, 94
importance in formation of humic matter, 95
structure of, 99
Carbohydrate-protein complexes, 31, 95, 120
Carbon cycle
definition of, 268
importance of humic matter in, 268
importance of humin in, 74
relation with humification, 75
Carbon/Nitrogen, C/N, ratio, 128, 165
of crop residue, 131, 272
of humic acids, 128-129, 155
of fulvic acids, 128-129, 132
of legumes, 272
of wheat straw, 272
relation with humification rate, 132
Carbon reserve
in humic matter, 268-269
Carbonization, 145
Carbonyl group, 146
Carbon-13 nuclear magnetic resonance, $^{13}CNMR$, spectroscopy, 192
characteristic spectral regions in, 149, 194, 195-196
chemical shift, δ, in, 193
cross polarization magic angle spinning, CP-MAS, 194
magic angle in, 194
resonance frequencies in, 192
resonance position in, 192
standard solvents in, 193, 197
spectrum of fulvic acids, 149, 194, 195-196
spectrum of humic acids, 149, 194, 195-196
standards used in, 193
Carboxyl groups
complexation constant of, 241
contents in humic substances, 143-144, 145, 150
content in relation with molecular weights, 145
contribution on negative charges, 214, 215
definition and function of, 143
dissociation of, 212, 215, 236, 257
effect on cation exchange, 216-217
effect of humification on, 214
effect of pH and pK_a on dissociation of, 219, 236, 243
functioning as electron donors, 266
methods of determination of, 144, 150
relation with total acidity, 215

steric hindrance in analysis of, 144
Carboxypeptidase, 276
Carotenoids, 101
Cation exchange capacity, CEC,
  definition of, 235
  due to permanent charges, CEC_p, 236
  due to variable charges, CEC_v, 236
  equation, 235
  maximum, CEC_m, 236
  measured, 235
  potential, 235
  relation with total acidity. 258
Cation exchange reactions
  importance of functional groups in, 216-217, 257
  models, 237
  significance of total acidity in, 215
Cation exchange resins
  in extraction, 44
  for purification of humic matter, 49
Cellogel, 43
Cellulose, 96
  as precursor in humic matter formation, 76
  in gel chromatography of humic substances, 43
Cenozoic rocks
  humic acid in, 27
Centrifugation
  in extraction procedures, 48
  in fractioning procedures, 50
Characterization of humic substances. 169
  chemical, 169
Charge characteristic
  in fractioning humic matter, 50
Chelates
  cation exchange by, 297
  humo-metal, 258, 259, 304
  importance of saturation of, 296
  metal, 240
  significance of metal-humic acid ratio of, 296
  pH effect on, 286
  role as nutrient carrier, 286, 297
  sink for toxic metals, 259
  stability constant of, 241
Chelation, 44,
  deactivation of metals by, 259, 273
  decomposition of soil minerals by, 247
  definition, 238-239
  effect of pH and pK_a on, 243
  effect on clay mineral formation, 248
  effect on soil genesis, 247
  in metal dissolution, 295-296
  of metal ions, 217, 240, 273
  role in weathering, 248
  significance of functional groups in, 216-217
  statistical modeling of, 249
Chemical composition

Index

issue of, 11, 13
of humic matter, 127
Chemical formulas
  controversy in, 11, 12
  of humic substances, 12
Chemical methods
  for fractioning humic matter, 50
Chemical reactions or interactions of humic matter, 225
  importance in soil and environment of, 225
  maximum, 244
  metal/ligand ratio in, 244
  types of, 225-226
Chemisorption, 233
Chitin, 98
  structure of, 99
Chitosan, 98
  structure of, 100
Chlorophyl, 99, 101
Cholesterol, 102
Chromoprotein, 99
Cinnamic acid, 86
City agglomeration, 20
Classification
  of humic matter, 29
Clostridium, 94
Co-adsorption, 45
  of metal and phosphate ions, 225, 284
  relation with bridging reactions, 225
Coal, 20, 22, 26
  as marker, tracer, for organic matter production, 27

sapropelic, 26
subbituminous, 26
Coalification, 26
  effect on COOH group content, 145-146
Color ratio
  definition, 122, 123, 177
  relation with molecular weight of humic substances, 177-178
  use as index of degree of condensation, 123
  values of fulvic acid, 177, 178
  values of humic acid, 123, 177, 178
Cometabolism
  by aquatic microflora, 278-279
  definition of, 278
Complex reaction or complexation capacity, CC, 242
  constant of Fe, 241
  definition of, 234, 238-239
  humo-Ca, 281
  of humic acid with a cation, 217, 228
  metal-organo, 230, 240
  significance of COOH on, 241
  significance of pK_a in, 243
  statistical modeling of, 249
Composition
  on an atomic percentage basis, 133-135
  C, H, O, N, S, P, 128
  elemental, 127-126
  factors affecting, 130

group compound, 142
 of reference humic acid, 130, 134
 on a weight basis, 128-129
Condensation, 26
 abiotic, 76
 phenol and polyphenol, 76
 quinone, 76
 sugar-amino, 76, 95, 112
 theory in formation of humic matter, 76, 118
Coniferaceae, 101
Coniferous lignin, 30
Coniferyl alcohol, 30, 78, 113, 157
Coordination
 bonding, 234, 258
 compound, 239
 number, 239
 reaction, 234, 242
Copropel, 17, 22, 31, 64
Corn belt, 293
Coumaryl alcohol, 30, 78
Covalent bonding, 242
Crenic acid, 6, 14, 15, 152
 formula weight of, 11
Cycles
 biochemical, 293
 carbon, 293
 nitrogen, 271-272, 292
Cytokinin
 in humate, 308

Dauerhumus, 3
Deamination
 aerobic, 93
 anaerobic, 93
 definition, 93-94
 nonoxidative, 93
 oxidative, 93
Decalcifying calcareous soils, 39
Decarboxylase
 amino acid, 94
 histidine, 94
 tyrosine, 94
Decarboxylation, 116
 definition of, 93-94
 of amino acid, 93-94
 of gallic acid, 87-88
 of orsellinic acid, 87-88
Decomposition
 aerobic, 269
 anaerobic, 269
 characteristics of, 130
 of carbohydrates, 96-97
 of lignin, 82-84
 of protein, 92-94
 products of, 269
Degradation
 biological soil, 299
 biophysical, 299
 chemical soil, 299
 of humic matter, see decomposition
 soil ecosystem, 293, 294, 299
 physical soil, 299
Dehydration theory
 in formation of lignin, 79
Dehydrogenase
 amino acid, 93
Delta, Δ, log K, 18
Demethylation

Index

of lignin, 84, 85
of orsellinic acid, 87
Density
　in fractioning humic
　　matter, 50
Depolycondensation
　of humic acids, 281
Depolymerization
　of humic acids, 281, 282
Detoxification
　biological, 297
　of heavy metals, 273, 297
　of radioactive elements, 297
　of xenobiotics, 297
Detritus, 17
Diagenesis, 26, 32
Diagenetic changes, 25
　of humic matter, 115. 136, 214
Dialysis
　as a purification technique, 49
Diffuse-ion swarm, 226, 228
Dimension of
　macro-molecular colloids, 42
Dimer concept, 10, 161-162
Dimethylformamide, DMF
　changes in composition of
　　humic substances by, 41
　effectiveness in extraction
　　of humic substances, 41
Dimethylsulfoxide, DMSO
　effectiveness in extraction
　　of humic substances, 42
Dispersion force, δ_p
　requirement for ideal
　　organic solvent, 42

Dissolved organic carbon,
　DOC, 23
　acidic, 24
　basic, 24
　content in ground water,
　　surface water, and
　　lakes, 24, 59
　content in foam, 24
　hydrophobic, 23
　hydrophilic, 23
　neutral, 24
Dissolved organic matter,
　DOM, 23
　acidic, 24
　basic, 24
　content in ground water,
　　surface water, and
　　lakes, 24
　content in foam, 24
　hydrophobic, 23
　hydrophilic, 23
　neutral, 24
Distribution of humic matter, 19
Distrophic environment, 25
Dopplerite, 17, 64
Dowex SAR, 44
Dowex SBR, 44
Duckweed
　growth as affected by humic
　　matter, 281
Duolite, 44
Dy, 17, 64
Dynamic system of humic
　substances, 12
Dystrophic lakes, 17

Echte huminsäuren, 15
Ectomycorrhiza
   in formation of humic substances, 87
Effective surface, 230
Electric charges
   origin and types of, 210
Electric double layer
   concepts, 223
   suppression of, 224
Electrochemical potential, 261, 262, 263
   formulation of, 261
   of a redox reaction, 261
Electron
   activity, 261, 262
   availability, 261
   poor compound, 260
   rich compound, 260
Electron donor-acceptor theory,
   in biodegradation, 109
   in humic acid reactions, 89, 109
   in triazine transformation, 110
Electron microscopy of humic matter, see transmission and scanning electron microscopy
Electron pair sharing, 242
Electron paramagnetic resonance spectroscopy, EPR, 189
   antisymmetrical resonance in, 190
   g value in, 189-190
   isotropic resonance in, 190
   spectra of fulvic acid, 191
   spectra of humic acid, 191
Electron spin resonance spectroscopy, ESR, see EPR
Electron transfer
   in half-cell reactions, 263
   intermolecular, 89
   microbially, 266
   in soil redox reactions, 260
Electrophoresis
   in fractioning humic substances, 61
Electrostatic attraction, 45, 242
   in cation exchange, 216, 217
Electrostatic factor
   of DMF, 42
   of DMSO, 42
Elemental composition, 10
Energy dispersive analysis by x-rays, EDAX,
   of plant cells, 289
Entropic action
   of soil, 288
Environment
   effect of humic acid on, 292
Enzymes
   co-, 275
   constitutive, 275
   definition of, 274
   denatured, 275
   effect of humic acids on, 274
   fixation of, 275
   free, 275
   induced, 275
   nomenclature of, 275
   proteolytic, 275, 276

Index 367

Epicoccum
  phenol formation by, 85
Epimer, 98
Equilibrium
  constant. 262-263
  law of, 230
Erosion
  accelerated, 300
  natural, 300
Escherichia coli, 79
Ethanol
  in fractioning humic acids, 52
Ethyl alcohol
  production by decomposition of saccharides, 97
Ethylenediamine, EDA
  Chemical changes induced by, 41
  in extraction of humic substances, 41
Euratium
  in phenol formation, 85
Eutrophic condition, 17, 22, 25
Exchange constant, K_{eq}, 238
Extinction
  coefficient, 177
  in color ratio, 122
Extractants
  criteria for ideal, 35
  inorganic solvents, 36
  organic solvents, 36, 39
  search of, 34
Extraction,
  advantages of NaOH, 36
  flow sheet for, 47
  in N_2 gas atmosphere, 46, 47, 48
  IHSS and SSSA procedures of, 46-47
  methods of, 36, 45
  of aquatic humic matter, 36
  of soil and peat humic matter, 34, 45
  pretreatments for, 46
  procedures, 46-47
  soil:extractant ratio, 46
  with acids, 39
  with dilute bases, 39

Fertilizer
  definition of, 308
  humate as, 307-308
  humic as, 307
  organic, 308
Fibric fraction, 2, 3
Filters
  common, 51, 52
  depth, 55
  fiber, 55
  flatbed membrane, 55, 56, 62
  hollow fiber, 62
  merits of different types of, 55
  micro-, 52
  ultra-, 58
Filtration
  according to linear sizes, 55
  according to molecular weight, 55-56
  in fractioning humic substances, 50
  issue and controversy in,

62-63
  micro-, 55, 60
  ultra-, 55, 62
  with activated charcoal, 57
Fixation
  of N, 283
  of NH_3, 302
  of P, 284-285
F-layer, 2
Fluorescence spectrophotometry, 179
Forces and mechanisms of adsorption
  chemical, 233
  electrostatic, 232
  hydrogen bonding, 233
  physical, 232
Formic acid
  formation of, 269
Formula weight, 170
  of humic substances, 11, 133, 154, 163, 165
  methods of determination of, 152-153
Förna, 17
Fossil fuel, 22, 26
Fractionation
  by anion exchange, 54
  by different solubilities, 51
  by dissolution, 52
  by filtration, 52, 55, 61
  by gel chromatography, 52, 61
  by precipitation. 61
  by ultracentrifugation, 61
  by ultrafiltration, 61
  definition of, 50
  merits of different methods of, 50-51
  methods of, 50
  of aquatic humic matter, 61
  of fulvic acid, 57
  of humic acid, 50
  of humic substances, 34, 50
  reasons for, 50
Freundlich equation, 230, 231
Frictional coefficient, 174
Frictional ratio, 174
  values of, 175
  versus molecular weight, 175
  versus particle shape, 174-175
Fructose, 95
Fulvic acid, 10, 14, 15
  A-, B-, C-, and D-types of, 18, 57
  definition of, 65
  fractions, 66-67
  generic, 49, 67
  inhibitory effect on crystallization of soil minerals, 249
  physical properties of, 66
  polymerization of, 66
  polysaccharide in, 95
Functional group
  acidic, 143
  charge transfer complexes formed by, 241
  composition, 10, 142
  definition of, 143
  oxygen containing, 143
  major types of, 143
Fused double layer concept, 224

Index

Galactosamine, 98
Gallic acid, 87
    by aromatization of shikimic acid, 87
Gapon equation, 237
    derivation of, 238
Gaussian Distribution model
    competitive, 249
Gelbstoff, 24, 25
    in black water, 24
Gel chromatography
    determination of molecular weight by, 172
    of humic substances, 43
    types of reagents in, 43
    XAD resins in, 43
Generic fulvic acid, 18, 67
    high molecular weight, 67
    low molecular weight, 67
Generic humic substances, 18
Geologic humic matter, 29
    diagenesis of, 32
    fulvic and humic acid content in, 32
    humin content in, 32
    types of, 32
Geologic deposits
    coal, 26
    fossil fuel, 26
    humic matter in, 26
    lignite, 26
    oilshale, 26
Gibberella fujikuroi, 105
Gibberellic acid, 105
Gibberellin, 105
    in humate, 309
Glucoprotein, 13
Glucosamine, 98, 162
    epimer of, 98
    structure of, 100
Glucose, 96
    amine reaction, 121
    content as affected by humic acids, 288
    glysine reaction, 121
    starting material in the acetate-malonate pathway, 87
    structure of, 99, 100
Glucosidic bond, 96
Glucosylamine, 162
    structure of, 163
Glycolysis, 286
Glycoprotein, 98
Glycoside
    role in plant of, 307
Glysine, 120
Gouy – Chapman equation, 222
    in determination of σ_v, 222
Granulation, 301
Graphite, 26
Grasses or bamboo lignin, 30
Grauhuminsäuren, 16
Gray humic acid fraction, 16, 52, 70
Green humic acid, 18, 71
Group compound
    definition of, 148
    methods of determination of, 148
Growth promoting substances, 104, 309
    types of, 104
Gymnosperm, 30

Index

Half-cell reaction
 standard, 263-264
Harbor agglomeration, 20
Hardwood lignin, 30
H-layer, 3
Hemicellulose, 96
 as precursor in humic
  matter formation, 76
Hemic fraction, 2, 3
Henderson-Hasselbalch
 equation, 213
 in ionization of functional
  groups, 211-212
Hendersonula
 in phenol formation, 85
Heterotrophic
 facultative, 282
 food source, 278
 organism, 283
Hexane
 in extraction of humic
  matter, 41
Histamine, 94
Histidine, 94
Histosols, 22
Hollow fiber ultrafilters
 merits of, 62
 polarization with, 62
Hormone, 99
 artificial, 105
 content in humate, 309
 Darwin's discovery of, 105
 effect of humic acid, 104, 290
 functions of, 104-105
 plant, 104, 308
Humate
 agricultural and industrial
  importance of commercial, 304-309
 ammonium, NH_4-, 307-308
 bio-, 307
 controversies as fertilizer, 307-308
 definition of commercial, 305
 foliar application of, 307
 mixed with microorganisms, 306-307
 origin and types of commercial, 305-306
 -P issue, 310-311
 potassium, K-, 307-308
 rate of application of commercial, 308-309
Humic acid, 5, 10, 14, 15, 115
 A, B, R_p and P, 70-71
 C/N ratios, 127-128
 definition, 65, 68
 elemental composition in
  different soils, 128-129
 fractions, 69-70
 green, 18, 71
 high molecular weight, 70
 hormone-like behavior of, 105, 111
 low molecular weight, 70
 properties of, 68-69
 role in mineral dissolution, 248
Humic coal, 26
Humic matter, 1, 3, 5, see also
 humic substances,
 agronomic importance of, 254
 allochthonous, 31

Index

anthropogenic, 33, 64
aquatic, 31, 64
autochthonous, 31
composition of, 5, 6
definition of, 5, 6
electrochemical properties of, 210, 211
genesis of, 75
geologic, 31, 64
historical concept of, 14
in soils, 20
issue of, 5
modern concept of, 17, 19
paleontologic, 27
peat, 32, 64
use of SOM for, 5
utilization as plant food, 281
terrestrial, 29-30, 64
Humic substances, 9
characterization of, 169
dynamic nature of
grass and bamboo, 30, 64
hardwood, 30, 64
issue of artifact, 6, 7
issue of chemical composition of, 10
issue of real compounds of, 8, 9
issue of operational compounds of, 7
softwood, 30, 64
types of, 63-64, 65
Humic-N
acid insoluble, 282
content, 282
forms of, 282
hydrolyzable unknown, 282
uptake by microflora, 283
uptake by plants, 283
Humification, 4
carbon fixation or sequestration during, 130
definition of, 75
energy gain in, 165-166
indexes, 122, 177
major pathways of, 75
maximum rate model, 126
model, 124-125
statistical modeling of, 121-122
theories of, 111-112
Humified fraction, 4, 5
Humified muck, 23
Humin, 5, 14, 15, 32, 113, 115, 120
aerobic, 72-73
anaerobic, 72-73
composition of, 71
definition, 71
extraction methods of, 72
fractions. 74
marine, 73
peat, 73
sapropelic, 64
types of, 72-73
Humoligninsäuren, 15
Humus, 1, 2, 4
composition of, 2
definition of, 2, 5
early concept of, 2
modern concept of, 3
use of SOM for, 4, 5
versus humic matter, 5
Humus acid, 5

Humus coal, 15
Humussäure, 5
HUN, 104, 282
Hydration
    of protoplasma by humic acids, 288
    shell, 250
Hydrogen bonding
    complex, 251
    definition of, 233
    force (δ_H) of, 42
Hydrolysis
    amino acid, 283
    enzymatic, 270
    of humic substances, 269-270, 282
    of protein by proteinase and peptidase, 93
Hydrophobic, 101
    bonding, 234
    compounds, 259
Hydroxycarboxylic acid
    polymeric, 24
Hydroxyhydroquinone, 79
Hydroxyl group
    alcoholic, 146
    phenolic, 146
    reactivity of, 146-147
    total, 146
Hydroxyphenols, 84
Hydroxyphenylperuvic acid, 79, 157
Hymatomelanic acid, 5, 14, 15, 52, 61, 69
Hypolimnetic areas, 17

Imogolite, 294
    humo-Al-, 256
    humo-Si-, 294
Industrial waste
    humic matter in, 27
Industry
    importance of humic acids
        in agricultural, 301
        in ceramic, 301
        in drilling fluids, 301
        in pharmaceutical, 301
        in staining, paints and ink, 301
Infrared spectroscopy
    absorption bands of humic substances, 180, 182
    characteristic spectra of major humic substances and lignin, 183
    classification of spectra of, 185
    fingerprint region in, 161, 181, 182
    group frequency region in, 161, 181, 182
    spectrum of fulvic acid, 181, 186. 187
    spectrum of hematomelanic acid, 184
    spectrum of humic acid, 184, 186, 188
    spectrum of humin, 184
    spectrum of lignin, 183, 185
    types I, II and III spectra, 185-186
Inner-sphere
    complexes, 226, 234
    surface, 226, 228, 239

Index

Internal oxidation, ω
  definition, 136, 137
  formulation of, 138
  importance of, 135-136
  limits in aerobic and anaerobic conditions, 138
  of carbohydrate, 137
  of methane, 137
  of water, 136
  values of humic substances, 134-135
Inorganic reagents
  effectiveness in extraction of humic matter, 37
  issue of artifacts by, 37
  types of, 38
Inositol, 107
Ionac series, 44
Ionization constant, 211, 213
Iron, Fe-, fulvate
  in ultisols and oxisols, 21
Isoelectric point, 219, 242

Kelp, 25
Kerogen, 31, 68, 115, 120
  paraffin in, 101
  types I, II, and III, 64
Kerr's equation, 237
  derivation of, 237
Ketose, 96

Laccase, 85, 89
Langmuir equation, 230-231
Lemna major or *minor*, see duckweed (Latin names)
Leonardite, 20, 26, 32
Lewis acid and base, 234
Ligand
  definition of, 234, 239
  exchange, 233, 234
  metal ratio, 244
Lignification
  definition of, 82
  functions of, 82
Lignin, 4, 7
  as precursor in humic matter formation, 76
  content in plants, 77-79
  coumaryl, 80
  decomposition of, 82-83
  definition, 77
  grass and bamboo, 79
  hardwood, 79
  methoxyl content of, 78
  monomers, 30, 78, 112, 116
  properties of, 82
  polymer, 78, 79
  -polysaccharide complex, 77
  softwood, 79
  structure of, 83
  theory in humification, 113
Lignite, 20, 26, 32
  commercial grade, 27
  fulvic acid content in, 27
  humic acid content, 27
Lignohuminsäuren, 15
Lignolitic fungi, 83
  types of, 83-84
Ligno-protein theory, 15, 30, 80, 91, 112-113
  of Flaig, 76, 113

structure of humic acid according to, 114, 157-158
Lipid, 4, 99
    definition of, 101
    derived, 102
    effect of, 101
    in microbial tissue, 101
    sources of, 101
Lipoidic acids, 101
    effect on plant nutrition, 101
    effect on rock weathering, 101
Litter, 2
    in nutrient cycling, 2

Magic angle spinning technique, see ^{13}C NMR
Maillard's reaction, 95, 99, 112, 119-120, 162
Markhoff model, 124-125
Mass action law, 230
    equations of CEC, 237
Mean residence time, MRT,
    effect of anaerobic and aerobic conditions on, 269
    of humic matter, 259, 268-269
    of pesticides as affected by humic acids, 259
    of xenobiotics as affected by humic acids, 259
Membrane
    effect of humic acid on, 278, 281, 287

penetration of root, 282
permeability of Diptera and Crustacea, 279
Melanic epipedon, 120
Melanins, 71, 86, 99, 120
Melanoidin, 120
    pathway, 99, 112, 119
Methoxyl group
    in lignin, 78
    decomposition by brown-rot, 83
Mesotrophic systems, 25
Mesozoic rocks
    humic acid in, 27
Metabolite
    effect of humic acids on plant, 287
Metamorphism, 26
Methane, 96
    formation of, 269
Methoxyl, OCH_3-, group
    in humic acid, 146
    in lignin, 78
Methylene chloride
    in extraction of humic substances, 41
Methyl isobutyl ketone, MIBK
    in isolation of humin, 72
Methylsulfoxide, MSO
    effectiveness in extraction of humic substances, 41
Microbial
    fermentation, 97
    phenols, 87
Mild humus, 2
Mineralization, 93
Mineral soil
    definition of, 23

Index

Minimum molecular weight, 13
Mires, 22
Mobilization of pesticides and xenobiotics by humic matter, 259
Model V in chelation, 249
Molar coefficient of absorption, 179
Molecular fractionation, 52
Molecular size
   of humic substances, 173
   versus molecular weight, 173
Molecular shape
   effect of molecular weight on, 175
   effect of pH on, 176
   of humic substances, 174
Molecular structure
   computer modeling of, 10, 162-163, 164-165, 166, 167
   Flaig's lignin monomer concept of, 157, 158
   ligno-protein concept of, 157
   methods of determination of, 155-156
   models of aquatic fulvic acid, 160
   Schnitzer's and Orlov's phenol concept of, 159
   Schulten's phenol concept of, 10, 159, 164
   Steelink's lignin tetramer concept of, 157, 158
   Stevenson's dimer model of, 161
   sugar-amine concept of, 162-163
Molecular weight, 169, 175
   characterization of humic matter by, 171
   in fractioning humic matter, 50
   methods of determination of, 152-153, 171-172
   minimum, 11, 152
   number-average, 65, 170
   of humic substances, 13, 133
   size versus linear size, 56
   values of, 171-172
   weight average, 171
   z-average, 171
Mollic epipedon, 296
Mollisols, 20
   fulvic acid/humic acid ratio in, 20
   organic C content, 20, 293
   effect of humic acids on properties of, 256, 293
   properties of, 293, 300
Monodentate, 240
Montmorillonite, 293
Mor, 2
Mucopolysaccharide, 98
Muck, 22, 23, 26, 32
Mull, 2
Municipal waste
   humic matter in, 27
Myrtaceae, 101

Nährhumus, 3

Negative adsorption
  effect of pH and pK_a on, 227
Negative charge
  effect on CEC, 211
  effect of ionization of functional groups on, 213-214
  functional groups responsible for, 210-211
  issue of COOH group on, 214
  origin of, 210, 212
  pH dependent, 211, 257
  pK_a and degree of, 215
  relation with total acidity, 215
  variable, 211, 257
Neutralization fraction, 16
Neutral salt solutions
  in fractioning humic substances, 16, 52, 70
Niacin, 107
NICA, noncompetitive adsorption model in chelation, 249
  or NICA-Donnan model, 250
Nicotine, 107
Nicotinic acid, 107
Nitrification, 272, 273
Nitrogen content
  correlation with organic carbon, 295
Nitrogen cycle
  definition, 271
  diagram, 272
  immobilization in, 271, 272
  importance of humic matter in, 268, 283
  inner cycle of, 271, 272
  outer cycle of, 271, 272
Nitrogen-15 nuclear magnetic resonance, ^{15}N NMR,
  analyis of N in humic substances, 197
  determination of N-functional groups by, 199
  spectrum of compost, 200
  spectrum of IHSS fulvic acid, 200
  standard solvents in, 199
Nitrogen fixation, 271
  effect on C/N ratio, 273
Nonhumic matter, 3
Nonhumified fraction, 4, 5
Nuclear magnetic resonance spectroscopy, see ^{13}C, ^{15}N, and ^{31}P NMR
  analysis of humic matter, 149, 192-194
  spectral regions, 149, 194
  types of, 189
Nucleic acid, 4, 13, 99
  composition of, 102
  definition of, 102
  deoxy, 102
  types of, 102
Nucleoprotein, 13
Nucleotides
  mono, 102
Nutrient cycling, 295

Oil shale, 20, 26
  organic C reserve in, 22-23
OM, acronym for organic

Index

matter, 4
Operational compounds, 7, 112
Optical density, 122, 177
   see also absorbance, extinction, 89
   demethylation of, 89-90
   oxidation of, 90
Organic acids
   in extraction of humic substances, 41
   types of, 40
Organic chelating agents
   in extraction of humic substances, 41
   types of, 40, 41
Organic C reserve, 22
Organic cycle, 271
Organic deposits, 22
Organic reagents
   chemical changes by, 41
   creation of artifacts by, 41
   effectiveness of, 40-41
   in extraction of humic substances, 39
   interaction with humic substances, 41
   properties for ideal, 42
   reason for use of, 39, 41
   types of, 40
Organic soil
   definition of, 23
Orsellinic acid, 86
   decarboxylation of, 89-90
Osmotic pressure
   as affected by humic acid, 279
   in plant cells, 288
   role of sugar content on, 288
   effect on wilting point, 287
Outer-sphere
   complexes, 226, 234
   surface, 226, 228, 239
Oxidase, 93
Oxidized humic substances, 264, 265-266, 267
   as electron acceptors, 267
Oxisols, 21
   organic C content in, 21

Paleontologic humic matter, 27, 32
Pantothenic acid, 107
Paraffinic compounds, 71, 101
   in humin, 73, 102
Paramagnetism
   in humic acids, 190
   use in EPR analysis, 190
Particle size
   in terms of linear sizes, 51
   in terms of molecular dimension, 51
   of humic substances, 50, 51-52
   of soil colloids, 51
pe
   definition of, 261-262
   negative, 261, 267
   positive, 261
Peat, 16, 20, 22, 26
   baths, 102
   bogs, 17
   distribution of, 22
   fulvic acid content, 23
   humic acid content, 23

medicinal properties of, 102
therapeutic effect of, 309
types of, 32
Peat lands, 22
Pelogoea, 17, 64
Penicillium
   in phenol formation, 85
Peptide, 90
   bond, 91
Peruvic acid
   hydroxy, 80
   hydroxyphenyl, 79
   phenyl, 79
Pesticides
   bio-, 304
   bioactivity of, 290
   biodetoxification of, 109, 290, 298-299
   persistence, 111, 290
Petroleum
   as marker, tracer, for organic matter production, 27
pH-dependent charge, see negative charges
Pharmaceuticals
   humic acid derived
      anti-cancer, 309
      anti-coagulatory, 309
      anti-inflammatory, 309
      antiviral, 309
      for heavy metal poisoning, 310
   importance of humic acids as, 309
Phenolase, 89
Phenols
   autooxidation of, 89
   definition of, 84
   formation of, 85, 86
   lignin derived, 84
   microbial, 85, 86
   poly, 84
   properties of, 84
   structure of, 84
   theory in humification, 112, 118
Phenolic glycosides, 68
Phenolic-hydroxyl, 77
   contents in humic substances, 145
   contribution to negative charge, 214-215
   dissociation of, 212, 236, 257
   effect of humification on, 214
   effect of pH and pK_a on dissociation of, 219-220, 236-237, 243
   factors affecting dissociation of, 219
   functioning as electron donor, 266
   in fulvic acid, 148
   in humic acid, 148
   importance of, 146
   protonation of, 218
   relation with total acidity, 215
   significance in chelation, 216, 217
Phenyl propane, 157, 159
Phenol-protein theory, 119, 159, 159-160
Phenoloxidase, 85

Index

Phenylalanine, 80
Phloroglucinol, 79
Phosphates
   biological fixation of, 284, 285, 302, 304
   bio super-, 304
   humo-, 304
Phosphorus–31 nuclear magnetic resonance, ^{31}P NMR,
   detection of orthophosphate by, 202, 284.
   purpose of, 201
   spectrum of $H_2PO_4^-$, 202
   spectrum of a humophosphate, 202
Photochemical
   breakdown by humic acids, 264, 295
   breakdown of xenobiotics, 295
   dissolution, 265
   reaction, 241, 264
Photolysis, 288
Photoreduction
   biological importance of, 267
   of Fe, 264-265
   quantum yield of, 265
   significance of crystallinity on, 265
Photosynthesis
   effect of humic acids on, 287
Physical methods
   for concentrating aquatic humic matter, 42
   for fractioning humic matter, 50
Pinus elliotti, slash pine,
   callus culture of, 288, 289
Pisolithus tinctorius, 307
   effect of fulvic acid on, 277-278, 282
   formation of humic substances by, 87, 277
pK_a
   definition of, 212, 213
   in determination degree of negative charges, 213-215
   in differentiating strong and weak acids, 213
   and rate of ionization of functional groups, 213-214
   significance in ionization, 219
pK_b
   formulation of, 219-220
   relation with pK_a, 221
   significance in protonation, 221
pK_{HB}
   definition of, 42
   of DMF, 42
   of DMSO, 42
Plankton, 17, 25
   marine, 267
Plantation agriculture, 294
Plant growth
   direct effect of humic acid on, 280-281, 282, 288
   indirect effect of humic acid on, 280-281
Plant nutrition
   effect of humic matter on, 280

Plant physiology,
    effect of humic acids on, 286
Plant respiration
    effect of humic matter on, 286
Plant species
    effect of humic acid on, 291
Plasmalemna, see protoplasm
Polyamides
    in gel chromatography of humic substances, 43
Polymerization
    abiotic, 114
    theory in formation of humic matter, 76, 114, 115, 118
Polyphenol theory, 113
Polysaccharide, 7, 68
    as precursor in humic matter formation, 77
    as reagent in gel chromatography of humic substances, 43
    hetero, 96
    homo, 96
    in hymatomelanic acid, 95
    in humin, 71, 73
    muco, 98
    structure of, 97
Polystyrene
    in gel chromatography of humic substances, 43
Polyuronides, 68
Pore spaces
    effect of aggregation on, 300
    macro-, 301
Positive adsorption
    effect of pH and pK_a on, 227

Positive charges
    origin of, 217
Poultry litter
    as waste, 28
    as source of humic matter, 28
    types of, 28
Precambrian rocks
    humic acid in, 27
Precursors of humic matter, definition, 76
    major types of, 76-77
    miscellaneous, 99
Protein, 4, 7, 90
    as precursor in humic matter formation, 77
    conjugated, 91
    decomposition of, 92-93
    definition of, 91
    gluco, 91
    glyco, 98
    lipo, 91
    muco,
    refluxing of, 91-92
    types of, 92
Protonation, 110
    of amino acid, 218-219
    effect of pH and pK_a on, 219-220
    of phenolic-OH, 219
Proton acceptor
    force (δ_b) of, 42
Proton pump
    electrogenic, 288
Protoplasma
    hydration of, 288
    permeability as affected by humic acids, 288

Index

viscosity, 288
P-type humic acid, 18
   P_b fraction of, 18
   P_g fraction of, 18
Purification
   of fulvic acid, 49
   of humic acid, 48
   use of HCl+HF mixtures in, 48
   methods of, 48-49
Purine, 103
   structure of, 103
Pyrimidine, 103
   structure of, 103
Pyrogallol, 79
   formation and structure of, 87-88, 89

Quinone, 159
   anthra, 86
   definition of, 89
   formation according to electron donor-acceptor theory, 89
   formation by white-rot, 84
   functioning as electron acceptors, 267
   properties of, 89
   structure of, 90
   units in humic molecules, 267

Radical
   ethyl, C_2H_5, 302
   methyl, CH_3, 302
   OH^-, 264
Reagents
   for extraction of soil humic substances, 37-38, 39-40
   for collection and concentrating of aquatic humic substances, 42-43
Redox, reduction–oxidation
   effect of humic acids on, 295
   in paddy soils, 295
   in photochemical breakdown, 295
   reactions of metals, 264
   role in mobilization and immobilization of soil elements, 283
   in spodosols, 283
   status of C, H, N, O, S, and other elements, 264
Redox potential, E_H, 261, 263
   formulation of, 261
   relationship with pe. 262
Reduced humic substances, 264, 265-266, 267
   as electron donors, 266
Reductive cleavage
   of aromatic bonds, 269, 270
   Na-amalgam method of, 270
Residue effect, 111
Resorcinol
   bioformation of, 88
Refractory nature of humic substances, 10, 12
Rhizobium sp., 308
Rhizopus sp., 105
Rhizosphere

soil, 281
Riboflavin, 107
   function in plants, 107
Root exudates
   effect on depolymerization of humic acids, 281
   types of, 281
Rotteprodukte, 15
R_p type humic acid, 18

Saccharides, 95
   decomposition of, 96
   definition of, 95-96
   di-, 96
   intermicellar adsorption of, 96
   mono-, 96-97
   oligo-, 96-97
   poly-, 97-98
   soil, 97
   types of, 96
Salt balance
   role of humic acids in, 285
Saponin
   in humic acid, 307
   in poisoning fish, 307
Sapric fraction, 3
Sapropel, 17, 22, 26, 31, 73
Sapropelic peat, 32
Sapros, 3
Scanning electron microscopy, SEM,
   advantage over TEM, 203
   freon-liquid-N technique in, 204
   macromolecular structure determination by, 209
   micrographs of
      aquatic fulvic acid, 207
      aquatic humic acid, 207
      soil fulvic acid, 206
      soil humic acids, 206, 208
   purpose of, 204
   ultra-rapid freezing technique in, 205
Schiff base, 120
Sedimentary rocks,
   humic acid in, 27
   prehistoric, 27
Sephagel, 43
Sephadex, 43, 52, 66. 67
   filtration of fulvic acid, 57, 62
   filtration of humic acid, 53-54
   types of, 54
Sewage sludge
   domestic, 273
   heavy metal content in, 273
   industrial, 273
   as a source of humic matter, 28
   as waste, 28
Shikimic acid, 117
   pathway, 117
      in lignin formation, 78, 79, 81, 87, 116
      in formation of phenols and quinones, 118
      in formation of pyrogallol, 87-88
Silicone
   properties of, 302

Index

types of, 302, 303
importance of, 302
Siloxane surface, 256
Sinapyl alcohol, 30
Smectite, 294
Sodium hydroxide, NaOH, extraction, 37
   chemical changes of humic substances by, 37
   historical significance of, 37
   issue of auto-oxidation by, 37
   issue of N_2 gas atmosphere in, 37
Sodium pyrophosphate, $Na_4P_2O_7$, extraction, 37
   chelation of sesquioxides, Al, and Fe by, 39
   effectiveness in extraction of humic substances, 38-39
Soft-rot fungi, 83
Softwood lignin, 30
Soil aggregation, 99
Soil amendment, 28
   poultry litter, 28
   sewage sludge, 28
Soil biological properties
   direct effect of humic matter on, 267, 268
   indirect effect of humic matter on, 267, 268
Soil biomass, 1, 282
Sequestration
   of soil carbon, 271
Soil crust, 255
Soil fatigue, 101
Soil humic matter, 29

Soil organic matter, SOM, 4, 5
   composition of, 1-2
   concept of, 1
   dead fraction of, 1-2
   life fraction of, 1
   preservation of, 281, 293
   in soil taxonomy, 2
Soil organisms
   absorption of humic matter by, 276-277
   decomposition of humic matter by, 276-277
   effect of humic matter on, 276
   formation of humic matter by, 277
   types of, 276
Soil physical properties
   andosols, 255-256
   effect of humic matter on, 254
Soil polysaccharides
   sources/types, 97
   structure of, 97
Soil redox system
   effect of humic matter on, 259, 264
   electron donor-acceptor concept in, 260
   significance in accumulation of humic matter, 259
Soil structure, 99, 286
   cementation effect on, 255
   significance of, 255
   types of, 255
SOM, acronym for soil organic matter, 4

Specific surface, 221
  definition of, 235
Spectrophotometry, see UV and visible light and infrared spectrophotometry
Sphagnum peat, 32
Spodic horizon, 21, 296
Spodosol, 21
  fulvic acid/humic acid ratio in, 21, 296
Stability coefficient of humus
  definition, 123
  of humic acid, 123-124
  relation with C/N, 124
Stability constant, log K_{eq},
  definition and derivation of, 244-245
  of chelates, 243
  importance in solubility of chelates and complexes, 243-245
  use as index of metal affinity, 246
  values of, 246
Starch, 96
Stereos (Gr), 102
Steroids, 102
  medicinal properties of, 102
  phyto, 102
  plant sources of, 102
Sterols, 101, 102
Structures of humic substances
  effect of humic acid concencentration on, 205-206
  effect of pH on, 205
  effect of soil differences on, 205-209

TEM and SEM of macromolecular, 204-205
  types of macromolecular, 206-209
Subcellular
  effect of humic acid, 278
Sugars, 68, 91
  amino, 95
  content as affected by humic acids, 288
  relationship with carbohydrates of, 95-96
Sugar-amine
  complexes, 95
  condensation theory, 118, 119-120, 162
Surface charge density, σ_s,
  calculation of, 222, 235
  formulation of, 221
  permanent, σ_p, 222
  units of, 221, 222
  variable, σ_v, 222
Surface potential, ψ, 222
Surface tension, 226, 227
  effect of saponin on, 308

Tannins, 68
Target organisms
  non-, 290
  persistence of, 290
Terpenoids, 101
Terrestrial humic material, 18, 25
  fulvic and humic acid content in, 30
  sources of, 30

Index

types of, 30
Terrigenous humic matter, see terrestrial humic matter,
Thiamin, 107
TOM, acronym for total organic matter, 4
Torfot
   humic acid derived, 310
Total acidity
   determination of, 147
   effect of COOH content on, 148
   effect of phenolic-OH on, 148
   of humic and fulvic acids, 145, 148
   of humin, 74
   relation with CEC, significance in negative charges, 215
Transamination, 79
   formation of humic matter through, 116
Transient nature of humic substances, 12
Transition metals, 230, 234
Transmission electron microscopy, TEM,
   purpose of, 203
   replica technique in, 204
   types of macromolecular structures determined by, 204
Triazine
   as electron donors, 109, 110
   effect of redox reactions on, 109
True fulvic acid, 18
Two-four-D (2,4-D)
   functions of, 109
   in *Hevea brasiliensis* culture, 109
   in latex production, 110
   structure of, 110
Tyrosinase, 85
Tyrosine, 80, 116

Ulmic acid, 14
Ulmin, 14
Ultisols, 21
   organic C content in, 21
Ultrafiltration
   of humic substances, 51
Uronic acid, 87, 92
UV and visible light spectrophotometry, 176
   derivative spectrum, 179
   importance in characterization of humic substances, 176
   spectrum of fulvic acid, 176
   spectrum of humic acid, 176
   use in quantitative determination of humic acid, 179-180

Van't Hoff
   law of, 226, 228
Variable charge, see negative charges
Vasoconstrictor, 94
Vitamins, 99
   B complex, 107

function and importance of, 106
-D, 102
in humic acids, 99-107, 309
types of, 106
functions of, 107
relation to nicotine, 107
sources of, 107
Vitamin C, 107
molecular structure of, 107
sources of, 107
role as redox agent, 107

Waste
agricultural, 27, 29
humic matter in, 27
industrial, 27, 29
in polluted ditches, 29
issue of contamination of groundwater by, 29
municipal, 27
Water
as extraction reagent, 42
Water bridging, 44
Water repellant soil, 101
Wax, 4, 101
Wetland humic matter, 22, 29, 31
Wheat belt, 281, 293
White-rot fungi, 83-84
demethylation by, 84
Wilting point
effect of osmotic pressure on, 287

XAD amberlite resins
extraction of aquatic humic substances by, 58-59
hydrophobic, 44
hydrophilic, 44
in gel chromatography of humic substances, 43
issue of bleeding with, 44
types of, 43
Xenobiotics, 259
definition of, 109
detoxification of, 259
effect of humic acid redox properties on, 109
effect in soils, 109
major types of, 109
Xylose, 96

Zea mays, corn, 281
germination of, 290
effect of humic acids on, 290
Zero point of charge, ZPC, 219, 242
Zwitter ion, 219